T0295882

Becoming a Supply Chain Leader

Becoming a Supply Chain Leader

Mastering and Executing the Fundamentals

Edited by

Sourya Datta, BE, MBA
Sudip Das, MBA, PhD
Debasis Bagchi, PhD, MACN, CNS, CFFS, MAIChE

Routledge
Taylor & Francis Group
A PRODUCTIVITY PRESS BOOK

First published 2022
by Routledge
6000 Broken Sound Parkway #300, Boca Raton FL, 33487

and by Routledge
2 Park Square, Milton Park, Abingdon, Oxon, OX14 4RN

Routledge is an imprint of the Taylor & Francis Group, an informa business

ISBN: 978-0-367-22081-5 (hbk)
ISBN: 978-1-032-05316-5 (pbk)
ISBN: 978-0-429-27315-5 (ebk)

DOI: 10.4324/9780429273155

Typeset in Minion Pro
by codeMantra

Dedicated to My Beloved Parents, Mrs. Manju and Mr. Asish Datta,

"Rekha", my beloved Wife, and My Darling Daughter "Tara" for

Continued Inspiration, Motivation, Love and Sincere Blessings

Sourya Datta

Dedicated to "Reema" and others, who like her indulge the restless.

Sudip Das

Dedicated to My One and Only Beloved Sweet Daughter "Dipanjali Bagchi"

Debasis Bagchi

Contents

Preface xi
Editors xv
Contributors... xix

SECTION 1   All Links of the Chain

Chapter 1   The Original Supply Chain Leader: Marco Polo.................. 3
Sudip Das and Ishaan Das

Chapter 2   Evolution of Modern Supply Chain 13
Sourya Datta

Chapter 3   The Solutions that Business Wants....................................... 35
Sourya Datta

Chapter 4   The Strategic Value Proponent – Six Sigma Approach 65
Sourya Datta

SECTION 2   The Supply Chain Leader

Chapter 5   Roles, Responsibilities and an Industry Overview 91
Sudip Das and Ishaan Das

Chapter 6   The Six Pillars of Supply Chain 123
Sudip Das and Ishaan Das

Chapter 7   Making Your Mark as a Leader.. 153
Sudip Das and Ishaan Das

SECTION 3   Practices, Perspectives and Leadership in Different Industries and Their Key Nuances

Chapter 8 Changing Landscape in the Printer Industry 179
Ashok Murthy

Chapter 9 Saving Lives through Medical Devices 199
Upayan Sengupta

Chapter 10 Indirect Procurement in Information Technology 217
Adhiraj Kohli

Chapter 11 Powering the Future of the Energy Industry 245
Bradley Andrews

Chapter 12 Driving Automotive Long-Haul Growth 263
Michael Cupo

Chapter 13 Managing Hyper Growth in the Cloud 279
Prasad Sabada

Chapter 14 The Logistics of Cooperative Supply Chain in Dairy 299
Andrew Bawden

Chapter 15 Data Analytics in Supply Chain Management 311
Tony Chen

Chapter 16 Longevity Challenges in Avionics 331
Ashok Das

Chapter 17 Bridging Innovation and Governance in Biotech 349
David Passmore

Chapter 18 Global Trendsetting in Apparel 361

Anupama Kapoor

Chapter 19 Bootstrapping Supply Chain in a Startup 377

John Jacobson

SECTION 4   Case Studies in Supply Chain Management

Chapter 20 Case Studies in Supply Chain Management....................399

Asish Datta, Sanjib Sengupta, Debasis Bagchi, and Sudip Gupta

Chapter 21 Dabbawala: A Hundred-Year-Old Six Sigma
Logistics Concept..413

Debasis Bagchi, Asish Datta, Manashi Bagchi, and Sourya Datta

Chapter 22 Conclusion...425

Ishaan Das

Commentary: A Path to Cognitive Leadership 429
Sourya Datta, Sudip Das, Ishaan Das, and Debasis Bagchi

Index..433

Preface

Becoming a Supply Chain Leader: Mastering and Executing the Fundamentals is a book on leadership in supply chain management. As the modern world is fueled and evolving through the supply chain, its study has become mainstream in academia all over the world. It has evolved into multiple specializations such as global logistics, industry-specific, and international supply chain management, vendor management, supply chain data analytics, supply chain finance, forecasting, and demand planning. This book complements the effort of academia as it focuses on developing leadership skills in the supply chain. Every year a large number of graduates of these disciplines join the supply chain workforce along with people trained in other domains. This book will prepare them to understand their roles, responsibilities, who they need to collaborate with and create values to ramp up as they grow.

This book is intended for academics and beyond. This is a book written for supply chain professionals which will be a companion for their entire professional journey. From an entry-level supply chain professional to a mid-level manager or a senior executive, everybody looks for opportunities to learn and grow and continue to make an impact. While there are plenty of books on supply chain theories and practices, this book is intended to focus on two key areas: how to demonstrate leadership and grow in this profession, and how to develop skills that are transferable within different supply chain domains and industries in order to foster personal growth. In the eyes of our group of authors, this book is a success if our audience can implement and use these ideas to create value within their organization at any level, find a map of their organization and stakeholder teams to collaborate with, and communicate to them with clarity and purpose as leaders. If readers can be inspired by our preeminent contributors' perspectives and develop motivation to grow to the highest possible leadership role in any supply chain industry they choose, then we will have done our jobs. Supply chain management is a highly transferable skill and this book provides a framework which may enable our audience to transition from one supply chain vertical to another without difficulty.

The book provides perspective into the intricate aspects of the supply chain through four major sections comprising 22 individual chapters. Supply chain and business leaders and academicians from a diverse set of industries such as enterprise and consumer electronics, energy, biopharmaceuticals, automotive, avionics and apparel have contributed to this book, giving glimpses into the specific challenges and unique aspects of each vertical. Supply chain management has been a strategic and vital business tool around the world for procurement and delivery of goods and services since ancient times.

The first section discusses the fundamentals of supply chain. In this section, Chapters 1 and 2 explore how and why the modern supply chain evolved. The two chapters discusses the types of business challenges supply chain solves and how supply chain principles are used to solve various challenges. Business challenges in a modern corporate environment are not solved in silos; they are solved through cross functional teamwork, aligning common goals, and building trusting relationships. Ultimately supply chain management is about the people involved, and Chapter 3 discusses the business partners of supply chain. After *people* comes *process*, and Chapter 4, the final chapter of Section 1, discusses Six Sigma, which is an often used method to improve supply chain processes through some examples.

Ultimately this book is not just about supply chain but about the people involved. Section 2 focuses on the people who run, manage, and lead supply chains. Chapter 5 discusses the roles, responsibilities, and opportunities for supply chain professionals. The perspectives are presented through three different angles, so that it's equally useful for an entry-level supply chain manager, a mid-level, and a senior-level supply chain leader. Chapter 6 presents a framework of six pillars of supply chain knowledge and the rest of the book revolves around this concept. The six pillars are used as a dynamic frame of reference to learn, execute, and demonstrate leadership and be successful in the supply chain roles at any level, in any industry vertical. Chapter 7 lists opportunities to move the needle in a supply chain organization. It suggests different initiatives to be successful and distinguish oneself at various organization levels.

Section 3 uses the principles and pillars of the supply chain knowledge framework to review different industry verticals. Ten leaders offered their perspectives on supply chain in their respective verticals, namely commercial electronic printers, medical devices, IT procurement, energy, automotive, cloud, dairy, avionics, biopharmaceuticals, and apparel. For each of the verticals, the industry outlook and the macro trends shaping its business and vertical-specific supply chain nuances are discussed. A supply chain professional interested in joining or already working in these verticals can learn about the critical supply chain skills required to thrive and succeed in these sectors. Finally, the contributors share their perspectives on leadership in these verticals. Two additional chapters in this section discuss application of data analytics in supply chain management and scaling a startup in the technology industry. The former presents various analytical methods and their potential applications, while the latter discusses a journey of a supply chain leader and his organization and its scaling up from formation to maturity.

Finally, in Section 4 we dedicate two chapters for case studies. Chapter 20 discusses a set of case studies which showcases various supply chain solutions such as consignment, technology-based supply chain disruption, cost management, and process integration. In Chapter 21 we explore the facets of a high performing logistics organization which is more than a hundred-year-old brand and known for its on time delivery accuracy. It analyzes the success of this organization as it uses a very limited amount of tools but has an incredibly well-managed team of people and simple yet robust processes.

The final chapter comprises a conclusion and a commentary from the editor's desk which wraps up the conceptual theme of all chapters presented.

Our sincere gratitude and thanks to all our eminent contributors, as well as CRC Press/Taylor & Francis editorial team members including Michael Sinocchi, Samantha Dalton, and Katherine Kadian for their continued support, cooperation, and assistance. Finally, the editors extend their sincere gratitude and special thanks to Ishaan Das for his continued support and efforts to complete this book.

Sourya Datta, MBA
Strategic Deals Manager
San Jose, California

Sudip Das, PhD, MBA
Senior Supply Chain Manager
Pleasanton, California

Debasis Bagchi, PhD, MACN, CNS, CFFS, MAIChE
College of Pharmacy and Health Sciences,
Texas Southern University, Houston, Texas, &
Research & Development, VNI Inc., Bonita Springs, Florida

Editors

Sourya Datta is a part of the Strategic Deals team at Apple where he is responsible for managing the procurement of a number of components for Apple. Prior to joining Apple, he managed the finances for eBay's North American division. This included managing P&L, budgeting, forecasting, and closing the month and quarters for eBay. He was also responsible for leading the analytics for Latam and Canada for eBay. Prior to that, Sourya was the supply chain manager in the Business Operations and Strategy Group at eBay. He received his Master's in Business Administration from the University of Pittsburgh, Pennsylvania, in Operations and his Bachelor of Engineering from Anna University, India, and worked in various consulting firms before doing his MBA. Sourya was featured in Poets and Quants as one of the best MBAs in 2015 and has won a number of awards in his tenure at eBay.

He is an avid sports fan and has completed a number of marathons and Spartan races since 2007. He has been swimming since he was seven years old and has participated in and won numerous state-level competitions. He has also taken active part in mountaineering and trekking since he was in elementary school. He has participated in various marathons in California and has been intricately associated with nutrition and functional food requirements for sports performance. He is also the author of two chapters in *Developing New Functional Food and Nutraceutical Products* that was published by Elsevier/Academic Press in October 2016.

He is also the author of a chapter in the second edition of *Nutrition and Enhanced Sports Performance* published by Elsevier/Academic Press in September 2018.

Sudip Das works at Amazon AWS supply chain where he is a senior leader for server component supply chain management. Prior to joining AWS, Sudip was a senior leader at LinkedIn, responsible for the data center server supply chain and hardware and software asset management. Prior to that Sudip was a director at eBay where he managed both infrastructure supply chain management, business operations, and supply chain analytics development including developing a market place for retired assets. Sudip started his supply chain journey at Cisco Systems where he was a global supply chain manager for network semiconductor components and network appliances. Sudip started his career as hardware engineer designing servers and networking products. He received his undergraduate engineering degree in Electronics and Electrical Communication Engineering from the Indian Institute of Technology, Kharagpur, and PhD in Electrical Engineering from the University of Kentucky, Lexington, Kentucky. He did his MBA from the Pepperdine University, Graziadio School of business, Malibu, California.

Outside academics, Sudip enjoys amateur astronomy and calligraphy. He is an avid fan of college basketball and loves soccer and cricket. Sudip is an avid traveler who enjoys visiting ancient ruins.

Debasis Bagchi received his PhD in Medicinal Chemistry in 1982. He is the director of scientific affairs at Victory Nutrition International, Inc., Bonita Springs, Florida,; an adjunct faculty member in the College of Pharmacy and Health Sciences, Texas Southern University, Houston, Texas; and an ex-professor in the Department of Pharmacological and Pharmaceutical Sciences at the University of Houston College of Pharmacy, Houston, Texas. He served as the senior vice president of Research & Development of InterHealth Nutraceuticals Inc., Benicia, California, from 1998 until February 2011, and then as the director of innovation and clinical affairs of Iovate Health Sciences, Oakville, Ontario, until June 2013, and then the chief scientific officer at Cepham, Inc., Somerset, New Jersey, from June 2013 until December 2018. Dr. Bagchi received the Master of American College of Nutrition Award in October 2010. He is the past chairman of the

International Society of Nutraceuticals and Functional Foods (ISNFF), past president of the American College of Nutrition, Clearwater, Florida, and past chair of the Nutraceuticals and Functional Foods Division of Institute of Food Technologists (IFT), Chicago, Illinois. He is currently serving as a distinguished advisor on the Japanese Institute for Health Food Standards (JIHFS), Tokyo, Japan. Dr. Bagchi is a member of the Study Section and Peer Review Committee of the National Institutes of Health (NIH), Bethesda, Maryland. He has authored 375 peer-reviewed publications, 43 books, and 20 patents. He is also a member of the Society of Toxicology, member of the New York Academy of Sciences, fellow of the Nutrition Research Academy, and member of the TCE stakeholder committee of the Wright Patterson Air Force Base, Ohio. Dr. Bagchi is an associate editor of the *Journal of Functional Foods*, *Journal of the American College of Nutrition*, and *Archives of Medical and Biomedical Research*, and an editorial board member of numerous peer-reviewed journals, including *Antioxidants & Redox Signaling*, *Cancer Letters*, *Food & Nutrition Research*, *Toxicology Mechanisms and Methods*, and *The Original Internist*.

Contributors

Bradley Andrews
Executive, Energy and Resources
  Industry
London, England

Debasis Bagchi
College of Pharmacy and Health
  Sciences
Texas Southern University,
  Houston, Texas
and
Director of Scientific Affairs, VNI
  Inc.
Bonita Springs, Florida

Manashi Bagchi
Director, Dr. Herbs LLC
Concord, California

Andrew Bawden
Sourcing Account Manager, Dairy
  Industry
Amsterdam, The Netherlands

Tony Chen
Supply Chain Data Scientist
Fremont, California

Michael Cupo
Senior Supply Chain Manager
Seattle, Washington

Ashok Das
Supply Chain Executive, Avionics
Grand Rapids, Michigan

Ishaan Das
Sales Development, Cloud security
New York City, New York

Sudip Das
Senior Supply Chain Leader, IT
  hardware and software
Pleasanton, California

Asish Datta
Prior-SVP, Berger Paints India
  Limited
Kolkata, India

Sourya Datta
Strategic Deals Manager
San Jose, California

Sudip Gupta
Prior-Divisional Head of Supply
  Chain, Nippon Paints
Chennai, India

John Jacobson
Senior Supply Chain Executive,
  Electronics
Santa Barbara, California

Anupama Kapoor
Global Procurement Executive,
  Apparel
Palo Alto, California

Adhiraj Kohli
Infrastructure Business Operations
  Manager, IT
Sunnyvale, California

Ashok Murthy
CEO Devices-Unlimited Corp.
Santa Ana, California

David Passmore
Head of US Business Development,
  YUMAB Inc
San Francisco, California

Prasad Sabada
Senior Supply Chain Executive,
  Infrastructure
Mountain View, California

Sanjib Sengupta
Prior-SVP, INDAL
Kolkata, India

Upayan Sengupta
VP & GM UL Life and Health
  Sciences
Austin, Texas

Section 1

All Links of the Chain

1

The Original Supply Chain Leader: Marco Polo

Sudip Das

Ishaan Das

CONTENTS

1.1 Silk Road 3
1.2 Explore and Prepare 4
1.3 Significance of the Silk Road 5
    1.3.1 Product 5
    1.3.2 Demand 6
    1.3.3 Supplier..6
    1.3.4 Value Creation 7
    1.3.5 Relationship 7
    1.3.6 Execution...8
1.4 Marco Polo, the Supply Chain Leader 9
1.5 Concluding Remarks..11
About the Contributor...12
References..12

1.1 SILK ROAD

It could be argued that Marco Polo (Polo and Latham 1958) was one of the well-known supply chain leaders in the history of the world. To prove it, let us first understand what is a supply chain and what does a supply chain leader do.

Simply put, a supply chain enables customers to get their desired products. It connects the producers to customers in a continuous manner.

DOI: 10.4324/9780429273155-2

By that definition, one of the oldest known supply chains was the Silk Road. Silk was the product and the Silk Road connected European customers in the west to the Chinese manufacturers of silk in the east (Whitfield 2015). The Silk Road was essentially a trade route. It started during the Han dynasty in the second century BC, many centuries before Marco Polo traveled from Venice to Kublai Khan's capital in China in 1269. So definitely there were supply chain leaders before him; however, due to his chronicles he had become perhaps the most famous one.

The Silk Road itself was actually a network of trade routes between Asia and Europe. Various types of goods changed hands as traders of each region exchanged their products. Spices like black pepper would find their way from Southeast Asia to Italy where it would be used for preserving meat. The finest silks would be traded multiple times before they finally got sold to the nobles in different parts of ancient and medieval Asia and Europe. Horses got traded for royal armies throughout many countries in the route. The Silk Road satisfied the function of the supply chain for many such items during ancient times.

1.2 EXPLORE AND PREPARE

In this chapter we will explore what it took to deliver Asian products to the European market in ancient times. This gives us an idea about what were the products, who wanted them, who made them, how the market grew, who were all the traders, and finally how this series of continuous trading got managed and executed for centuries. Ultimately, this book is not about what needs to be done but about who does it. We will do an excercise in this chapter. We will consider an example which we will explore through the lens of some logical question to baseline our generic understanding of supply chain. We will, however, dive deeper into these questions in the subsequent chapters in this book.

This chapter will start to scratch the surface of supply chain leadership which will be followed in greater detail throughout the rest of this book. For the moment, we will do an armchair analysis about the roles and responsibilities of the traders who ultimately made the European Kings and Queens look chic in great Chinese silk or enabled them to enjoy the finest cured meats when they threw opulent parties.

1.3 SIGNIFICANCE OF THE SILK ROAD

Let us dive deep into the Silk Road as a supply chain, using a list of questions. Let us go back in history and imagine ourselves as a young Venetian like Marco Polo in the thirteenth century. Let us put ourselves in the shoes of Romeo, an imaginary young Venetian youth, belonging to one of the noble trading families of Venice. The family might have expected young Romeo to step up and stop being a player and instead getting into the family business. The young man did not know much about the business and had vaguely heard that it would involve going to hazardous places far away from the creature comfort of home and bringing back amazing things like silk and spices. People in Venice were crazy about those goods and therefore as soon as he would roll into the town with bags full of exotics, people would offer large sums of money and would make him rich beyond imagination. Let us put ourselves in Romeo's place and come up with a framework or a list of things that he had to be prepared for (Haksöz et al. 2011).

1.3.1 Product

First, Romeo needed to understand what were the products that he was supposed to buy. He had to buy at least two categories of items which were spice and silk. There were also various types of spices in the Far East. The two primary spices that could get most value for money were pepper and cinnamon. All natural products have different grades or qualities, price point, shelf lives, local preferences, etc., so learning the differences between different grades was must. He had to learn how to negotiate the price of pepper and cinnamon, what were the locations in the Far East where these two spices would be available, and what would be the exchange rate against gold or some other currencies. Then there would be shipping logistics as both had to be brought in large volume to cover the travel cost. How much cinnamon or pepper could a single horse carry? Similarly for silk, the logical list of questions would be what were the different levels of quality of silk and how to recognize them and negotiate their prices. Silk could be available as rolls which could be cut and tailored or could be available as finished and beautifully embroidered dresses. Minerals were used to dye the silk and price could vary based on color. Romeo needed to understand what type of silk he would have to procure and their prices and how to bring them back or ship them.

1.3.2 Demand

After understanding the product the next critical item was to develop an understanding of the demand of these items in the Venetian market. What would happen when he brought in pepper, if another supplier also brought in pepper? Would that make the price of pepper fall and impact the profit margin? Would Romeo need some statistics like what was the annual amount of peppercorn sold in Venice and were there many traders with back-end suppliers or was the Venetian market always undersupplied and the demand always stayed high? What were the shelf lives of the peppercorn and cinnamon bark so that they would survive the journey plus the amount of time it would take to sell them? Similarly how much silk should he buy? Should he buy only premium silk or buy a mix of top and medium-grade silk? Should he buy equal worth of these three items? The young Silk Road supply chain rookie needed to understand what were the demands of the market to balance his target portfolio of products. There could be additional options like procuring some precious stones which were a lot less bulky with high return on investment and also in high demand.

1.3.3 Supplier

The first two sets of information were available in Venice itself. Walking around the Venetian market or by mingling with the Venetian high society, Romeo could develop the necessary understanding of the product categories and their sources of demand. However, identifying the sources was the next challenge. One way of procuring would be going to the kingdom of origin and buying the products at their lowest cost. For example, silk would be available inside China where the weavers had set up silk weaving looms next to mulberry farms for the silk worms. Cinnamon on the other hand came from Ceylon or Indonesia and both were far away from China. So as a supplier manager, young Romeo had to understand the strategy, whether to procure from the manufacturer at the lowest cost or from an intermediary like a wholesaler at an outpost on the Silk Route. There could be traders in Kerman, Persia, who got supply of peppercorn and cinnamon but the price would be higher than Ceylon but still relatively low compared to Venice. On the other hand there could be traders at outposts all along the Silk

Routes who might be selling silk but in Khashgarh at the border of China; Afghanistan could be the biggest wholesale market. Silk from different parts of China was available in one spot. The next spot for buying silk would be Khanbaliq or modern day Beijing which could offer maximum choice at the best cost but would also add months of distance in both directions. There were suppliers in each of these locations and there were different kingdoms and different rules of trade and different selling prices. Understanding and being aware of political situations, possibilities of wars breaking out or possibilities of seasonal disruption like flood, etc. were important as these could cause disruption in the Silk Route.

1.3.4 Value Creation

The goal of a supply chain was not necessarily to go to sources for the lowest cost with a target of maximizing profit. After understanding what to procure, how much to procure, or from where to procure, Romeo must decide how he could be the best. However, that would beg the question – best in what? What would the young supplier manager desired to be known for? Would he like to be known for being the best source of silk in Venice, as the purveyor to the Doge of Venice? Or would he like to be remembered as the peppercorn king of Europe? The bottom line is there should be something unique or special like finding a new source of silk which was better than what was available before. Or it could be about discovering a sustainable supply chain so that peppercorn and cinnamon could arrive on time year after year. There could be opportunities to create value like bringing in extremely rare jewels from Ceylon or bringing in exotic porcelain or jade figurines from Khanbaliq. It was not about being another trader in the Silk Route but doing something innovative or unique. For example, Romeo might want to be the fashion trendsetter of Venice by discovering a new silk supplier who had created new dyes and produced silk in colors which were never seen before in Venice.

1.3.5 Relationship

Ultimately success of all supply chain activity boiled down to people and knowing the right set of people. From being able to finance the whole journey, to being able to travel to distant land and procure highest quality silk and spices, to being able to safely bring them back to Venice and getting permission to sell, every step involved working with an enormous number of people. So getting their trust and cooperation were pivotal. For example, the banker would be investing a large amount of money which had not only an extremely high risk of return but also a very long waiting period. Similarly, the outpost where the spices and silk were available had kings and lords who could make their lives easy and provide access to the bazaars and safe passage instead of robbing them. Even coming into Venice, the political situation could change in Venice or it could be at war with another town. Selling the goods would need contacts and permissions. Relationships were built based on trust and track record. Being young, Romeo had to use his family's track records and connections built by his father and uncles. The young supply chain manager had to make a list of people whom he should meet before leaving Venice and the key contacts made by his family in previous journeys. He had to understand the depth of relationship, which one was strategic and which one was transactional as he would be leveraging those and would be investing and building new ones as well. He would need to understand how personal the existing relationships were and how were they forged and how it had grown. For example, working with a gemstone trader in a distant outpost would require not only trust but extreme security and could be an exclusive relationship with a gemstone trading family. On the other hand there could be more transactional relationships with silk merchants as there were many. For silk, it would be more important to develop a deep relationship with the emperor who would give him wider autonomy to do business under the royal sponsorship and that was what Marco Polo did. He built a strategic win-win relationship with emperor Kublai Khan who trusted him and appointed him to be his foreign emissary. Marco built such a relationship with the emperor, that as part of his last service to the emperor, he took the perilous responsibility of escorting Princess Kokochin from China to Persia for marriage with the Persian ruler. He became family. Romeo would really need to step up in this category.

1.3.6 Execution

The culminating step before commencing the journey would of course be execution. In spite of doing great in the five items above, things could fail anytime. For example, the first step of the journey would be going to Asia from Venice by crossing the Mediterranean. A shipwreck or an attack at the port of Acre in Palestine could end the journey right there. Arranging logistics throughout the journey and safety and security were the primary execution requirements. Being healthy in faraway land, not to attract undue attention from opportunistic local lords or jealous traders, scammers, or robbers, would require both luck and flawless execution in every step and planning and anticipating challenges ahead of time. One example of an executive decision that the Polos made was to follow through Kublai Khan's quest to build a relationship with the Roman Pope. However, the previous Pope had died and they waited for 2 years and a new Pope was still not announced. After the Polos arrived in Palestine, they heard the news that the new Pope had been announced. They went back to get Pope's response to Kublai Khan's request. This ensured access to the Mongol court and audience with the emperor which later helped Marco immensely. In every step there could be the possibility of being robbed or taxed, so converting profits into gemstones of various sizes and gold coins would be equivalent to money management during the journey. For example, Marco got massively taxed by Turkey on his return trip and had to give away four thousand gold coins. Young Romeo must reach out to other traders, understand these execution challenges, and be aware that if faced similar situation he could think on his feet.

This is a thinking exercise which could be done in greater detail if desired. However, this presented a broad framework of supply chain management during the ancient time.

1.4 MARCO POLO, THE SUPPLY CHAIN LEADER

Though Marco Polo was not the first European to go to China, his name has been uniquely connected with the Silk Road. If we consider the Silk Road as an ancient supply chain, then Marco should be considered as one of its supply chain leaders. Now, how did he become *the* supply chain leader on the Silk Road and revolutionize the trading practices?

Marco Polo traveled from Venice to Asia between 1271 and 1295. He met Kublai Khan, a Mongol leader in China (Man 2016) and went on many diplomatic missions as Khan's foreign emissary. Though he was not the first European who went to China, he traveled extensively to various parts of China, Southeast Asia, India, and Persia. He converted his fortune into precious gemstones and returned to Venice. His extended stay in Asia gave him a unique perspective into the culture, wealth, and the workings of the

Mongol Empire and China. He chronicled his travels in *The Travels of Marco Polo*, a book that gave the first detailed view of China, Persia, India, Japan, and other Asian countries to European traders for the first time. It captured the culture, customs, geography, and also the products of the orient like porcelain, coal, gunpowder, paper money, and exotic Asian plants and animals. His travel book inspired many travelers and also influenced European cartography and led to the creation of the Fra Mauro map which could be considered as one of the earliest versions of modern world map.

Now would it make sense for Romeo, the young Venetian novice who wanted to get into the family business, to hit the family library and get a copy of *The Travels of Marco Polo*? Would it make sense to read the book cover to cover and then perhaps follow in his footsteps? Seeing as he wrote the book on Asia and talked about the trading business in great detail including the geography and cities en route both via land and by sea, as well as being an emissary for Kublai Khan and coming back home as a successful merchant, wouldn't Romeo want to have Marco's wisdom? Now let us unpack these achievements through the previous six sets of questions and figure out if this book would be a must read.

First of all Marco wrote a book (Polo and Latham 1958) to share the information and did not keep it a secret. It got translated in many European languages and was a bestseller. A leader is also a coach who shares and inspires others to become leaders. The book helped other traders to understand the Silk Road and the business and political environments of the various countries on the road. It helped traders to chart their courses, understand distance between places, and take alternate routes to avoid a certain region to mitigate risk. The book also helped them to prepare by familiarizing with customs and cultures of the various countries to trade effectively. It also helped them to find out where the different merchandise were available. In a way the book would help you to create a network of merchandise and their sources. Upon reading the book, Romeo would get to know what were the needs of Venetian people and which parts of the Middle East, Central Asia and East Asia would produce them. It would probably help Romeo to prepare for the journey by assessing the security risks, ports to avoid, protection and logistics needed to avoid the bandits, or the amount of supply needed to be carried between cities and the type of transportation needed.

In order to be successful one would need to be able to sell their merchandise and purchase the merchandise that you would be bringing back home.

The ten thousand-mile round trip distance between Venice and Khanbaliq (Beijing) would be perilous. Plus the traders should be willing to do trade. Marco built a relationship with the Mongol King himself and Marco even liaised between the Vatican and Kublai Khan. That implied the relationship and influencing power of Marco and contracts between two great powers. As a trader of the Silk Route Romeo would need to build relationships with heads of trading guilds, heads of cities and countries to get safe passage, and access to the goods, transportation, foods and other logistics like facilities to store merchandise, local guides, and local help. Relationships were to be viewed as contracts between oneself and the facilitators or makers of the products that would be procured. The most important aspect would be safety but not losing investment and keeping cost manageable were also key. So the book would help Romeo to understand the markets from the Middle East to China. It would give ideas about the people of different countries and their products. Following the book would also help in understanding what kind of relationships need to be forged with the network of people on the Road.

Finally, Marco converted his profits into gemstones which would help Romeo to understand how to maintain financial liquidity when needed. Trading bulky items which requires complex logistics like slow moving donkey carts that were prone to bandits attack and converting them to light gemstones seemed to be the right hard currency for international trade. Also the price of gems would never diminish in comparison to merchandise which could be perishable or have wavering demand, causing prices to fall. The book would give ideas on how to balance liquidity between merchandise and cash.

1.5 CONCLUDING REMARKS

In summary Marco Polo was the first supply chain leader because he definitely excelled at all the fundamentals which made him successful and went above and beyond in becoming Kublai Khan's emissary and a successful Venetian merchant. And of course, the greatest legacy he left behind was his book where he shared the wealth of his knowledge. He inspired and educated traders all over the world, and for Romeo, Marco was definitely a supply chain leader to aspire for.

ABOUT THE CONTRIBUTOR

Ishaan Das is currently working in technology sales development for Box, a cloud storage company, in New York City. He graduated with honors from the University of Michigan, Ann Arbor with a degree in Economics and Entrepreneurship. He has worked in a multitude of tech startups primarily focusing on growth in sales development and customer acquisition as well as organization management. He also has experience working in supply chain growth management, developing delivery cost models and securing venture capital funding. He has been instrumental in shaping the organizational flow of this book and bringing in a student's perspective from creating and organizing content and coordinating among authors and contributors.

REFERENCES

Haksöz, C., Seshadri, S., and Iyer, A. 2011. *Managing Supply Chains on the Silk Road: Strategy, Performance, and Risk*. Boca Raton, FL. CRC Press/Taylor & Francis.

Man, J. 2016. *The Mongol Empire: Genghis Khan, His Heirs and the Founding of Modern China*. London. Transworld Publishers Limited.

Polo, M., and Latham, R. (Translator). 1958. *The Travels of Marco Polo*. London. Penguin Books.

Whitfield, S. 2015. *Life along the Silk Road*. Oakland, CA: University of California Press.

2

Evolution of Modern Supply Chain

Sourya Datta

CONTENTS

2.1   Objective...14
2.2   Dabbawala (Mumbai Dabbawala 2019) 14
2.3   History of Supply Chain 15
    2.3.1   Timeline of Supply Chain: Mid-1900s–2000s
        (Institute of Supply Chain Management 2018) 16
2.4   Various School of Thoughts on Why We Need Supply Chain 17
    2.4.1   Lead Time Focus 17
    2.4.2   Cost Focus..18
    2.4.3   Accuracy and Quality Focus 18
    2.4.4   Customer Focus...18
    2.4.5   All Round Efficiency Focus 19
2.5   Types of Supply Chain Management – Production Methods
    (Anderson 2008)...20
    2.5.1   Make to Stock 20
    2.5.2   Build to Order..21
    2.5.3   Constant Replenishment or Continuous Replenishment........21
    2.5.4   Channel Assembly 22
    2.5.5   Real-Time Change of Supply Sources......................................23
    2.5.6   Turnkey Contract or Order to Make..23
2.6   Principles of Supply Chain Management 24
    2.6.1   Focus on Customer Need..24
    2.6.2   System Thinking in Supply Chain..25
    2.6.3   Disruptive Innovation in Supply Chain...................................25
    2.6.4   Collaboration between Verticals within Organization 26
    2.6.5   Flexibility of Supply Chain 26
    2.6.6   Corporate Social Responsibility for Manufacturers 27
    2.6.7   Technology-Based Transformation in Supply Chain..............28

DOI: 10.4324/9780429273155-3

2.6.8   Communication and Information Sharing
          within Organization 28
2.6.9   Analytics Used in Supply Chain 28
2.6.10   Disruptive Innovations in Product Technology 29
2.6.11   Warehouse Management for Storing 29
2.6.12   3D Printing Process for Prototype Development....................29
2.6.13   Global Perspective of Supply Chain Process and Systems 29
2.6.14   Impact of Economic Conditions of Various Countries.......... 30
2.6.15   Risk Management and Mitigation Strategies for Buyers.........31
2.6.16   Political Conditions and Navigating Through
          Political Changes...31
2.6.17   Visibility of End-to-End Supply Chain and
          Catching Problems Early 32
2.6.18   Value Creation within Supply Chain 32
2.7 Concluding Remarks..32
References...33

2.1 OBJECTIVE

In the previous chapter we discussed the ancient Silk Road which took silk from China to Europe. We talked about how the supply chain connects the producers to the customers in a continuous manner. We will focus on the continuity and the process aspects in this chapter and trace its development through the history. We will look at the key concepts and few examples to understand the concepts along with a few strategies that are employed within the supply chain to make the supply chain more efficient.

2.2 DABBAWALA (MUMBAI DABBAWALA 2019)

Continuity implies discipline and a process. Silk road (Mark 2018) traders like Marco Polo (Biography.com 2021) went back and forth to China to get silk, and weavers in China produced silk year after year to meet the demand. This repetition required discipline. It would imply weavers spun silk thread and weaved the finest silk, traders took risk and made the perilous trips to purchase silk and then returned to Europe to sell them. All of these activities happened continuously in a cyclical manner with a predictable timeline which created a predictable demand for the traders to go back year after year. In other words it was a process which spanned over many nations and involved many people. Their discipline to manufacture, purchase, transport and sell made it a continuous process. Therefore the product and process both went hand in hand to make the Silk Road a success.

To highlight the discipline and how it leads to developing a process that is reliable and suitable for commerce, we will highlight a more recent example. This example is about a food delivery service in Mumbai, India, which is run to perfection by an army of minimum wage laborers. This organization and the team of food pick and delivery service providers are known as the 'Dabbawalas' (Mumbai Dabbawala 2019). Their job is to deliver home-cooked lunch to office workers during the day time. Currently about 5000 Dabbawalas pick up and deliver about 200,000 lunches daily for less than a dollar per delivery. We will have a case study on Dabbawalas (Chapter 21) to explain the details more but this point to point pickup and delivery process, which has its roots in the late 1800s, evolved long before the internet-based food delivery services that we are so familiar with.

Overall this is an example where a team of people who have formed an organization and have been successfully executing a process to deliver lunch on time. People, process and product are three key ingredients of the supply chain. This book is all about these and the following chapters will provide more details on these.

2.3 HISTORY OF SUPPLY CHAIN

The root of the supply chain is from logistics. There were a few key events that led to the growth of the supply chain. Even though the term 'supply chain' was not coined during the 1900s but in 1911 Frederick Taylor came out with a book *The Principles of Scientific Management* (Hyde 2019) which had a number of principles of supply chain embedded in it. The industrial engineering and operations research had roots in logistics and the popularity of operations research increased tremendously due to World War II in 1940. Then came the integration of industrial engineering and operations research, which led to the emergence of supply chain management. Here is a brief timeline of events which played a role in the evolution of modern day supply chain such as automation in warehousing logistics, transportation of goods, using data for tracking, and making decisions using computers.

2.3.1 Timeline of Supply Chain: Mid-1900s–2000s (Institute of Supply Chain Management 2018)

1940: Machines such as forklifts and pallet lifts were introduced, which started revolutionizing stocking and shipping of goods swiftly. Previously these were mostly done by humans.

1950: A growth in transportation management started. Containers became more popular for freight along with an increase in the use of trucks, trains and ships.

1960: While Europe continued on its dependence on freight movement by train, there was a shift from train to truck transportation in the United States. This made shipment faster and more nimble. Physical distribution became a norm across countries and continents.

1970: Computers were introduced in operations management. Logistics planning and warehousing methods started getting automated. Apart from computerization, material handling and transportation were also developed by introducing conveyor belts and silo systems. In case of liquid or gas items, the distribution started to be done through pipeline.

1980: Introduction of true supply chain management happened in the 1980s. Flexibility of multiple people doing analysis using spreadsheets started gaining popularity. Optimization of models to bring in analytics into supply chain management gained traction.

1990: Executives of companies started giving more importance to the supply chain and operations as a way to improve bottom line. Not only did chief operating officers in the companies get involved but also chief financial officers became an integral part of decision making in operations.

2000: Enterprise resource planning (ERP) systems were implemented, which led the move to a decentralized supply chain. Supply chain became more digitized. In Japan just in time (also known as JIT) was introduced to reduce inventory cost and utilize space for storing materials.

2010: The entire 2000 including the 2010 saw continued outsourcing for lower cost. There were few efforts to bring manufacturing in-house but it could not financially compete with the outsourced models.

2020: COVID-19 happened, which caused havoc in the supply chain industry. A push toward in-house manufacturing gained some steam. US and China tensions in trade caused some industries to give a hard look at supply chain

2.4 VARIOUS SCHOOL OF THOUGHTS ON WHY WE NEED SUPPLY CHAIN

Organizations face unrelenting pressure to find new and effective ways to bring their products and services from concept to delivery on to the hands of their customers. The demand of today's fiercely competitive environment has made it imperative that companies connect the end to end process in a most efficient way. Each company endeavors to approach and manage this in a way that it gives them a competitive edge. They focus on a particular pain point and try to create value for their customers by becoming best in class in minimizing that challenge. The section below lists some of the common focus areas chosen by modern enterprises. Organizations try to remove every ounce of inefficiency related to their focus area that exists within their system of delivery. They also redefine and reengineer how supply chains should function to be the best in the eyes of their customer and build their brands. The goal is to establish an error-free, totally efficient network from original supply to final consumption. Several schools of thought have developed on how to accomplish this purpose.

2.4.1 Lead Time Focus

One group of technologists seek maximum efficiency with the shortest amount of cycle time for production. For this group, such a system will foster a stream of innovative products that can be brought to market faster than any competitive network. Unfortunately, the track record for many of these new introductions is less than exciting. Fewer than a hundred new products survive from the thousands introduced each year. But the consumer's desire for a stream of innovation must be satisfied and this group is in the forefront on that aspect. Let's take a slightly different example to explain the concept with one of the most renowned companies of the world, Google. It is one of the most successful companies of the world. We could take the example of the innovative side of Google and Google's mantra in delivering the best product to its customers.

Google brings numerous software products to the market very fast. Sometimes the goal is to bring products to the customers even though there could be improvements in the future. The goal is not to have a final, 'not-to-be-modified' product but the goal is how fast can the products be brought to the market for customer consumption and then iterate based on customer feedback.

2.4.2 Cost Focus

Another school of thought is solely focused on cost and how to maximize efficiency to minimize cost. This school advocates extracting every possible savings to reduce the already attractive 'everyday low price' (EDLP) (Hyde 2019) that exists at certain retail outlets. If they do this, they can tell consumers that there is no need for special discounts or promotions. The everyday pricing will be as low as possible at all times. For this group, the intention is to maximize market position. However, for the consumer to become loyal to EDLP companies, these groups must face unyielding pressure for cost improvement by quickly delivering the fruits of their efficiency to the store shelves. Walmart is a great example for this scenario. Walmart's focus is to provide quality products at 'everyday low prices' and Walmart tries to have stores within 50 miles of a customer so the inventory management is much simpler and Walmart can reach customers at a lower price point than another competing organization.

2.4.3 Accuracy and Quality Focus

A third group would use leading-edge information technology to link the members of a supply chain to the network, which would then become a model of effectiveness. This network would eliminate all duplicated effort and reduce paperwork to an absolute minimum in a virtually errorless system. For this group, success comes from knowing that their network cannot be matched in terms of efficiency. Low cost is not only presumed to exist in this type of supply chain but becomes an integral part. A great example of this model is the car manufacturing company Toyota. Toyota is in the forefront of accurate and high-quality manufacturing and has an extremely well-managed process without any duplication in effort.

2.4.4 Customer Focus

A fourth group seeks the means to turn the traditional 'push' of the product toward the retail customer into one that would 'pull' the product through the network in response to consumer purchases. For many years, supply chains were oriented around the concept of manufacturing goods in demand, at the lowest unit cost, for movement (push) toward consumers. Improvements were generally focused on faster speeds or on the unit-to-time ratio, so item costs would steadily decline. As profits were generated, a portion always went to increasing capacity, causing more units to be produced at ever-lower costs. The sales concept was to push this volume toward the consumer. Inventories that were produced between manufacturing and consumption were conveniently stored in warehouses or distribution centers so that the push would be uninterrupted.

In the case mentioned above, the engineering concept is to eliminate the significant costs embedded in those inventories and warehouses by developing closer links between actual consumption and production. Using ever-increasing databases containing information on what has been purchased by specific categorized customers, it is now possible to pull products and services in direct response to what has been consumed. This network will function with the lowest inventories, damage and obsolescence. Quick response is the motto for this group, which strives to build a system that pulls through only the products that are in actual demand, using a methodology that concentrates on the transactions taking place at the cash registers and checkout counters.

A good example in this area would be the drugstore and pharmacy company CVS. Even though CVS is a giant organization, inventory systems are so efficiently built that CVS is able to replenish inventory to all its stores based on customer demand which is a 'pull' instead of 'push'. CVS also tries to give extremely high importance to fulfilling customer's needs in time even though it follows a 'pull' process from the customers. The above example of CVS is in the pharmacy specifically and not necessarily in the store. The store model needs to have inventory on shelves.

2.4.5 All Round Efficiency Focus

Growing armies of specialists chasing supply chain improvement are considering, testing and developing these and many more approaches, focusing attention on the manufacturing, storage and logistic distribution of products and services. A great opportunity is seen to combine the energy of all the groups discussed above into one single unified effort that seeks what each group ultimately craves for: an optimized supply chain network that generates shared savings for all constituents while establishing a competitive advantage in the market. The savings extracted from such a system could be shared in part with the ultimate consumers, but a portion would create the funding to develop further network enhancements on a continuous basis.

Only in this way will the network keep pace with the ever-changing patterns of consumption to sustain the coveted competitive advantage. To develop such an optimized scenario, it is needed first to understand the supply chain and the factors that are affecting the drive for improvement. This knowledge begins with a consideration of exactly what constitutes a supply chain.

There are a number of organizations who have made efficiency a focus of the supply chain. General Motors, Ford and Toyota are great examples where they see end-to-end supply chain and efficiency as a big part in running the operations.

2.5 TYPES OF SUPPLY CHAIN MANAGEMENT – PRODUCTION METHODS (ANDERSON 2008)

In an organization production of goods and services or manufacturing and its supply chain goes hand in hand. Production and supply chain both are under the umbrella of operations but the goal of production is to 'make products efficiently' and the goal of supply chain is to bring the manufactured product successfully to the hands of the customer in the most efficient possible way.

We listed the strategic focus of a company in Section 2.4. The section below lists the different methods of supply and production. In order to meet the strategic focus an organization refines these production and supply chain processes to perfection.

Let's look at the various types of production strategies. The types of production strategies are important to understand the various steps that organization can take for effective supply chain management and reducing cost.

2.5.1 Make to Stock

Make to stock also called MTS is a type of production method which is used to match inventory with customer demand. Traditionally a company sets a production level and post production process to sell goods to customers. The MTS concept is slightly different. MTS estimates how many orders will be generated and then produces enough to meet the orders. MTS only manufactures products based on demand forecasts.

MTS is one of the most efficient models in the supply chain. As soon as the replenishment of finished goods are completed, new stocks come into the inventory so that the next set of products are ready as soon as the demand from the customers come in. This process keeps happening. Everything is tied together in a process where the direct procurement team (direct procurement is the team responsible for goods or services directly associated for production of goods and services by the organization) is notified when to buy raw materials and how many to buy. It is very important to predict the demand requirement in the market. Amazon and Starbucks are great examples where they have perfected this system.

MTS is often used when there is high focus on inventory management control and ultimately there is a high focus on cost management. This is possible due to the development in technology and tools which captures customers' needs and can send the details from customer requirement system to production management system immediately.

2.5.2 Build to Order

Built to order is a production method where products are built only after an order is confirmed. In this case once the final customer orders the products and determines the time they need the product, the organization starts building the products.

As soon as the customer order comes in, the company starts assembling. Inventory management and control could be extremely difficult in this case and one way to avoid that problem is to use as many common components and subcomponents as possible. Dell and NetApp are great examples of business-to-business transaction (which includes servers and storage) companies who use this method.

Build to order is used when the focus is on customer need and timing is important. A good example of this scenario is the business-to-business division of Dell, where Dell is able to send laptops to the schools only when the schools need laptops during the beginning of the new session at schools. This is possible due to communication and information sharing within organizations where the sales team in Dell and supply chain teams are in close contact with each other.

2.5.3 Constant Replenishment or Continuous Replenishment

This is a supply chain strategy where replenishment happens continuously. In order to maintain better supply chain flow, constant replenishment takes place from supplier to distributor to retailer. Oftentimes the inventory is managed by the vendor (vendor managed inventory, VMI) and the decision of when and how much inventory to replenish comes from vendor and not necessarily the distributor or the retailer. This process needs agreement between all parties and replenishment is very frequent.

It's a supply chain strategy in which frequent replenishment takes place from the supplier to the retailer or distributor in order to maintain better flow in the supply chain and minimize bullwhip effect. It's a kind of VMI system where the decision of quantity and time to replenish lies with the supplier and not the retailer. But such moves need agreement between supplier and retailer. Also continuous replenishment is a misnomer; it is not continuous but frequent replenishment.

Constant replenishment takes place when inventory is constantly replenished based on the demand. This requires a very close relationship between suppliers, company's internal subgroups and systems. A good example of this would be prescription medicine providers like CVS.

Continuous replenishment takes place when there is a focus to increase all round efficiency. Constant replenishment is possible due to flexibility within the supply chain. CVS is a master of this where they have facilities built with the ability of increasing or decreasing production faster than many other organizations.

2.5.4 Channel Assembly

Channel assembly is the process where products are accumulated as the items travel around the various channels. A good example of this process is items first getting accumulated at courier companies FedEx or UPS and once it reaches a critical mass, they are sent to the final destination.

Channel assembly could be seen as an extension to the 'Build to Order' process. The individual components are gathered and assembled at various steps of the distribution process. It could mean physical assembly or partnerships with third party logistics firms. An example of this process is Tesla where they buy different components from different suppliers and final assembling happens when all individual components come together.

Channel assembly is used when there is a cost focus and is possible due to the partnership of external and internal players within the supply chain. Amazon is known for managing customer delivery where they often use a partner to gather all the purchases and carriers like UPS or FedEx deliver all products on the same day. Amazon also gives an option to send all the products to customers one day a week, and if the customers agree to do that instead of getting separately, the channel assembly method is utilized in those scenarios.

2.5.5 Real-Time Change of Supply Sources

Based on assessing future demand and potential future release, a company sometimes builds an ability to increase capacity or have an alternative supply source to retain its own or increase market shares. A number of retail companies like Target and Walmart have built this capacity to satisfy increased demand during specific retail events.

Let's take an example of how retail stores like Walmart and Target use the holiday shipping season and hire temporary workers to increase the sales from the higher demand. Each retail store has specific strategy and based on the customer demand, stores have a variable workforce to meet the demand and the workforce is managed easily since the temporary high selling season is managed by temporary workers. This example relates to managing cost and companies like Walmart and Target are able to manage well due to the analytics or data that they have collected in the past years and determined the trends for the future.

2.5.6 Turnkey Contract or Order to Make

For execution of a big contract job or a supply of big machinery and equipment, a bill of materials (BOM) is prepared first and then critical path analysis (CPM[*]) is created for assessing the requirement of materials and spare parts that need to be used.

The next procedure and order is initiated after that. Then the assembly exercise or execution of the jobs takes place in the right sequence. The supply chain procedure for this type of industry is very different from normal regular supply chain activities.

Turnkey contracts are popular when there is a high focus on lead time management. One of the principles that's used in this is technology-based improvement in supply chain. This process is very commonly used in oil industries when they are building a new oil line. Companies such as Shell first create BOM and determine CPM. Once the companies know how much

---

[*] CPM is a project management technique that maps key tasks that are required to complete a project. This process identifies the amount of time necessary to finish each activity and the dependencies of each activity on others. CPM is heavily used in supply chain processes to understand the dependencies between tasks.

the total cost will be, they take the next step. They get partial payment from the customers and then start working on the project.

2.6 PRINCIPLES OF SUPPLY CHAIN MANAGEMENT

In Section 2.4 we discussed the focus areas and in Section 2.5 we introduced the production and supply chain strategies. Focus is on value addition or unique differentiator, and in order to achieve it inefficiencies need to be removed from the execution of the production strategies. This section discusses the methods of achieving that goal. The methods are referred to as principles of supply chain. Focus of the supply chain is the value addition, and principles are 'ways' or 'methods' to deliver the value addition in the production and supply chain processes. We described various types of supply chain and production processes above and the principles are various steps that organizations take to get the desired results.

We will be reviewing the supply chain principles which are useful both in a production-based and a service-based organization.

The principles described below are developed from past supply chain learnings, success of other supply chain methods and organizations that have built successful supply chain principles and processes. We will try to explain the principles with examples so that the reader can understand the principles and the application. Another way of thinking about this section is that these principles are guidelines and could vary from one organization to another.

2.6.1 Focus on Customer Need

Supply chain management starts with understanding who the customers are and why they are buying the product or service. Whenever customers buy, they're either solving a problem or filling a need. Supply chain managers must understand the customer's problem or make sure that their companies can satisfy customers' needs better and cheaper than any competitors can. Customer focus covers a lot of ground and can be different based on customers. Companies have also realized sometimes customers don't know what they want and organizations can provide that added benefit to customers.

A good example in this regard is the car manufacturing company Tesla. Tesla manufactures cars where a car gets enhanced functionality with software updates, and it's a very unique model which other car manufacturing companies didn't think of (at least initially). Tesla CEO uses social media heavily to interact with customers and drive customer focus and value with software updates which makes customers surprised and happy. Another advantage with this model is the fact that if customers don't like some features, the next software update can actually make tweaks or even scale back on the previously released features.

2.6.2 System Thinking in Supply Chain

Supply chain management requires an understanding of end-to-end systems – the combination of people, process and technologies that must work together so that product or service can be duly provided. System thinking involves an appreciation for the series of cause-and-effect relationships that occur within a supply chain. Since they are complex systems, supply chains often behave in unpredictable ways and small changes in one part of the system can have major effects somewhere else.

Let's explain with an example. In a complicated supply chain system where there are a number of suppliers responsible for providing components separately, one change from one supplier or delay can jeopardize the entire system. Let's take automobile manufacturing as an example. The delay in setup or shortage of one part can cause the entire supply chain to fall through, so it's important to see the entire picture end to end and have all components in place.

2.6.3 Disruptive Innovation in Supply Chain

The world of business is changing quickly and supply chains need to keep up the innovation to keep pace with competitors.

Continuous process improvement is not always enough to really bring in change. In comparison a new technological advancement can really disrupt an industry in a major way. This effect is called disruptive innovation. When a new solution for a customer's needs emerges and becomes accepted, this solution becomes the new dominant paradigm.

Let's understand the above concept with an example. A good example is that of video streaming. In the US and all over the world, cable and satellite TV along with video rental stores were in existence for a long time. Video streaming services were at the initial stage of development in the late 1990s but rose very quickly and disrupted the way people watch shows and movies. Netflix has become the largest subscription provider in the US now.

Companies like Blockbuster LLC, who were pioneers in the movie renting business, did a lot of improvement in processes, and the improvements did provide a better customer experience but disruptive forces like Netflix completely changed the way movies and videos are watched by customers.

2.6.4 Collaboration between Verticals within Organization

Supply chain management can't be done in a vacuum. People need to work across silos inside an organization and also with suppliers and customers outside the organization. A self-centered mindset leads to transactional relationships where people focus on short-term opportunities while ignoring the long-term results. This actually increases spend in the long run because it creates a lack of trust and an unwillingness to compromise among players in the supply chain. An environment in which organizations trust one another and collaborate for shared success is much more profitable for organizations than an environment in which each person or group is concerned only with their own success. Also, a collaborative type of environment makes working together a lot more enjoyable.

When the sales and marketing team of an organization promises the features and product functions to customers without discussing with the operations or when operations team tries to simplify operations without inputs from engineers, the organization suffers since either products don't meet customer expectations or leads to situations where quality takes a hit. These problems can be avoided if all groups work together with a common goal to provide the best in class products for the customers.

In the past, this has happened real time in many different sectors and we can take an example from the technology and marketing sector. There was no discussion between product and marketing departments. As soon as the product team made a customer-facing update both online and in-app without taking input from the marketing team of the brand, customers started rejecting the update and the product. This could have been avoided if the teams would have talked to each other.

2.6.5 Flexibility of Supply Chain

Surprises happen, and that's the reason supply chain processes need to be flexible. Flexibility is a measurement of how quickly the supply chain can respond to changes, such as an increase or decrease in sales (or change in supplies). This flexibility often comes in the form of extra capacity, multiple sources of supply and alternate forms of transportation. Usually, flexibility costs money but also has a value. The key is understanding when the cost of flexibility is a good investment.

In one scenario, let's say one company makes widgets and needs to buy one thousand raw materials per month. A more cost-effective option could be to buy all the raw materials from a single supplier, which would lower the supply chain costs due to economy of scale. But it might be a problem if the incumbent supplier experiences a natural disaster like flood or fire which affects the production in the factories, or files for bankruptcy and cannot make the raw materials for a while.

Even though the initial purchase price was lower for the widget making company, the risk was high and there was no flexibility. In other words, having a second supplier would have provided flexibility.

We can think of the extra cost paid to the second supplier would work as an insurance policy. By paying more up-front to have that insurance policy, the flexibility of the supply chain is increased.

2.6.6 Corporate Social Responsibility for Manufacturers

Demands for corporate social responsibility (CSR) in labor practices and environmental stewardship are also having an impact on manufacturers and producers. With instant worldwide communications, it's hard for any company to hide from what they or their suppliers are doing. The fallout from poor supplier practices in apparel and consumer electronics has damaged brand reputations and brought renewed calls for companies to improve their corporate supplier responsibility programs and supplier oversight responsibilities. This trend toward greater social awareness and demand for sustainable business practices is expected to grow. Greater emphasis will be given to the sourcing process and supplier relationship practices.

CSR is not just a buzzword anymore. Companies have seen returns in increased profit from revamping or changing their CSR models. A number of companies still think CSR is tied to increase in profit but the most successful companies do not think of CSR as a method to generate more profit but profit could be a by-product of CSR activities. A good example is that of FedEx where they introduced a robust CSR process that saw them going to alternate sources of energy which not only created a new dimension of advertisement for the company but helped them to save cost in the long run.

2.6.7 Technology-Based Transformation in Supply Chain

The rapid evolution of technology, for moving physical products and for processing information, has transformed the way that supply chains work. Even a few years ago, if an item was selected from a catalog and the order was placed along with the payment, goods were delivered after that. Right now, orders may be placed on phones or emails and the payment could be completed electronically and the goods may be received immediately thereafter. Supply chain management requires understanding how technologies work and how to use them to create value at each step in the supply chain.

The impact of technology on supply chain operations and management has already been enormous. Without the advances of the past two decades, vast global supply networks would not exist.

2.6.8 Communication and Information Sharing within Organization

Enabled by the internet, the communication and information sharing technologies will continue to grow. The new systems available today have changed the fundamentals of how organizations can be structured. Rather than employing full-time staff, companies can assemble available resources from anywhere in the world on an as-needed basis. Virtual teams will become a great option that would also include collaborators from supplier firms. The recent COVID situation has already pushed companies to embrace technology and determine how businesses can be run virtually.

2.6.9 Analytics Used in Supply Chain

Making sense of big data will improve supply chain and logistics problems. Current technology is capable of capturing and storing vast amounts of data and running machine learning models on the dataset. This includes structured data, such as point of sale (POS), records and retail loyalty-card transactions and unstructured data, such as that from social media and web search logs. However, data are meaningless until information can be extracted from it. Organizations, such as SAS (originally Statistical Analysis System) and IBM, along with many others, are actively competing in this space.

Data mining and use of predictive analytic tools have the potential to vastly improve the accuracy of forecasts and produce results in real time. Once meaningful information can be extracted about trends, market conditions and consumer behavior, supply chains can potentially be reinvented to be much more responsive.

2.6.10 Disruptive Innovations in Product Technology

Disruptive innovations cannot be predictive. They appear and may launch new markets or change the value proposition for existing technologies. The iPad and iPhone relied on improvements in existing microprocessors, miniaturized components, manufacturing and software, but their introduction altered the landscape for handheld devices.

The internet is now mobile and the ways in which this new customer interface will change business practice have only begun. The consequences will be felt throughout the supply chain as customers locate products, make orders and payments and demand delivery wherever they are.

2.6.11 Warehouse Management for Storing

Warehouse management will continue to become more efficient through the use of sophisticated software applications and automated storage and retrieval technologies. The increased use of robots and the further expansion of RFID into the supply chain have already and will increasingly transform physical distribution.

2.6.12 3D Printing Process for Prototype Development

Additive manufacturing or 3D printing has been used to make prototypes for over two decades. It is a process that can create a three-dimensional product from a computer image. While the replicators are not yet of Star Trek fame, these machines have the potential to replace some commercial production in the future. On the industrial scale, manufacturers may offer internet-based processes that allow customers to design and create truly individual products.

2.6.13 Global Perspective of Supply Chain Process and Systems

The ability to share information instantly and move products around the world with the least cost means every company today operates in a global marketplace. No matter what product or service is required, companies are global. Supply chain managers need to recognize how the business depends on global factors to supply inputs and drive demand for outputs and think globally about the competition. After all, the company's real competitive threat could be a company never heard before and on the other side of the planet.

Organizations are not located in one geography anymore – they are physically present in different parts of the world, source materials from suppliers who are at different parts of the world and sell products to customers who are at different parts of the world. As a result of this, the operations of the organization need to be different from an organization where the scope is limited to one geography or location. Natural disaster of one supplier could impact the organization's ability to build the product and thus the ability of sourcing from multiple suppliers is a 'must' now. A sudden change in geopolitical situations in the country where the majority of the customers are located could hamper the growth of the organization, so it is important for the organization to think about other sources of revenue from other countries or other sources of revenue from different products.

2.6.14 Impact of Economic Conditions of Various Countries

A number of nineteenth-century supply chain strategies were designed to drive cost down and many of those involved outsourcing to low-cost countries. Nevertheless, the global business environment has changed and will continue to do so. BRICS (Brazil, Russia, India, China and South Africa) are developing closer economic relationships with developed countries. With greater than 40% of the world's population in BRICS nations, supply chain managers will need to determine the past, current and the future impact that these nations have (will have) on global commerce.

On another side, the market opportunities are expanding in these developing economies, but at the same time the labor costs within the countries are rising. Strategies that have only embraced the use of low-cost labor will need to change. The rationale for where companies have production facilities will need to be determined by local demand, and companies will need to manufacture and source both domestically and offshore.

This is applicable in the medical industry where drug manufacturing companies source a lot of raw materials from China. However, the current US-China trade tensions have made organizations worry about future sourcing from China. Also the labor cost in China has been increasing and organizations are not getting the benefit of lower labor cost which they were able to enjoy before. Another worry is the safeguarding of intellectual property which is a big factor in determining the countries and locations where the organization wants to partner with.

2.6.15 Risk Management and Mitigation Strategies for Buyers

When a company combines high-performance requirements with complicated technologies and there is a very high dependency on global customers and suppliers, the risk increases. A lot of variables are introduced and a number of issues can happen. Even a small disturbance in the supply chain, such as delay in shipment, can lead to a series of problems further down the supply chain, such as stock outs, shutdown and penalties. Supply chain teams should be aware of risks, and implementation processes should be predefined to detect and mitigate the threats. Stability may be key to making supply chain work smoothly, but risk management is key for avoiding or minimizing the costs of dealing with surprises. If planned well, risk management can provide opportunities to capture value during times of uncertainty.

This goes back to the system thinking section and the example of automobile manufacturing companies where parts are very important. As an example, if the tires are provided before the main chassis, the tires will occupy warehouse space and provide additional cost for storage. This actually goes beyond just speed of the production but goes deeper into determining availability. In our example, if the chassis is not available or there is a delay, that causes the automobile organization further delay.

2.6.16 Political Conditions and Navigating
Through Political Changes

Anticipating risks and planning to manage risks must be an essential element in supply chain strategy development. Unfortunately, not all sources of risk can be predefined or known early. For instance, the consequences of political events and governmental decisions can alter the landscape of business overnight. This means that some companies will need to rethink and redesign their supply chain and operation to accommodate the unanticipated changes and emerging challenges. One possible unexpected adjustment identified by the World Economic Forum (WEF) could be nontariff measures. Another example could be a change in political situation where one country falls out of favor with respect to another country. Supply chain systems need to be built with the ability to change direction when such situations occur. This brings in sudden risk for the organization.

2.6.17 Visibility of End-to-End Supply Chain and Catching Problems Early

Visibility is an important factor in supply chain management. The success of operations depends on the vision of the supply chain manager where supply chain managers can foresee the problem(s) in advance and take due measure to protect the company by taking advanced adequate measures. The decision needs to be taken on data and facts and not on estimation or assumption. Having better visibility on the demand and supply will result in better managed inventory and reduce cost.

One of the requisites of a good supply chain manager is to catch problems early. Early detection of an issue saves the organization a lot of money. Detection actually is both for short term and long term – anything that might impact the organization. Certain disruptions are long term, for example, quality of teaching in developing nations. Some are time-scheduled events like new regulation coming in at certain times and some happen with short warnings such as floods and hurricanes, but certain natural calamities might happen without any warning at all like fires or earthquakes. Some of these are difficult to foresee but a supply manager shines if they are able to catch design defects or contaminations in products early in the production phase. This saves an organization a lot of money.

2.6.18 Value Creation within Supply Chain

Supply chain management is about creating value – meeting customers' needs at the right place, at the right time, at the right level of quality and for the lowest cost. This value is the heart of supply chain management. One principle to describe the whole process of supply chain management is value creation.

Supply chain value addition is heavily dependent on determining the right focus for the organization, its suppliers and its customers. Once focus is determined, production methods need to be analyzed which will help to achieve the desired focus and that's how the organization adds value.

2.7 CONCLUDING REMARKS

Supply chain has come a long way from where it started and supply chain processes and systems have to keep innovating to keep organizations relevant. Even though the focus areas are managed by the team under supply chain management, however, almost all the problems are business problems that the organization needs to solve. The implication of not solving the problems is much more broad than supply chain; it impacts even operations and the organization as a whole. Supply chain principles help to achieve the focus area goals, and after applying the underpinning principles on their production and supply chain processes, organizations will be able to achieve their focus area or the broader goal in creating successful, distinct, long-term products as desired by their customers.

REFERENCES

Mumbai Dabbawala; 2019. https://mumbaidabbawala.in/about-us/ (Accessed January 1, 2021)

Mark J; 2018. Silk Road. World History Encyclopedia. Horsham.

Biography.com Editors; Marco Polo; 2021. https://www.biography.com/explorer/marco-polo (Accessed January 4, 2021).

Institute of supply chain management; Supply Chain Timeline; 2018. Newcastle upon Tyne.

Hyde R; 2019. How Walmart Model Wins With "Everyday Low Prices". New York.

Anderson D; 2008. Build-to-Order & Mass Customization; The Ultimate Supply Chain Management and Lean Manufacturing Strategy for Low-Cost On-Demand Production without Forecasts or Inventory. 1st edition. C I M Pr.

3

The Solutions that Business Wants

Sourya Datta

CONTENTS

3.1 Objective...36
3.2 Significance of the Early Supply Chain System 37
3.3 What Did We Learn from Supply Chain Methods and Processes.......39
    3.3.1 Create a Long-Term Strategy – Business Problems
         and Opportunities 39
    3.3.2 Multiyear Roadmap 40
    3.3.3 Where and When to Make Trade-Offs...................................... 40
    3.3.4 Flexibility...41
3.4 Lean Supply Chain (Packowski 2013) 42
3.5 Partners in Modern Supply Chain 45
3.6 Customers – Internal and External... 46
    3.6.1 Internal Supply Chain Management Functions 47
    3.6.2 Marketing (Roylance 2008) 47
    3.6.3 Finance 49
    3.6.4 Production (Operations in Service Companies)........................51
    3.6.5 Logistics...51
    3.6.6 Information Technology 51
    3.6.7 Engineering..52
    3.6.8 Supply Management 52
    3.6.9 Other Partners – Procurement Subgroups and
         Collaboration with Legal 53
         3.6.9.1 Supplier Selection 53
         3.6.9.2 Make-versus-Buy and Outsourcing Decisions........... 54
         3.6.9.3 Global Sourcing..55
         3.6.9.4 Supplier Relationship Management.............................. 56
    3.6.10 Legal and Role in Supply Chain..57

DOI: 10.4324/9780429273155-4

3.7 External Supply Chain Members..58
    3.7.1 External Manufacturer or Producers..58
    3.7.2 Suppliers 59
    3.7.3 Transporters (Transportation Service Providers) 59
    3.7.4 Wholesalers and Distributors...59
    3.7.5 Retailers..61
    3.7.6 Customers 61
3.8 The Strategic (Preferred) (Segel and Shay 2003) and
    Transactional Suppliers..62
3.9 Concluding Remarks...63
References.. 64

3.1 OBJECTIVE

In the previous chapter we explored why we need a supply chain. It is needed not only for producing goods and services that an organization develops for its customer but also for creating unique value for its customers, by making the production process efficient. An organization achieves this efficiency goal by creating processes based on the supply chain principles. However, the supply chain is not narrowly used for just fixing the inefficiencies. Instead, most of the problems that the supply chain organization solves are often general business problems and that's where the supply chain team shines. It's not always specific to operations and supply chain-specific issues but more generic business issues. We will take the readers through a few examples to explain these business challenges in this chapter.

These solutions that the supply chain brings to the table are not work of an individual or done by a single team in isolation. It takes a village to implement and manage the supply chain processes and create solutions from end to end. This chapter talks about that village or in other words the people in the supply chain ecosystem. At the end of the day modern enterprises are a collection of people. The ecosystem consists of all the various groups of people that the supply chain team members work with. This chapter will cover the external partners in the supply chain process along with internal collaborators within the enterprise. We will look at the various functional organizations, teams and people in the ecosystem. This chapter will also touch on the concept of designing lean supply chain as a method of solving business challenges together with the ecosystem partners.

In order to understand the people involved we need to understand the business challenges that supply chain solves. In that vein, let's re-inspect the familiar Silk Route supply chain one more time and understand the business challenges it faced. How the ancient traders applied the supply chain principles that we explored in the previous chapter. Understanding the business challenges the supply chain addressed for them would help us understand the stakeholders they solved it *for* and solved it *with*.

3.2 SIGNIFICANCE OF THE EARLY SUPPLY CHAIN SYSTEM

Whether its Marco Polo or Silk Route or the Dabbawalas or any form of supply chain that occurred in the past, the basic principles of inventory management, reducing inventory carrying cost and delivering the right products to the customers in the most streamlined fashion, haven't changed over time. The focus on lead time, cost, accuracy, quality and efficiency were applied to the supply chains mentioned before and applicable to any modern day supply chain as well.

Supply chain management is the planning and coordination of all people, process and technology involved in creating value for the company. Managing a supply effectively involves coordinating all of the work inside the company with everything that is happening outside the company. In other words, it means looking at the business as a single link in a long, end-to-end chain that supplies something of value to the customers.

Following the supply chain of the past, some of the concepts have been incorporated in the modern supply chain as well. Some of the disruptive innovations are now part of the modern supply chain. Flexibility has been improved and so has collaboration, but before we go to the modern supply chain let's look at the significance or impact that early supply chain had. Below are a few important aspects of the ancient supply chains.

a. Streamlined multiple components in the supply chain: There was a distinct advantage of supply chain with streamlining trading posts, markets and transportation. Exchanges, distribution and storage of goods were simplified in the existing supply chain. The distribution was planned in such a manner the goods were off loaded on a point-to-point basis. Alternately, the parts of the total consignment were offloaded sequentially to avoid haphazard and zigzag movements. Let's relate this to a current scenario. Walmart is a very good example of making this process very efficient. Walmart trucks deliver goods to the Walmart stores and find the most optimal path for delivering goods to stores. If there is a possibility of delivering to stores beyond the normal route, another vehicle takes up that delivery process. Not only does Walmart deliver goods to stores in a streamlined fashion, but Walmart has built most of their stores very close to the freeway which keeps zigzag movements of delivery trucks to minimum.

b. Distance covered: The supply chain of the past covered a lot of areas. From ancient times, Greece, Rome to modern Iraq were part of the earlier trade route. Also routes covered Iran, Turkmenistan to modern Afghanistan and to Mongolia and further to China. The vast length was unique which helped in designing the modern sea and land transportation routes. In the past, when the companies had export orders to various other countries, the route was decided after getting entire information from shipping companies. The company used to follow the same route for offloading the export consignments and informed clients in advance about the tentative schedule of arrival of consignments at the respective place.

c. Items traded: There were a number of items that were traded. Early supply chain saw a number of items such as vegetables, fruits, livestock, grain, leather, tools, artworks, stones, religious objects, metals, paper, even gunpowder and everything for daily livelihood transported via the road. Damages or losses were avoided due to frequent shipment of the goods instead of one bulk shipment.

d. International traders supplying materials to all over the world also set up various stock points to cater prompt and efficient service from these stock points to the nearby prospective customers to reduce inventory carrying time. This helped the end user to get the materials in a short time and simultaneously build up confidence and dependency on the supplier.

e. Consignment stocking system: It is surprising but at that early stage, businesses took advantage of consignment stocking concepts where materials were stored temporarily when there were options to store items cheaply. This concept was also utilized when the price of the products would vary over a period of time. Buying and storing goods could be cheaper when there was an abundance of goods as compared to the time when there was limited availability. These goods were taken out from storage and sold when there was a higher demand.

3.3 WHAT DID WE LEARN FROM SUPPLY CHAIN METHODS AND PROCESSES

The early supply chain taught us a number of methods even though they were not called such. Most supply chain systems were following those principles already. The expectation from the supply chain is to provide the right products to the customers at the right time with lowest cost.

In most organizations today, approximately more than 10% of the total cost arises from the supply chain (Kristensen 2020). A very high percentage of revenue for a manufacturing company is also dependent on how the supply chain is managed. Keeping all of the parts of the supply chain aligned is key to running the business successfully.

A number of factors could impact supply chain efficiency in the short and in the long run as well. This made long-term planning virtually impossible. The factors such as commodity price swings, natural disasters and financial meltdowns due to wars definitely had direct impact. Prior logistics, supply chain helped in reducing the inventory carrying time and cost by being strategic in the transportation of goods. Supply chain ecosystem was mainly based around on in the earlier times, and early methods of transportation showed the importance of key concepts. The key concepts defined below are applicable in today's supply chain as well.

3.3.1 Create a Long-Term Strategy – Business Problems and Opportunities

In order to be successful, it is important to think about the long-term strategy and create a model based on that. The focus needs to be on the customer's needs and wants and if that focus is maintained, then most of the time it will be a successful supply chain and operations.

Frequent product changes will disturb the set up. The early Silk Route understood that the logistics model developed will be applicable for a long time and it is easy to recover the modern day 'depreciation cost' and recover return on product research and other investment costs. Let's explain the above concept. Right now when an organization builds factories or warehouses, they invest money, and over the years there is depreciation of the factories and warehouses and the value decreases. It's the same concept of that of a car. When a new car is bought, it is much more expensive than the same car in 5 years since there is depreciation associated with it. The older car will start having defects and will need regular maintenance, and that is the reason the value drops. Similarly every year there is a certain cost that is associated with the depreciation of the factory. Now organizations sell products and the revenue they make is higher than all cost, and depreciation cost is one of them. As long as the revenue is higher than the total cost, the organization makes a profit. The Silk Route was also based on the same concept. The process made enough money which they were able to cover. Obviously current organizations need to spend on other aspects such as research and development and product development, but the goal is to make more money than all costs added together, and the early supply chain was able to do that effectively.

Any good profit generating company falls in this criteria.

3.3.2 Multiyear Roadmap

In order to develop a supply chain ecosystem which can sustain for a long period of time, it's important to look ahead and plan ahead. The Silk Road was an example of building a road system which can sustain for many years. Any system that works for a long time has a distinct advantage that the work will flow in its own way and inertia will help to continue the process longer. There will be always scope of more improvement but that improvement can happen while the current system is utilized.

A number of companies in the US had made long-term strategic initiatives and roadmaps with production in China. The current US trade regulation has caused a number of companies to think about their future strategy. As trade relationships change, companies who had alternate sources of either acquiring raw material or distributing final products are facing less trouble than companies who are sole sourced from China. This is very evident in the medical industry where US companies acquire raw materials from China.

3.3.3 Where and When to Make Trade-Offs

Supply chain is very complex and there will always be scenarios where trade-offs will be needed. It's important to understand which elements are more important. To corelate to the modern world, the current supply chain should also determine the value and question if the existing process that is working well needs to change.

This concept is evident in manufacturing companies such as Cisco where the company is extremely careful to change any process in the manufacturing if there is a lot of risk or uncertainty in the proposed modification in production and logistics process. Cisco likes to do small improvements and capture the benefits instead of changing a streamlined process.

3.3.4 Flexibility

Higher flexibility in the supply chain ecosystem increases cost. The additional cost is generally borne by the party who is requesting the flexibility. This is mostly applicable in the case when the customer needs the product. Oftentimes in the supply chain ecosystem, the cost is borne by both parties – seller and buyer – and the split percentage is determined between the two parties.

This is evident when the company tries to bring in an alternate source of supplier to increase the flexibility. The company needs to spend more money in order to bring an alternate supplier.

Let's take an early supply chain example and a current supply chain example to drive the point. At present, company A is only sourcing raw material from one supplier (supplier B), and it pays $10,000 for service every month. That service could be just for support of internal infrastructure for company A. The support consists of supporting the entire infrastructure software in our example (even though one supplier supporting entire infrastructure software is very difficult). With the present model company A is getting support for the entire infrastructure base at a fixed cost. Company A is worried about the performance of supplier B and wants to bring in supplier C for supporting a small subset of the infrastructure software. When company A does this, it doesn't necessarily decrease cost from supplier B since they were already supporting enterprise software at fixed cost. But now company A has to spend additional money on supplier C since they need to be compensated for the support work. This is a classic example of how the additional cost is paid by a company who wants to bring in flexibility, and in this case flexibility is defined as an option between vendors B and C. The early supply chain also followed the same model. When the service is requested from two vendors, it increases the overall cost and the party asking for flexibility is responsible to pay for that.

Apart from the above, in addition to the ancient, supply chain, we also learnt a few more concepts even though they were not exactly termed or called as such.

For better supply chain management and reduction on cost, substantial stock reduction will have a direct impact on reduction of working capital and inventory. This doesn't stop here and the release of working capital also helps to improve the bottom line. Modern term for this process is **reduction of working capital and interest cost.**

A good supply chain ecosystem will try to minimize the import duty and try to pay the import duty in various parts or phases. Another method is to import bulk consignments which hold the price at a discount. This process has been used from the Silk Road times (even though it wasn't called import duty). The concept was still used where bulk consignment was moved from one location to another. This is actually an **effective storage of items.**

Market survey helps the company get an indication of requirements which provides the details behind the stocking plan. This also helps in budgeting and creating a roadmap for the company as well. Market surveys have stayed constant and an important part of the supply chain ecosystem. Current method will be called **a market survey.**

During the early days of Silk Road, inventory was managed manually but with the advent of scanners and robots, managing factories has become much easier. Any good supply chain ecosystem has all the technical capabilities to manage inventory. However, the technology development didn't happen in one day. As we discussed before, there are negligible errors in the Dabbawala system due to a process of numbering and color coding introduced for efficiency. The current technology development is an extension of that. This will fall under **introduction of technology.**

3.4 LEAN SUPPLY CHAIN (PACKOWSKI 2013)

Another evolutionary step in modern supply chain management is the introduction of a concept called lean. Lean supply chain is a principle and can be used by any company that wants to streamline processes and eliminate waste.

There has been a lot of discussion on lean and the origin of lean. It is believed that lean manufacturing was first used by car manufacturing company Toyota in 1930. Toyota coined 'Toyota Production system' which was followed around the world. However, it is also believed that the history of lean started in the 1400s and the first company who truly integrated lean was Henry Ford. In 1913 Henry Ford modified the flow of production by interchanging different parts.

Let's look at an example to get a sense of how the principles of lean evolved. We will take the example of Nike (Distelhorst 2016) and how Nike used lean principle to modify the organization to where it is now.

Nike was facing a lot of issues with labor problems in the 1990s. Nike decided that board member Jill Ker Conway will travel to the contract factories in Southeast Asia to experience and tackle the labor issues first hand. Jill Ker Conway soon realized that there were a lot of 'wastes' in the factories and the conditions of the factories needed to be improved. The relationship between workers and management were at all-time low. So she first determined the 'flow of goods in the factories first', and in this case, the flow of goods was actually workers and their relationship with management. She understood the issue and was open in her discussion with both management and the workers. Basically she performed 'Show Customer's usage data and pattern to all supply chain partners'. In this case customer's usage data are the management's view of the workers.

She created a partnership between the International Youth foundation so that they could survey sixty-seven thousand female workers in the Nike contract manufacturing factories. As a result of this, a lot of data were gathered and it showed the variation in supply chain (Goal: 'Reduction of variation in the supply chain'). She also realized that there is no 'Collaboration' and 'discipline' in the process, and due to a lot of dissent, Nike was spending a lot of money which it didn't need to spend ('Important to minimize "Total cost" of supply chain'). She also realized that there were few bad players on both workers and management sides which could be seen as non-value adds to the system ('Avoid long time inventory carrying costs. Eliminate if not adding any value but avoid stock out situation'). Conway decided to change the entire structure of Nike factories in Southeast Asia. Nike started appointing a smaller subset of managers to be responsible for one factory floor instead of the entire location so work was divided into zones and nonperforming team members were let go ('Foresee correct demand supply situation. Reduce nonmoving or obsolete stocks'). Nike created a training facility in Sri Lanka for training managers. Also the physical layout of the shop floor for various factories were changed ('Factory or warehouse layout is very important for movement and storing of goods). Nike started doing certification of stocks, increased salary for workers but created a strong system where any theft will be punished strongly and if caught will lead to termination of employment ('Proper codification for all stocks, e.g. Raw materials to finished goods with all stages'). This also saved Nike cost in the long run ('Reduction of working capital which ultimately helps improve bottom line') and helped in 'Improvement in logistic management'. Nike also created the plan where managers attending training will cross-train other managers in other parts of the world so that managers could understand the culture and be cross-trained for different geographies. ('Synchronizing the supply chain functions with other activities'). This improved the overall value in the supply chain ('Value creation'). The overall process resulted in 15% reduction in labor noncompliance. Brining in new methods and technology was an integral part of making the process successful ('Technology').

In order to have a lean supply chain in the ecosystem, a number of improvements in the supply chain processes need to take place. Modern lean supply chain system has the following features which makes lean successful.

- Eliminate all wastes to make sure only 'true value' remains
- Determine flow of goods to each concerned party/parties
- Show customer's usage data and pattern to all supply chain partners
- Lead time reduction
- Use of 'Pull system'
- Increase in velocity and throughput
- Reduction of variation in the supply chain
- Collaboration is not an option but key to be successful
- Use discipline in process
- Important to minimize 'Total cost' of supply chain
- Avoid long time inventory carrying costs. Eliminate if not adding any value but avoid stock out situation
- Foresee correct demand supply situation. Reduce nonmoving or obsolete stocks
- Factory or warehouse layout is very important for movement and storing of goods
- Proper codification for all stocks, e.g. raw materials to finished goods with all stages
- Reduction of working capital which ultimately helps improve bottom line
- Synchronizing the supply chain functions with other activities
- Improvement in logistic management
- Value creation
- Improvements in technology to improve supply chain

There are a number of benefits of a lean supply chain (Myerson 2012). Designing supply chain is very challenging and if designed incorrectly it hurts the organization by increasing the cost.

Generally the design will vary from company to company and place to place. Lean supply chain helps both the organization responsible for creating the supply chain and the customer. When designed correctly, there are advantages to the organization side, stock view on both organization and customer side and the supplier and buyer side.

3.5 PARTNERS IN MODERN SUPPLY CHAIN

So far we tracked the evolution of the supply chain. We discussed how supply chain principles are applied and supply chain objectives are defined to meet overarching business goals.

We learnt the business problems and how supply chain principles could be applied to solve the problems. We also explained the concept of lean with a real-life example. Let's now focus on the models of supply chain. Since the models of supply chain have evolved, it is important to understand the various models before we go to one of the most important parts of the supply chain – people. In this section, we will analyze three models in their order of complexity and then we will analyze who are the internal and external stakeholders of the process.

3.5.1 Supply chain systems started with a very simple structure as below. The process started with a single supplier, an organization who took delivery of raw materials from the single supplier and then delivered the final product to the customer. This was a one directional flow where the organization just sourced raw materials, built a product from the raw materials and finally delivered it to the customer.

A good example is a small shop that is using only one vendor for raw material and then selling the final product to one customer.

3.5.2 The supply chain system slowly extended and started expanding where the levels of suppliers increased on one side and the end customers expanded on the other side.

Supply chain started bringing more players in the domain. The level of suppliers increased from just one supplier to supplier's supplier on one side and on the other side, it also extended to customer's customers. If you see in the below chart, oftentimes, underlying service by a service company is provided to all levels in the supply chain.

A good example of this model is Dell. Dell is getting components for laptops and uses a number of suppliers for raw materials and the final laptops are sold to retail chains such as Walmart who are one set of customers and at the same time customers can directly buy from dell website.

3.5.3 The supply chain did become more complex and complicated and started involving more entities.

In the above example, if we look closely the model is very complicated. Lot of teams have been added now in the supply chain process. Logistics, Finance, Legal and Sales from not only the buyer but also the supplier(s) and final customers are involved now. Buyer is doing the design in-house and the implementation of the design is done by the supplier(s). Buyer is doing the market research and sometimes is also responsible for feasibility testing along with the suppliers. Suppliers are not just one entity anymore but now consist of initial manufacturers along with distributors.

A good example of this type of supply chain process is medical device companies such as Intuitive. Intuitive is designing the robotic arms and surgical systems and the implementation is done by its suppliers. Intuitive is doing the market research to determine which products they should focus on. Extended support teams such as legal and finance are involved in Intuitive's suppliers, Intuitive itself and Intuitive's customers. Intuitive is also doing studies with patients and hospitals directly instead of leaving that to the suppliers or end customers, which in this case could be the hospitals.

3.6 CUSTOMERS – INTERNAL AND EXTERNAL

Customers in the supply chain vary widely depending on the type of company – a manufacturing company has a different set of customers compared to a software company. The set and type of customer also vary based on the size of the company.

In the case of a software company, the purchasing department has to worry less about external customers as customers are internal. It is different from a manufacturing-based company. The internal customers could be indirect purchasing, broader operations, engineering, sales, marketing, finance, legal and even administration.

Supply chain teams have become broader, and it's not just that team alone but a number of others who help the organizations manage the supply chain functionality. Let's discuss the selected few internal supply chain management functions. The goal is not to cover all the internal functions but a select few which supply chain team will interact with.

3.6.1 Internal Supply Chain Management Functions

The size of an organization and the industry sector in which it operates are major factors in determining who the internal stakeholders are. Small manufacturing operations may have one team or even a set of people who wears many hats and has functional responsibilities for several supply chain areas. In larger organizations, the duties in these areas would each be performed by separate groups that include marketing, supply management, finance, production, logistics and information technology (IT). Other internal teams, such as the legal, human resources and customer service department, are very important as well.

It's important to understand the functions of each department and how they interact with the supply chain division.

The activities of some functions are mentioned herein below:

3.6.2 Marketing (Roylance 2008)

The *Practice of Management* legendary management guru Peter Drucker wrote: 'Because the purpose of business is to create and keep a customer, the business enterprise has two and only two, basic functions: marketing and innovation. Marketing and innovation produce results; all the rest are costs. Marketing is the distinguishing, unique function of the business'. According to Drucker, marketing's basic function is 'to attract and retain customers at a profit'.

To accomplish that goal, marketing personnel manage four elements in a marketing mix – product, price, place and promotion. These '4Ps' have been expanded over the years and now often include others, such as people, process and physical evidence (packaging and placement).

1. Product: Whatever satisfies the needs or wants of a customer is a product. This may be tangible good or an intangible service.
2. Price: Expressed as an amount of money, this is whatever the buyer and seller agree upon to make the exchange for a product.
3. Place: This is primarily concerned with distribution. Successful marketing requires getting the right product to the right place for customers.
4. Promotion: This is communication; marketing must inform and persuade. The many ways in which marketing gets the message out about its products and attempts to encourage a customer's desire for them include public relations, press releases, advertising, direct marketing and personal selling (word of mouth).

5. People: This refers to all the people involved in the exchange of products and services and includes the customers and the internal employees whose customer service and relationship-developing skills help build customer loyalty and differentiate a company.
6. Process: All the tasks, procedures and the policies that assist in getting a product or service to the customer are part of the process. Process also includes the methods in place for providing customer service. Communicating and interacting with customers, receiving orders and handling complaints are just a few of the many customer-related processes in marketing.
7. Physical evidence (packaging and placement): Packaging not only encloses and protects a product as it travels through the chain of supply, it also influences the customer's buying decisions. Therefore, the size, shape, color, graphics and many more visual attributes of packaging need to be addressed. In retail operations, the 'physical' aspects also apply to the layout of the store and the positioning of displays.

Let's take an example of how collaboration happens with supply chain and packaging.

There are different levels and types of packaging that require attention from a supply chain perspective. At a basic level, packaging is whatever is used to enclose and protect until it gets to the final consumer. It can add unacceptable cost and little customer value if not properly designed.

Retail-ready packaging is a rather recent European import to the United States. Retail-ready packaging serves the protection purposes of conventional packaging but allows a retailer to use it as part of shelf display. For instance, candy bars located at the grocery checkout counter are displayed in retail-ready packaging. The individual candy bars arrive in boxes that may be opened and immediately displayed.

Many factors need to be coordinated among supply chain members from the producer to the final customers before this becomes a 'great' idea.

From the producer, shipper and warehousing perspectives, issues such as the quality, size and weight of the packages to be shipped are important.

From the retailer's perspective, the ease of use, shelf-space requirements, simplicity of product identification and the quantity of waste to be disposed of or recycled are important concerns.

Customers just want it easy. They want to see it, be able to grab it and be able to put it back on the shelf without any hassle.

It is a good idea, but there's no way of meeting everyone's needs without a high degree of collaboration up and down the supply chain.

3.6.3 Finance

The finance department manages the company's money. These are the people who write the checks. Finance has three critical functions: (1) provide support for business and operational planning, (2) provide operational support to the company's other functions and (3) produce required internal and external reports. Overall finance responsibilities usually incorporate both financial management and accounting services.

Financial management includes the following activities:

1. Planning: Financial strategies, budgeting, revenue and expense projections and estimating the cost of capital projects.
2. Compliance: Monitoring expenditure activity; assuring compliance with federal, state and local regulations; assuring tax abatement methods and internal auditing to guarantee that the company policies are met and safeguarding against fraudulent activities.
3. Guidance: Finance must provide management with quantitative data and well-supported recommendations to help them make decisions about projects, capital purchases and investments.
4. Risk management: Financial risks are serious concerns for an organization. To prevent damage to the company, finance professionals work to identify and manage any risks. This may include managing risks through the use of insurance or other financial measures.

Accounting services include the following activities:

5. Tracking expenditures and revenue: The accounting group is responsible for measuring and recording all business transactions. Every dollar that is spent or received must be tracked. Accounting is responsible to see that expense and revenue records are correct and accurate. They are also responsible for billing customers.
6. Making payments: Invoices from vendors and suppliers must be paid. Timely payments are often a requirement in order to get a discount from suppliers.
7. Creating reports for external use: Includes reporting to the Internal Revenue Service (IRS), the Securities and Exchange Commission

(SEC) and to shareholders. These reporting documents include balance sheets, profit and loss statements and cash flow statements that are created according to generally accepted accounting principles (GAAP)

Let's look at the role of finance in a manufacturing company. The role of finance expands to a lot of other functions. Even the current role of finance in the supply chain is much broader than what it was even a few years before. While the supply chain team is negotiating with suppliers, finance acts as a valuable partner to determine whether there is enough cash flow for the organization to meet the payment commitment. Procurement works very closely with finance to determine the best model for transportation/delivery cost, inventory holding cost, apart from the regular cost of raw material. Role of finance can be thought more as band master in a musical troupe [3].

In a manufacturing set up, finance plays the following roles:

1. Creating a baseline: First step is to effectively run the supply chain function and gather baseline information. The baseline information will contain current production cost details. Once the baseline data are captured, the finance team needs to understand and model finances based on improvements of current product and introduction of new products.

2. Demand planning: Based on forecasts and other factors affecting demand, supply chain works closely with finance to create demand planning models. Multiple related factors could impact demand which could be geography, locations, price and other promotions.

3. Supply planning: Supply planning follows closely with demand planning. Understand the capacity requirement, and based on the capacity requirements, the supply plan is created by the supply chain team working in close partnership with finance.

4. Balance between demand planning and supply planning: An organization needs to reconcile between demand and supply in order to understand various constraints. The constraints could be factory floor, machines, employees, warehousing facilities. This task is mainly carried out by the finance team with the input from the supply chain.

5. Planning for revenue, cost, margin and profitability as well planning for cost reduction: This is a very important step for understanding the projection for revenue and profit. This will also include projection of cash flow statements to understand funds needed or funds available.

6. The finance team also works with the supply chain to ascertain the ideal payment terms which will help the company to take decisions in a timely manner.

3.6.4 Production (Operations in Service Companies)

The Institute for Supply Management (ISM) defines operations as: 'the planning, scheduling and control of activities that transform inputs into products or services'. However, the term production most often refers to what is done by manufacturers and producers. In service companies, such as banking, insurance and healthcare, operations refers to the activities required to support the firm itself and the services it provides.

3.6.5 Logistics

The basic function of logistics is to manage the means by which an organization moves and stores products and materials from the extraction of raw commodities through the chain of supply to customers. Currently, the term logistics is associated more with a set of functions and activities that happen inside a company rather than with the larger completion of the process. Supply chain management has become the accepted term for describing the management of all the activities, including logistics, which are required to bring a product to customer's hands from the start of manufacturing.

The companies may have a department responsible for all their logistics activities or they may have individual departments responsible for the various logistics functions.

3.6.6 Information Technology

Generally referred to as IT, this term is applied to infrastructure (could be departments within the engineering organization). Infrastructure refers to the electronic equipment (mainly computers), the systems used to store, retrieve and analyze data and communication equipment used to transmit. The internal supply chain members are the personnel responsible for managing the infrastructure. The IT group's basic function is to manage the means by which an organization collects, stores, processes and communicates information.

The IT department serves many purposes. These professionals are in charge of assessing and making recommendations about equipment, working with service suppliers, determining system capacity requirements and maintaining all the electronic communication and data processing assets of the company. This includes both the hardware and the software required to run a business's applications, such as the ERP system, communication services and desktop programs.

3.6.7 Engineering

Engineering is one of the most critical functions who the supply chain team will be in close connection with. Engineering team is responsible for specs, design and initial and final feedback on products. In certain companies, the engineering team defines the roadmap for the future as well.

Let's look at the supply chain management function to understand how they interact.

3.6.8 Supply Management

The primary function of supply management is to identify and obtain all the external resources necessary to support the activities of the company. This is equally true for both companies offering physical products or services. In either case, supply management must ensure that everything necessary for the company to produce products or provide services and support operations is available when and where required.

There are four terms that organizations frequently use to identify the business department responsible for buying materials and services. They could be: purchasing, procurement, sourcing and supply management. The term used is 'supply management', which the ISM defines as, 'The identification, acquisition, access, positioning, management of resources and related capabilities the organization needs or potentially needs in the attainment of its strategic objectives'. It is the most inclusive term and, therefore, the most appropriate definition of what function it performs today.

Supply management has both strategic and tactical responsibilities. Strategic duties are those that involve greater long-term and financial consequences for a company. One of these is supplier relationship management (SRM). SRM is a vital activity and one of the most important duties in supply chain management. Tactical duties are those that deal with the daily operational aspects of acquiring the goods and services that keep the wheels turning in a company.

3.6.9 Other Partners – Procurement Subgroups and Collaboration with Legal

The primary function of supply management is to purchase all the external resources necessary to support the company's activities. Sourcing is a part of the supply management process used to identify and select potential suppliers. There are four essential ways that supply management supports supply chain network performance and enhances its value to customers: (1) supplier management, (2) negotiating and contracting, (3) cost control and (4) innovation and new product development (NPD). Supply chain group in the company have multiple partners. Oftentimes, these groups act as extensions of procurement and work very closely with finance and legal.

After the preliminary pricing negotiation with the outsourcing vendor the Supply Chain Manager works with the finance team to finalize the payment terms and frequency of the payment to the vendor. The supply chain manager works in parallel with legal to determine the contract terms and conditions.

For sourcing of materials the Supplier Management team needs to go through the following steps:

a. Suppliers selection
b. Make-versus-buy outsourcing decisions
c. Global sourcing
d. Supplier relationship management (SRM) (O'Brien 2018)

3.6.9.1 Supplier Selection

The first step in supplier management is selecting the right suppliers. Suppliers are critical for supplying the raw materials for final production. The method a company uses to identify, evaluate, qualify, select and manage suppliers is critical. A rigorous process is involved for identifying suppliers till the approval of the suppliers.

Process steps to identify a supplier:

Step 1: The first step in the process is the development of a supply management strategy. The supply chain team prepares a guideline regarding the sourcing strategy.

Step 2: The preliminary selection criteria gets modified a number of times. The strategy involves determining, cost, location, proximity of suppliers location to buyers and financial health of the supplier(s).

Step 3: Depending on the industry, the process of determining suppliers varies. For a technology industry, most times, the first list of potential suppliers is determined by engineering. It also depends on its existing technology or new technology. In most existing technology, the suppliers are already determined. The supply chain team works very closely with the engineering team to determine the supplier list. Information is also gathered from trade shows, publications and research.

Steps 4 and 5: These involve detailed research to determine a list of suppliers that are capable of providing raw materials. These two steps involve meetings between all the stakeholders to determine the final list of suppliers.

Step 6: This step continues with further in-depth investigation. Starts with request for information (RFI), and is followed by request for quote (RFQ). In parallel with completing the RFQ, the buyer looks into the financials of suppliers along with supplier's past performance and reputation. The buyer might also complete on-site visits to evaluate potential supplier's capabilities with respect to supplier's capacity and quality assurance.

Steps 7 and 8: Once steps one through six are completed, a detailed assessment is made of all the information gathered and finalized and the supplier(s) that meets all the required criteria would then be added to the approved supplier list. Buyers always like to have backups for unseen situations that may arise in the future. A potential supplier who didn't make the final cut is sometimes asked to make some adjustments or improvements to be considered in the future.

Steps 9 and 10: There are few steps post selection of supplier(s). Certain tax forms and documents need to be finalized between the buyer and the supplier(s).

3.6.9.2 Make-versus-Buy and Outsourcing Decisions

The choice of whether to do something internally or have an external company provide that product is always a hard decision. The term outsourcing is frequently applied to any type of activity that was or could be performed internally but is done by an outside provider instead. In other words, any make-versus-buy decision that results in contracting with an external provider is called outsourcing.

There are many reasons why a company may want to go outside for the manufacturing of products or for service providers.

Companies will not have expertise for manufacturing all raw materials and would go to individual suppliers for raw materials. If the expertise lies outside the company, it's often better to use supplier's expertise. This is a common practice across any industry. Take the example of automobile manufacturing for electric cars. The battery is often sourced from suppliers rather than manufacturing on their own. There could be multiple reasons why the products or the raw materials are not built in-house. Some of the common reasons are as follows:

i. Lack of in-house expertise
ii. Shifting to supplier with better capabilities reduces buyer's risk
iii. A sole source supplier is the only one capable of providing a product or service
iv. Technology is unavailable in-house
v. Temporary requirements do not justify doing it within organization
vi. Customers have required the use of specific service providers in other aspects
vii. Total cost is improved because an external provider is more efficient and effective
viii. Buyer's core competencies do not include doing this in-house
ix. The lead time available is too short for internal employees or tools to complete

There is a difference between out-tasking and outsourcing. Out-tasking is done for something relatively minor, whereas outsourcing is a strategic decision. Out-tasking is a subset of outsourcing. However, such decisions rarely have the sort of strategic impact that the outsourcing of an entire company function or process would have.

3.6.9.3 *Global Sourcing*

Where in the world are source suppliers? As global sourcing has become more important over the past two decades, supply management's role has gotten more complex and challenging. If it's difficult to manage suppliers locally, it's even harder when the buying and supplying companies are thousands of miles apart.

Sourcing typically refers to the process by which a company identifies and selects sources of supply for the products and services it requires. (In the supply chain operations reference [SCOR] model, the term sourcing embraces all the activities associated with the purchasing process.) Global sourcing is the search for and selection of suppliers in the world. This type of supplier selections elevates the process to an even higher level of importance. The identification, evaluation, qualification and selection of suppliers is challenging for domestic sourcing. Doing it globally is much more difficult, time-consuming and costly.

Supply chain management's responsibilities for sourcing outside the country are really no different than when they are finding domestic suppliers; however, there are additional considerations that must be addressed in areas such as category management, SRM and total cost of ownership (TCO). The concept of TCO is defined as the 'sum of all costs associated with acquisition, use, ownership and disposal of any organizational purchase'. The application of this technique is particularly valuable when a company is making offshore sourcing and buying decisions.

The primary reason for the shift from domestic to foreign sourcing has been cost reduction. This is why an up-front TCO analysis is important— to validate or invalidate the assumption of cost savings. Offshore buying does not always result in the expected cost savings. The geographic location, infrastructure, characteristics of the labor force, culture and political or legal environment will all influence the comparative costs of sourcing around the world. Each must be factored into a TCO analysis.

The elements of TCO provide a quick analysis and decision-making structure.

3.6.9.4 Supplier Relationship Management

One of the major changes in the way suppliers are managed is a move away from an arm's length, adversarial relationship between buying and selling companies to a more collaborative perspective. SRM efforts reflect that change. SRM is not a software package. Software is a tool that facilitates but does not substitute for a well-structured and practical method of managing supplier relationships. SRM includes guidance in how that relationship is to be managed and the processes by which this is accomplished. There are no universal ways to manage suppliers and every situation will be different. What SRM does is set up a means of establishing mutual expectations, facilitating regular communication and resolving issues.

How a company views suppliers makes a big difference in the way supplier relationships are handled. A strategic focus is long-term oriented and cost driven; a tactical focus is short-term oriented and price driven.

There are many types of supplier relationships. Different types of relationships require different SRM methods. The greatest benefits come from the strategic alliances set up with key or critical suppliers. These benefits include improved value to the supply chain through collaboration in the following ways: cost management, production development, inventory management, transportation management, product and process innovation.

The interdependencies and the supply chain linkages between buying companies and their supply chain network require a well-structured SRM process including the methods and the tools that companies use to develop and maintain collaborative working relationships with suppliers. For example, the key performance indicators (KPIs) are measurements used to monitor supplier performance in critical areas such as on-time delivery, quality standards and cost management. On-time delivery would be a KPI calculated as the percentage of a supplier's deliveries over a specified period that had been made on the date required.

SRM software is the tool that aids in the assembly, analysis and display of the data associated with these KPIs. Interpersonal communications are also a part of SRM. Members of senior-level management as well as buying personnel regularly interface with key suppliers' management. These are top-tier suppliers who are responsible for critical parts or materials and support of joint-product development activities. Senior buying personnel, such as category managers, purchasing managers or senior buyers, are the primary contacts within the buying company. They are the individuals responsible for the execution of the SRM processes.

Executives in such companies as major manufacturers place SRM high on their priority. Mutual long-term value, not short-term price connections, is their objective. Close internal collaboration may be required in order to achieve this objective and resolve conflicting departmental perspectives about what is important. For example, the quality assurance personnel might want to focus only on those suppliers with the best track records of conformance to quality standards. Engineering may want only the suppliers with the highest technical capabilities and innovative ideas. Supply management personnel are the ones who bring together the various perspectives and create a unified approach to supplier performance assessment.

3.6.10 Legal and Role in Supply Chain

Just like a lot of other functions, legal plays a very important role and has changed over the years. Apart from looking at all the terms and conditions, there are specific terms and conditions that are determined by the supply chain working with the legal team in an organization. The goal of working with legal is not just to have a contract that protects the organization when something goes wrong but also creates a structure which will govern the relationship between the supplier(s) and the buyer. Some of the key components will provide details on following:

1. Performance of the supplier
2. Acceptance of the product from supplier to buyer
3. Cancellation clause from buyer and seller
4. Support criteria from supplier
5. Warranty and indemnity
6. Contract termination criteria
7. Country-specific rules which will also include various laws across the globe that will be useful to conduct due diligence and conformity of various statutory norms within or outside
8. Current tax liability and future tax implications
9. Options for mitigating risks and can include supplier risk mitigation best practices
10. Settlement options if that is needed and might include settlement of disputes primarily through arbitration clause and/or finally in the court under jurisdictional authority

We will look into more detail on the various functions and how the divisions interact in more detail in Chapter 4.

Let us look at the external supply chain members now.

3.7 EXTERNAL SUPPLY CHAIN MEMBERS

In terms of external customers, it varies and there could be a number of external customers. It could be the second level of distributor, third party manufacturer, suppliers, direct customers, retailers including e-commerce retailers or organization's own retail outlet.

Some of external supply chain members and activities are described below.

3.7.1 External Manufacturer or Producers

Certain times an organization might outsource production of some items – either finished goods or any intermediary products. The manufacturers may be an individual or organization that transforms inputs to create products for final sales by the principal organization. The products delivered by the manufacturing company could be the intermediary products for the principal organization or the final end products that go to the customers directly. Oftentimes, the principal organization will decide the principal manufacturers and the principal manufacturers may decide on the raw material vendors for correct quality of the inputs.

3.7.2 Suppliers

Suppliers are the external organizations or individuals that provide the goods, materials and services that the companies need to create their own products or services. A supplier is often an agent between manufacturer and retailer but with the complexity in supply chain systems, the levels of suppliers have also expanded from one layer to multiple layers. Suppliers play an enormous role at all stages in the product life cycle. Suppliers are responsible not only to just provide raw materials but also make sure they are able to provide better options for raw materials. Suppliers are also responsible to help in ramp up production, improve quality in raw materials and provide best possible price.

3.7.3 Transporters (Transportation Service Providers)

The basic function of transporters is to move materials and goods from one location to another. The transportation could be an external organization whose responsibility is to deliver production to the right location from the supplier to the primary organization or primary customer. Let's look at this with an example. Suppliers of Walmart employ third party trucking companies to deliver products from suppliers warehouse to Walmart facilities or warehouses.

3.7.4 Wholesalers and Distributors

These two labels are frequently used interchangeably or sometimes even used together – sometimes referred to as wholesale distributor. Sometimes a wholesaler serves as an intermediary between the manufacturer and the retailer; however, that intermediator's definition varies.

One difference between wholesalers and distributors is that distributors may have direct business relationships with the manufacturer whose products they stock.

This sometimes includes the exclusive right to offer a manufacturer's products in certain geographic regions. In that scenario, wholesalers and retailers must work together with the distributor for the manufacturer's products in order to purchase products for resale to other customers.

Wholesalers (or whole distributors) are categorized by the type of products they sell:

1. Durable goods: These are things that don't wear out quickly. These items are expected to last longer than 3 years. Durable goods can be of two types: consumer or commercial. Consumer durables are products such as cars, jewelry, office furniture and washing machines. Commercial durables are products such as trucks, aircraft and other heavy equipment.
2. Nondurable goods: These are products that usually don't last long, such as clothing or things and are used quickly, such as cosmetics.
3. Food: The range of items in this category is enormous, but the word explains itself – items from carrots to coconuts, lettuce to lemons and doughnuts to dark chocolate.

However categorized, the primary purpose of both wholesalers and distributors is to buy in bulk and create smaller lots to be sold to their customers. In general, they purchase large quantities of products from manufacturers. These large lots are then stored, repacked and distributed to customers in smaller lots. These customers are generally not the final consumers. Instead, a wholesaler's customers are institutional, industrial and retail operations. As an example, a wholesaler's customer could be

An institutional buyer employed by a public organization such as a school
A commercial buyer employed by a company
Someone employed in merchandising or category management for a retailer

Usually, it's the retailers that sell individual items, for example, consumer goods or food, such as a carton of milk, a six-pack of beverages, a toaster. But wholesalers may also sell single items to industrial buyers. These sales are generally durable goods, such as large industrial equipment, tractors, cranes, forklifts, etc.

Depending upon the industry, wholesalers may not even be in the chain of supply. Apparel and other textile products are often sent directly to the retailer and don't pass through a wholesaler's warehouse. For example, John

Player is a major player in the garment industry. The company designs and markets clothing that is sold in thousands of John Player retail stores around the world. The manufacturing of these garments is done in countries such as Bangladesh, India, the Philippines and Pakistan.

John Player has International Sourcing operations that oversee the production of its clothing at numerous facilities worldwide. A few key suppliers who have a multitiered network of material suppliers and subcontractors make the John Player's clothing. Once completed, the garments produced are shipped through one of John Player's International Sourcing offices or directly to a John Player regional distribution center.

3.7.5 Retailers

An individual or business that sells goods or materials directly to consumers is a retailer. Major types of retail supply chain members can be categorized by the variety of products offered, level of service provided and price as applicable.

3.7.6 Customers

Merriam-Webster's Dictionary provides interesting definitions for the word customer.

1. one that purchases a commodity or service;
2. an individual usually having some specified distinctive trait.

The above two definitions are quite different from the dictionary's definition of a consumer, which is 'one that utilizes economic goods'. What the definitions offer is a very easy way to look at the distinctions between which supply chain members are customers and which are consumers.

When defined as one who purchases a commodity or service, the term customer includes everyone in the chain of supply. Is there really any difference between customers and consumers?

Customers are the ones who pay for something but they may not be ones who actually consume or use it.

Consumers are the ones who actually use the product or are the recipients of the benefit of a service.

In that sense, some supply chain members are both customers and consumers, e.g. suppliers, manufacturers or producers. Wholesalers and distributors are customers but not consumers because they do not actually use the products they buy. But why is this of any importance? It is useful information because the needs of both customers and consumers must be satisfied throughout the supply chain.

Supply chains create value by addressing customers' needs. Often organizations spend a lot of money to understand what customers are expecting and these could be

a. Analyzing industry trends
b. Conducting surveys
c. Interviewing customers
d. Attending conferences
e. Sponsoring focus groups

The moment the organization gains new insights about what the customers value is, that information is used to improve the metrics for supply chain, ensuring that supply is performing in terms of metrics that matter to key customers. This provides more value since this method is used to capture new customers, sell more to existing customers and introduce new products and services that address needs that aren't being currently met.

Let's analyze the above situation with an example and we will take Amazon with Amazon Web Services (AWS) strategy. The introduction of Amazon was very different when Jeff Bezos started with an online book store. As the main business started growing, Amazon quickly understood there is a need from a completely different set of customer base where small- to medium-sized businesses are looking for compute storage. The initial market study and analysis showed there might be a need which Amazon can solve even though the existing business model was completely different. Once a few businesses were involved, Amazon was able to show the value and more business became involved. From small- to medium-sized businesses, bigger organizations started using AWS and the cycle continues.

3.8 THE STRATEGIC (PREFERRED) (SEGEL AND SHAY 2003) AND TRANSACTIONAL SUPPLIERS

Organizations often make a distinction between strategic and transactional suppliers. The strategic suppliers are very important for buyer organizations and it's important to understand the difference (Table 3.1).

TABLE 3.1

Preferred Suppliers Versus Transactional Suppliers

Strategic Suppliers (Preferred Suppliers)	Transactional Suppliers
High level of involvement in current customer's business model	Lower level of involvement with customer or customer's business model
Customer dependency and expectancy of high support	Lower level of support expected and defined under service level agreement
High impact on core business	Low impact on customer's core business
Supplier taking enormous effort to expand relationship	Low and/or medium effort by supplier to grow the business
Mutually beneficial relationship to both customer and supplier	Supplier is looking at short-term gains
Future/potential partnership	Looked at individual deal
Supplier in the list of 'preferred suppliers list'	Supplier maybe even removed from the system after the transaction

TABLE 3.2

Strategic Versus Transactional Procurement

Strategic Procurement	Transactional Procurement
Requirement analysis is the main goal	Standardizing documents is the main goal
Responsible for source selection	Using current source and optimize source
Cost modeling is a key activity	Automatic matching of purchase order and invoicing
Negotiations is primary focus	Automate bid award process
Build sourcing initiatives	Use existing initiative to make process simpler

The idea of strategic and transactional suppliers also could be expanded internally into the concept of strategic and transactional procurement. Strategic procurement is the front end of sourcing while transactional procurement is at the back end but provides immense value to procurement and supply chain in general.

The difference between the strategic and transactional procurement is as below (Table 3.2):

3.9 CONCLUDING REMARKS

As we saw from the chapter, supply chain processes and steps are complex. The ultimate goal of an organization is to make profit. In order to make profit, managing cost is one of the most important criteria and that's where an effective supply chain can add a lot of value. Supply chains need both good processes and people to be effective. We have shown the importance of process and people in this chapter and it could be argued that people play even a greater role than process for supply chain to be successful. We talked about all the partners both external and internal who make the supply chain successful and also explained with a small case from Nike why it's important to focus on people.

REFERENCES

Kristensen, T. 2020. Five ways to cut supply chain costs while boosting customer satisfaction. https://www.pwc.com/us/en/industries/private-company-services/business-perspectives/supply-chain-costs.html (Accessed January 1, 2021).

Packowski, J. 2013. *LEAN Supply Chain Planning: The New Supply Chain Management Paradigm for Process Industries to Master Today's VUCA World*; 1st edition. Boca Raton, FL: CRC Press/Taylor & Francis.

Greg Distelhorst, G. 2016. HBR article: Can lean manufacturing put an end to sweatshops?

Myerson, P. 2012. *Lean Supply Chain and Logistics Management*; 1st edition. New York: McGraw-Hill Education.

Roylance, D. 2008. *Purchasing Performance; Measuring, Marketing and Selling the Purchasing Function*; 1st edition. Boca Raton, FL: CRC Press/Taylor & Francis.

O'Brien, J. 2018. *Supplier Relationship Management: Unlocking the Hidden Value in Your Supply Base*; 2nd edition. London, UK: Kogen Page.

Segel, R., and Shay, T. 2003. *How To Become The Preferred Vendor: 251 Strategies for Doing More Business with Retailers*. New York: Specific House.

4

The Strategic Value Proponent – Six Sigma Approach

Sourya Datta

CONTENTS

4.1 Objective..66
4.2 Six Sigma Overview...66
    4.2.1 What is Six Sigma? (Gitlow et al. 2015).....................................66
    4.2.2 DMAIC, Lean and Six Sigma and Interrelationship................67
4.3 Case Study for Implementing Six Sigma Concepts (Journal of
    Engineering and Science Research 2012) 69
    4.3.1 Case Problem Statement 69
    4.3.2 Case Study Solution Steps..70
    4.3.3 Define Phase – Determine Current Status 71
    4.3.4 Measure Phase 73
    4.3.5 Analyze Phase...76
    4.3.6 Conclusions/Summary...77
    4.3.7 Next Steps/Lessons Learned...78
4.4 Process Details of Six Sigma and Its Significance 79
4.5 Theory behind Six Sigma and How It Relates to Case of the
    Utility Company 80
    4.5.1 Step One: Identify Key Customers
        (Define Phase in Case Study) 81
    4.5.2 Step Two: Define Customer Requirements (Moving from
        Define Phase to Measure Phase in Case Study) 82
    4.5.3 Step Three: Measure Current Performance (Measure
        Phase in Case Study)...83
    4.5.4 Step Four: Prioritize, Analyze and Implement
        Improvements (Analyze Phase in Case Study) 85

DOI: 10.4324/9780429273155-5

     4.5.5   Step Five: Expand and Integrate the Six Sigma System
             (Conclusion Phase in Case Study) 85
4.6   Conclusion 87
References...87

4.1 OBJECTIVE

We will first define what is Six Sigma and what is a Six Sigma roadmap. We will go through each and every step under the roadmap. We start the chapter with a case study and take the readers through the rationale of the steps and explain the background. We will discuss the objectives and deliverables for each step and show how the Six Sigma processes can be applied. We cover the definitions of each concept when we go through the case and then cover the theoretical concepts to correlate the case with the theoretical aspects of Six Sigma. This chapter will help us understand, guide and provide a framework for tracking the steps for employees interested in pursuing green belt in Six Sigma as well. We will explain Six Sigma and discuss DMAIC (define, measure, analysis, improvement, control) steps in depth, which are detailed steps for implementing Six Sigma projects. The idea of this chapter is to provide guidance to all levels of the organization: executives, mid-level managers and junior employees.

4.2 SIX SIGMA OVERVIEW

4.2.1 What is Six Sigma? (Gitlow et al. 2015)

Six sigma was made popular by Jack Welch, the famous leader of General Electric who used Six Sigma processes heavily in General Electric during his time. It is believed that Bill Smith of Motorola was responsible for coining the term 'six sigma' in 1986.

Six sigma is a set of tools and techniques used for process improvement. Concept of lean (Packowski 2013) is embedded within Six Sigma in a way Six Sigma covers lean processes. Six sigma is a process where 99.99% of features in the product are expected to be defect free. Six sigma is important for organizations that are manufacturing products, and the most important aspect of Six Sigma is the opportunity to remove any defects and as a result reduce cost. Six sigma will be successful only if there is a high level of collaboration between all the stakeholder verticals within an organization.

4.2.2 DMAIC, Lean and Six Sigma and Interrelationship

DMAIC (Rasmusson 2006) is an acronym and steps under the Six sigma process. We will cover a case study showing the application of DMAIC and then relate that with the theory. Six sigma does process improvement by following the DMAIC process. DMAIC steps are as follows:

Step 1 – Define: First step is defining the problem and the impact on current process

Step 2 – Measure: Measure data. Supply chain team examines process in place and also identifies what is not working

Step 3 – Analysis: Analyze data to determine root cause of the problems

Step 4 – Improvement: Determine solutions, test the solutions and determine the improvements

Step 5 – Control: Continue making improvements over time. Keep refining the process

Even though the five steps mentioned above make the DMAIC process, depending on the project or problem statement, slight adjustments are made to the steps. In almost all cases, there are no changes in the first three steps. All projects need to define the project, measure the data and analyze the data. The Six Sigma processes in a manufacturing set up could be different from a company offering service, a project in operations might have different steps (steps 4 and 5: improvement and control) than a project involving marketing services. There are projects that will not have steps such as improvement and control but conclusions and next steps might be more applicable in certain scenarios. Not all Six Sigma projects will always find steps to improve; the process might already be functioning well and no improvement may be needed. Let's explain the concept with an example. In one case, a company is making widgets and there is a very long lead time to deliver final products to customers. The Six sigma project finds that warehouse locations are not suitably managed and there needs to be relocation of some existing warehouses and some existing warehouses need to be shut down. In this case, there is a definite improvement and control step. Let's take another example. A fortune 100 company is trying to target selling their products more to a certain age group. The marketing materials are very targeted, but the company is not successful in attracting the group of people belonging to the age group. When the six sigma project was implemented in this scenario, it was determined that this was a wrong choice from the company to choose this age group and the company needed to focus on a different age group since the product is more relatable to the group who the company was not focusing initially. In this case instead of improvement and control, fourth and fifth steps of the Six Sigma project could be 'conclusion' and 'next steps' should be used since the strategy needed to be changed and no changes are needed in the process. The company was trying to focus on college students for valuable coins but it was determined that coin collection is an expensive hobby and people who are earning will be more interested in collecting coins and can also afford to do that. We will go into a case study which will capture the above points as well. For certain projects, step four will be different and involve prioritization, analysis and implementing the improvements defined in the analysis phase, and the last step (step 5) is actually integration and expansion of Six Sigma processes. A good example of the above scenario will be a scenario managed by cruise ship companies where they are always struggling with determining the number of ships they would need to have and the number of trips that they will make in a year to effectively make the maximum profit. (Just making more trips does not necessarily mean the cruise company will make more profit.) The maximum profit will depend on the cost of ticket per passenger, the experience level of cruise employees and route taken. In the cruise ship example, step four in the six sigma process will be for prioritizing, analyzing and implementing improvements. This would be the case where the cruise ship company models other routes based on the success of one route. Step five could be used for expanding and integrating the Six Sigma system for all future planning of the cruise shop company. In the section after the case, we will again touch on the concepts.

Let's now go to the definition of lean and relationship between lean and Six Sigma.

Lean can be defined as a process of removing wastes. Six sigma is used to reduce defects. Even though lean and Six Sigma have the same goal, the approaches are slightly different where identification of root cause and wastes are determined differently. Sometimes organizations do both lean and Six Sigma together. Lean is used in production and Six Sigma is used to reduce error in production and non-production environments. Six sigma can be used in other departments of the organization as well, for example, in the marketing department of an organization while lean is used mainly in production.

4.3 CASE STUDY FOR IMPLEMENTING SIX SIGMA CONCEPTS (JOURNAL OF ENGINEERING AND SCIENCE RESEARCH 2012)

We will talk about a case in the utility industry and look at the roles and responsibilities of different levels within the organization. We will see how the utility company is faced with inaccurate lead times and how the organization comes up with a solution. The senior leadership of the company wanted inventory system lead time to improve since lack of control was directly impacting the bottom line of the company.

Whenever we talk about any inventory measurement issue, it is always difficult. Oftentimes, employees are less motivated to find a better solution to make the inventory management system better since the quality of data deteriorates very fast (within a year of actual physical count or sometimes even less than that) and that makes the employees redo the inventory count again. So the solution needs to be such that it solves a problem persisting for a long time and stops the same inventory issue from recurring in the future again.

In order to complete a Six Sigma project successfully, it is very important to define and follow the steps end to end. The steps for the Six Sigma project are generally created with different members and cross-functional teams within an organization. The Six Sigma project involves team members from supply chain, finance, marketing and sales. Each group is represented by executive leaders from the various verticals or divisions of the organization. Typically the heads of the verticals are the project sponsors, the customer (internal or external) becomes the main stakeholders and the employees going through the steps become the project initiators and a few senior members are chosen to lead the project. Just like the executive leaders, the leads are from various verticals in order to get feedback from different parts of the organization.

Before we start talking about the case and how different members worked to solve the case, let's define the problem statement in detail.

Note: *This case is not a real case and made up scenario to explain the concepts*

4.3.1 Case Problem Statement

A big utility company in California, USA, is experiencing inefficiencies in recording and receiving parts in their inventory system. This is causing inaccuracies in lead time calculation, thus resulting in inaccurate inventory levels. It is estimated that approximately 30% of the calculated inventory lead times are incorrect in the system. This has resulted in inaccurate reorder points and this has been occurring for the last 5 years.

Since the company is not able to accurately estimate the specific component or part (raw material in this case) delivery times with confidence, the company needs to buy excess inventory and pay expedited freight charges, and there are constant delays in repairs due to parts not in hand. Currently the utility company has 12 times more inventory than what they use each year. The problem has become worse since the last few years as the customers are looking at alternate sources of energy. Customers have started using solar energy and utility company's ability to make profit has decreased further as a result. On one hand the utility company is facing pressure on revenue and on the other hand, the company is facing an increased cost. In this case we will be looking at the increased cost aspect and how to solve the issue of the increased cost.

4.3.2 Case Study Solution Steps

In our case study, the employees of the utility company along with the senior management came up with the following set of phases (Six Sigma Steps (ASQ Press 2012)) to define and tackle the problem and show results:

1. Define phase
2. Measure phase
3. Analyze phase
4. Conclusions/summary
5. Next steps/lessons learned

All the employees of the organization are responsible to define the steps. Steps 1–3 are the result of work between different verticals. Define phase is the most critical phase where the problem needs to be defined accurately and the customer (either internal or external) needs to sign off. For a Six Sigma project to be successful, it is very important to get the sign off from the customer.

Supply chain team leads steps 1, 2 and 3 and works with cross-functional teams as described above. The team members assigned to the project complete steps 1, 2 and 3 and the team leads provide inputs along the process. Supply chain takes the lead on this process since that team is in the best position to understand the overall goal of the project and see the big picture. Also defining the right problem is critical for Six Sigma projects to be successful.

The conclusion and summary (step 4) is presented by the cross-functional team (different verticals) to the team leaders first and then to the executive leaders and the project sponsors.

The next steps/lessons learned are a joint discussion that the entire organization together does, and then steps are used to solve the current problem. The key to a successful solution is patience. Six sigma projects don't show results immediately and it takes some time. It is also important to continue doing minor changes in the future to keep improving on the solution.

Let's go through each of the phases in detail

4.3.3 Define Phase – Determine Current Status

Table 4.1 captures the task and the responsible parties.

The process starts with visiting the customer to understand the problem. The customer(s) is interviewed and asked a number of questions. Oftentimes, in order to understand the problem, the Six Sigma project team travels to the actual work location of the customer and follows the steps undertaken by the customer. We will have a table which will define the detailed steps and the people involved in this process. The goal or outcome of the first step is to determine the main problem statement.

Let's talk about various steps taken and the responsibilities under each step (Table 4.1).

In our case above, the most important step is the warehouse visit to interview various people. For other projects, however, there might be a need for site visit and separate interviews with various departments within

TABLE 4.1

DMAIC Define Phase: Determine Voice of Customer Phase

Task	Description	Responsibility
Site visit	This site visit could be the site for internal or external customers	All team members including executive leadership of the organization
Interview project sponsors	Interview project sponsors to determine end goal	Cross-functional meeting between various organizations
Warehouse site visit	On-site plant visit in factory/warehouse. Interview people responsible for inventory system updates and review processes	Core team along with team leaders. Again a cross-functional effort
Determine data needs	Once initial site visits and interviews are complete, the team begins to request data for analysis	Core team only (mainly driven by supply chain vertical). Team leads are provided updates

the organization. The table captures the common steps taken during the defining process of Six Sigma.

Define Phase – Action/Analysis (Voice of Customer: Data Analysis)

The next step after this is understanding and determining the voice of the customer which covers the data analysis.

This step involves analyzing the data for the period in question (when the issues have occurred) and focusing on the last 6–8 months' specific data (could be purchase orders in the above example). The purchase orders will show possible discrepancies in lead time data entries and the team looks for patterns in items with incorrect lead times (similar parts, vendors, etc.). The team also talks to the set of people who are responsible for the actual data.

In our example, the supply chain team needs to talk to warehouse personnel at the company to identify the actual steps that are taken by the warehouse employees to record the order and receive dates within the Inventory System

Define Phase – Action/Analysis: Supplier–Input–Process–Output–Customer Diagram

Once data are collected, the supply chain team (core team) creates a table showing the various steps. In the above problem statement, a SIPOC diagram is first created (supplier–input–process–output–customer). The SIPOC diagram helps achieving two goals:

a. The various members who are present at various steps
b. Exact steps followed by the members in the steps

Both 'a' and 'b' are critical to understand the exact problem. Let's show how a SIPOC looks like (Figure 4.1):

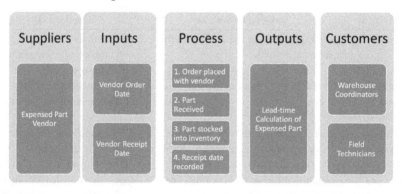

FIGURE 4.1
Define Phase – Action/Analysis: Supplier–Input–Process–Output–Customer.

Define Phase – Outcome/Results Key Findings

This is the final step in the define phase. The team needs to determine the findings, and in our example, findings include the following (the below details are just examples which will change based on the case but will give the reader information on what to collect and define at this stage).

The core team defined that there is a lack of defined process and purchases were made through various outlets – email, online, fax, website and even phone. Another observation was recording of purchase dates can be manipulated in the inventory system, which is not representative of actual order date. The team also determined that approximately 30% of lead times are incorrect and finally inventory on hand is 20 times the amount of spend on parts per year.

At this stage, the team gets the sign off from the customer. Since the team has a good handle definition of the problem, the team goes to the measure phase

4.3.4 Measure Phase

The first step in the measure phase is to create the measurement system analysis data flow.

An entire data flow is created at this stage which captures the order date, receive date, flow of the data to the inventory system of the customer and the determination of lead time. This process has input, measurement and then the output. The input is provided after discussion with the customer, the measurement system is the actual tool used by the customer (Purchase Order (PO) in the system) and the output is the measured value (lead time in this example).

It could look as shown in Figure 4.2 for the current project:

Next comes the data collection part

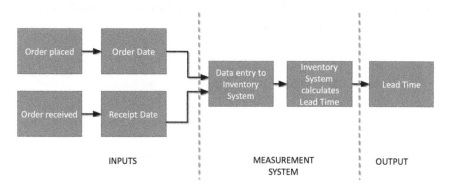

FIGURE 4.2
Input, measurement, and output.

Measure Phase – Purpose/Objectives (Data Collection)

Since the core team had collected purchase orders from the warehouse, they now select a random sample of a set of purchase orders. The purchase orders need to be from various locations, using a 95% confidence interval. Let's understand the details. An important aspect of the measurement phase is the one sigma (one standard deviation) plot above and below the average value on the normal distribution curve. This region would define 68% of data and two sigma above or below would determine 95% of data. At this stage of the project, a distribution curve is created as shown in Figure 4.3:

Measure Phase – Data Collection Plan and Key Inputs

Data collection plan and key inputs will include purchase order number, location of orders received, order date, receipt date or delivery date of products, delivery time of products and shipment type.

In this step, it is very important to have face-to-face interviews with the inventory management and invoicing team. The purchase order details are collected and random samples of purchase orders are done using a 95% confidence interval.

Measure Phase – Determine Defect Types

Defects from human error, system issue, invoice date, shop date and product availability are determined. Few examples of errors in the case study could be as follows (this will depend on the actual project but the below are examples to explain the point):

Invoice date: Invoice date before order date (Obviously, there is some error since an invoice can't happen before the part is ordered)

Ship date: Ship date outside of order and receipt dates (Again this is an error since part should be shipped after order date and before receipt date at warehouse. The ship date should be within order date and receipt date but that's not the case in the case study)

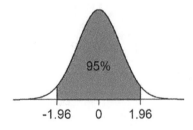

FIGURE 4.3
Distribution curve.

Shipment type: This is another error where there is a mistake in the system for both pickup and next day delivery. Pickup from the same location should ideally happen on the same day since the buyer and seller are located in the same city. Also next day shipping can't be more than 5 days.

'Pickup' in same location > 5 days
'next day' shipping > 5 days

Measure Phase

The last step for measure phase is creating a fishbone diagram (Bowen 2016). Typical fishbone diagram will need to have a number of 'causes'.

A very important step in the measure phase is creating a fishbone diagram. A typical fishbone diagram (also known as Ishikawa, after professor Kaoru Ishikawa at the Faculty of Engineering at The University of Tokyo) is a cause-effect diagram that shows various reasons for error, variation and defects. Since the diagram looks like a fish's skeleton, it's called a fishbone diagram. The problem statement is at the head of the fish and the causes are along the spine of the fish.

For the case study above, the fishbone will outline the various causes which the team gathered around the spine as described below. The causes for the problem are broken down under various separate components since there are various reasons for the error. The various causes are environment, measurement, material, method of capturing the data, human error and infrastructure or system error. Details under each of the causes are captured in this phase (Figure 4.4):

Let's go to the next step which is analyze phase.

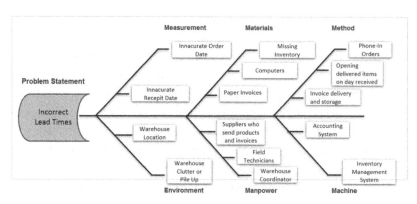

FIGURE 4.4

Fishbone diagram (measurement, materials, method, environment, manpower, and machine).

4.3.5 Analyze Phase

There are various steps in this step. This is the step where the data are analyzed and errors are determined. The steps will include analyzing data collected in the measure phase, identify the most viable causes for possible lead time defects and finally determine and prioritize the root causes, which result in inaccurate lead times.

Analyze Phase – Ship and Invoice Errors
Various errors could be determined from the system. These are errors like vendor shipped product before order date, vendor shipped part after receipt date in utility company's system and part shipped before order date where part sent and received on the same date. There could be other issues like zero lead time, but that is not possible since some part vendors are in a different locations and shipping and receiving goods from those vendors take time. Other errors like invoices received prior to ordering could be part of issues as well. In this case study, it is important to determine the error by each warehouse since it needs to be determined if the problem is specific.

Analyze Phase – Actions/Analysis
In this phase, a dashboard is created to show errors belonging to each of the four categories above and the total incorrect lead times is captured in terms of percentage

Analyze Phase – Return Back Fishbone Diagram to Determine Actual Causes
Based on the above determination, the fishbone diagram now needs to be updated with actual issues, and based on the findings, the outcome or result based on analysis is now determined and fishbone diagram is now updated.

Based on the findings, it was determined that the main problems or causes were under measurement, employee (human or manpower) and infrastructure (machine). Again, depending on the problem, it will vary and the right determination needs to be made specific to the case. Order date and receive date were incorrect, the person at the warehouse is responsible for incorrect method of data entry and the system itself shouldn't allow the way it allows the current data entry. The determination is also made whether this is a specific warehouse problem or it is common across all warehouses for the utility company. All the 'causes' are determined and captured under the fishbone diagram. The new diagram will look as in Figure 4.5 with all causes determined.

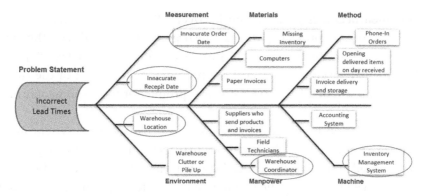

FIGURE 4.5
Fishbone diagram (output).

Analyze Phase – 5 Why Analysis (Roderich 2020)

The 5 why analysis is another good approach and can also be used to update with questions and answers for each of the questions. The goal of the 5 why process is to answer all the 5 'why' questions and if answers to any of the questions is not known, the team goes back to find the answers. A good example of our case study is shown in Table 4.2.

Once the analysis phase is complete, the team can go to the next step.

4.3.6 Conclusions/Summary

As described earlier, conclusion and summary is a joint meeting where the team first presents its findings and provides solutions. Before presentation to the executives of the company, the team presents information to team leaders for alignment and feedback. The team leaders provide feedback which is incorporated in the final presentation.

TABLE 4.2

DMAIC Analysis: 5 Why

Question	Answer
Why is lead time inconsistent?	This is the problem statement
Why are there inconsistencies in inventory level?	Incorrect lead time calculation
Why are there incorrect lead time calculations?	The incorrect order date/receipt date was entered
Why are the dates entered incorrectly?	Process and communication gap
Why are there process gaps?	Lack of controls on inventory system Free field entry

This then becomes a discussion between all the involved team members including the project sponsors. The organization decides on the next steps and action items. Based on our case study above, a few steps or action items that could be useful are provided below. Most of the times the solutions are in all aspects – process, product or tool, and people and protocol.

Process solution points to making inventory management system more functional. In order to make inventory management better, few steps need to be followed and this will include logging timestamps on every edit, computing the lead time using the moving average method, periodic audits, and placing constraints on order and receipt date (order should always be before receipt date).

Product/tool will have a few steps. Tool needs to continue to measure the process capability to see improvement between the voice of the customer and the expectation (specifically on pick up and next day shipments). The tool also needs to check the data for errors, observe trends between various months, specific warehouses and types of error and finally there needs to be a standard operating procedure (SOP) for all warehouses which will help establish a baseline to further investigate errors in the inventory system.

People and protocol will need to confirm the standardized protocol between all the steps performed. This will include stamping data on the packing slip when inventory is received. This stamp will serve as a reminder when the inventory is logged in the system. **The procurement agreements/ supplier partnerships standardizing sourcing helps remove errors and a performance metrics to evaluate warehouse managers needs to be introduced.**

4.3.7 Next Steps/Lessons Learned

Following all the processes in the Six Sigma, a broader meeting is often arranged where all the people responsible for the Six Sigma project meet and discuss the learnings. For a Six Sigma project to be successful, it is important not to have any assumptions and define results based on data. **It is almost impossible to fix an issue if there is a lack of quality of data and hence it is very important to have good quality of data.** Lessons learnt are often a very important step in a Six Sigma project.

4.4 PROCESS DETAILS OF SIX SIGMA AND ITS SIGNIFICANCE

Now that we have described the case in detail, let's focus on the theoretical aspect of Six Sigma. The below section will go through the core methods and process of doing a Six Sigma project. Based on the project and needs of the organization, Six Sigma steps will also vary. The following section will provide the step-by-step guideline of doing a Six Sigma project and what is needed for each step. The theoretical aspect of Six Sigma will also show how Six Sigma projects are joint efforts between different verticals of the organization.

The five steps outlined below are similar to the DMAIC process but organizations sometimes modify the DMAIC steps to make it shorter or longer depending on the criticality and time of the project. The below steps are the same variation of the steps from the utility case study and are actually the same DMAIC processes but with minor modifications. As explained before, the concepts of define, measure, analyze, improvement and control don't change but the steps are tweaked based on the project, the need of the organization and the project environment. Based on the DMAIC steps, we will go through the following steps in the theoretical discussion.

a. Identify existing processes and key customers
b. Define customer requirements
c. Measure current performance
d. Prioritize, analyze and implement improvements
e. Expand and integrate the six sigma system

In order to get the maximum output from the employees, the management team alone doesn't make the roadmap but employees at all levels come together to make the roadmap. Six sigma processes within the company is more successful when employees at all levels get involved and define the process. The successful implementation of Six Sigma processes benefits the organization tremendously and is also very useful for employees to grow within the company as well.

Following exact DMAIC steps may not be the only path to six sigma improvement, the process and steps will need to be adjusted depending on the current problem that the organization is looking to solve. What makes this path 'ideal', however, is if followed in the right order, these activities build up the essential foundation that will then support and sustain six sigma improvement. Specifically, the advantages of defined steps include:

1. A cleaner understanding of the business as an interconnected system of process and customers.
2. Better decisions and use of resources to get the greatest possible amount of benefit out the six sigma improvements.
3. Shorter improvement cycle times, thanks to better upfront data and selection of projects.
4. More accurate validation of six sigma gains – in cost savings, minimizing defects and improving customer satisfaction.
5. A stronger infrastructure to support change and sustain result.

These DMAIC steps are guaranteed to win a poll as the 'ideal' implementation approach among six sigma veterans, as well. Everyone who has been involved in a six sigma launch – executives, implementers and team members – should work together and needs to agree on the path that needs to be followed. Since this is not a push process (coming from management and down to the employees), success rate is often higher since everyone in the organization has worked together to determine the process and the steps. The defined path may need to be tweaked or modified and results may be achieved sooner or might take some time and patience is often key in Six Sigma process. (Toyota didn't become a success overnight with their Six Sigma implementation.) The success depends on proper and correct implementation of the various steps of six sigma, and once the assessment is done correctly and implementation is done sequentially, the results will come automatically.

The beauty of Six Sigma is the fact that this could be introduced at any level in the organization. As an example, shop floor employees could even start Six Sigma projects.

4.5 THEORY BEHIND SIX SIGMA AND HOW IT RELATES TO CASE OF THE UTILITY COMPANY

Now that we have gone through the case study, let's look at all the principles and tie them back to the case to explain the Six Sigma concepts. The concepts will be important to use in a Six Sigma project.

As business becomes ever more dispersed and global, customer segments more narrow and products and services more diverse, it gets more difficult to see the 'big picture' of how the work actually gets done. We will discuss various steps which will be useful for the core team within the organization to break down the big picture into clearer focus by defining the critical activities and getting a grasp of the broad structure of the business system.

Let's discuss all the steps involved and how to make the process successful at all levels in the organization.

4.5.1 Step One: Identify Key Customers (Define Phase in Case Study)

The first step in determining the core steps in the Six Sigma process is to identify the problem. Identifying the problem is not easy as it sounds primarily due to alignment of all employees with the leadership regarding the most critical problem to solve a number of problems or issues that the organization has at that particular time and identifying that Six Sigma is the right approach to solve the problem. As a result of the above, it's very important to identify the core processes and the customers who are in need and the project that will benefit the organization. This is particularly important since understanding the various activities within the organization becomes critical. Also another important factor to keep in mind: customers can be internal or external.

The objectives for step one are applicable to an entire organization or a segment within. Even a department or function that serves internal customers – human resources, information technology or facilities has its own 'core process' that delivers products, services and value to customers. Let's look at the objective and deliverable of step one below (Table 4.3).

TABLE 4.3

Objectives and Determine Deliverables for 'Identifying Key Customers'

Objectives	Deliverables
Create a clear, 'big-picture' understanding of the most critical cross-functional activities in the organization and interface points with external customers	A 'map' or inventory of value-delivering activities in organizations, driven by three questions 1. What is the core value-adding process? 2. What products and/or services to provide to customers? 3. How do processes 'flow' across the organizations?

The knowledge gained from step one is important as a prerequisite for providing the best value to customers. A more significant benefit of this is inventory management, but the actual goal is to have a clear understanding about organization as a whole.

If we look at our case, the first step of **Define phase – determine current status** along with the table that shows the involvement of various groups talks about this step.

4.5.2 Step Two: Define Customer Requirements (Moving from Define Phase to Measure Phase in Case Study)

One of the discoveries often admitted by business leaders and managers, after embarking on six sigma, is not knowing what customers really want. Quote from an executive, *'We really didn't understand the customers very well'*. Getting good customer input on a company's needs and requirements may be the most challenging aspect of the six sigma's approach. The value added to the company could be provided by employees and that creates a ladder for success for the employee(s) within the company. The objectives and deliverables for this step are shown in Table 4.4.

If the organization has no idea of customers' needs or requirements, it's impossible to give it to them. Moreover, in the context of achieving six sigma performance, meaningful development can't be measured until clear specific requirements are available. This is an opportunity for employees to create a niche for themselves in the organization. Creating a scope of work to determine what customers want or even extending to some ideas where

TABLE 4.4

Objectives and Determine Deliverables for Define Customer Requirements'

Objectives	Deliverables
1. Establish standards for performance that are based on actual customer input, so that process effectiveness can be accurately measured and customer satisfaction predicted 2. Enhance systems and strategies devoted to ongoing 'Voice of the Customer' data gathering	A clear, complete description of the factors that drive customer satisfaction for each output and process aka 'requirements' or 'specifications' in two key categories: a. 'Output Requirements' tied to the end product or service that make it work for the customer (what quality gurus have called 'fitness for use') b. 'Service Requirements' describing how the organization should interact with the customer

the customers didn't know what they want is very transformative for the employees and the organization.

This rationale goes even further which is that of 'attitude'. A number of companies, even entire industries have got into serious trouble in the past with 'we know what's best for the customer' mentality. Almost bad is the misguided belief that 'we are really tuned in to the needs of our market', when in fact the company is out of touch with changing demands. Arrogance and ignorance may have been tolerable 20 years ago, but in today's competitive environment either one is a sure predictor of failure.

In the twenty-first century, it will be the companies and employees that really listen to their customers that are most likely to see long-term survival and success. This is a process where organizations and suppliers need to closely work with each other and the onus is at all levels in an organization from an entry-level person to mid-manager to senior manager to the executives.

If we correlate to the case **Voice of customer** (Yang 2007) was the step that we took in the case followed by **Define Phase – Outcome/Results Key Findings**.

4.5.3 Step Three: Measure Current Performance (Measure Phase in Case Study)

While step two defines what customers want, step three looks into how well the organization is delivering on those requirements today and how likely they will do this in the future. On a broader level, performance measures focused on the customer serves as the starting point for establishing a more effective measurement system. Without properly measuring the data, it is impossible to analyze or come up with a solution. This is why this step is so important. The objectives and deliverables of this step are captured in Table 4.5.

TABLE 4.5

Objectives and Determine Deliverables for 'Measure Current Performance'

Objectives	Deliverables
To accurately evaluate each process's performance against defined customer requirements and establish a system for measuring key outputs and service features	i. Baseline measures quantified evaluations of current/recent process performance ii. Capability measures assesses ability of the current process or output to deliver on requirements. These include 'Sigma' scores for each process that allow comparison of very different processes iii. Measurement systems determine new or enhanced methods and resources for ongoing measurement against customer-focused performance standards

The measurements need to be determined correctly and the systems should also capture data on the efficiency of the processes: costs per output, energy or material consumption. Highly inefficient operation leads to only one option – an unprofitable option.

The need for an accurate 'grade' of performance against customer requirements should be obvious. There are several other benefits of step three, however, that make this much more valuable than just a report card:

a. Creating a measurement infrastructure: Measurement infrastructure gives the power to follow changes in performance (good or bad) and the ability to respond promptly to warning signs and opportunities. Over time, these data points become key inputs to the responsive, always-improving six sigma organization. Creating a measurement infrastructure and effectively using the infrastructure will help employees within the organization to define a growth path and is a very effective mechanism for both organization and the employees.

b. Setting priorities and focusing resources: Even in the short term, knowledge derived from those measures derives decisions to determine the most urgent and/or high-potential improvements. The impact is a higher return on investment for process design, redesign or improvement projects.

c. Selecting the best improvement strategies: Having accurate process capability measures allows organizations to gauge the real nature of performance issues. Organizations need to determine if it's just occasional problems or minor issues or situations implicitly demanding that an entire product line or process be revamped. Making this distinction is key in improving the supply chain in the organization.

d. Marketing commitments and capabilities: Employees from different organizations need to work together to make the final product successful. Sometimes we see sales and marketing are at odds with operations. Sales comes to operations asking to enable the feature set or product set that customer has asked and operations complain about 'impossible commitments' made by the sales. At the same time just better communication between employees from different departments will not resolve these disconnects. These differences are extremely challenging and very costly for the organization.

The ideal way to solve the problem is for employees to have added advantage of the knowledge gained through six sigma methods – both about what customers really want and what the organization can actually deliver.

Going back to the case, **data flow** was determined, **data collection** took place and **defect types** were determined.

4.5.4 Step Four: Prioritize, Analyze and Implement Improvements (Analyze Phase in Case Study)

From this step, employees can take advantage of six sigma. In this step, organizations don't only provide anecdotes and opinions but are equipped with facts and measures. The objectives and deliverables of this step are captured in Table 4.6.

A key to be successful in the six sigma system is to choose the improvement priorities carefully and not to 'overload' the organization and the employees with more activities than it can handle. The value of the improvement methods applied in step four is that they encompass the best techniques for driving out defects and improving process efficiency and capacity.

If we go back to the case, **Analyze phase – actions/analysis** was performed and then **Analyze phase – fishbone diagram** was created which depicts this phase.

4.5.5 Step Five: Expand and Integrate the Six Sigma System (Conclusion Phase in Case Study)

Real 'six sigma performance' will not come through a wave of improvement projects, it can be achieved only through a long-term commitment to the core themes and methods of six sigma. That's why it's important for

TABLE 4.6

Objectives and Determine Deliverables for 'Prioritize, Analyze and Implement Improvements'

Objectives	Deliverables
Identify high-potential improvement opportunities and develop process-oriented solutions supported by factual analysis and creative thinking Also, the goal of this step is to effectively implement new solutions and processes and provide measurable, sustainable gains	1. Improvement in priorities: Potential six sigma projects should be assessed based on their impact and feasibility 2. Process improvement involves solutions targeted to specific root causes ('continuous' or 'incremental' improvements) 3. New or redesigned process: New activities or workflows are created to meet new demands, incorporate new technologies or achieve dramatic increases in speed, accuracy and reduce cost called six sigma or business redesign

TABLE 4.7

Objectives and Determine deliverables for 'Expand and Integrate the Six Sigma System'

Objectives	Deliverables
Initiate ongoing business practices that drive improved performance and ensure constant measurement of steps, reexamine products, services, process and procedures, core	1. Process controls, measurements and monitoring to sustain performance improvement.
	2. Process ownership and management: Cross-functional oversight of support processes, with input from voice of customer, voice of market, voice of employee and process measurement systems.
This is the step where the employees within the organization work hard to achieve the vision set by the leadership – the vision of a six sigma organization	3. Response plans: Mechanisms to act based on key information to adapt strategies, products/services and processes
	4. Six sigma 'culture': Organization positioned for continuous renewal. Six sigma themes and tools form an essential part of the everyday business environment

employees between different departments to work closely and integrate Six Sigma processes (Table 4.7).

This step probably is the hardest for the organization and success comes after a long time.

Following features describe organizations with effective Six Sigma processes.

a. An accurate, well-channeled customer feedback system
b. Well-integrated process, with easy handoffs and information flow upstream and downstream
c. Timely measurement systems that track not just spend but also defects, changes in key activities and variations in key inputs like raw materials
d. Expertise in correcting problems and making improvements – either by fine-tuning processes or by creating entirely new processes, products or services to meet changing customer needs

Step 5 defined above becomes a key component in the success of employees within organizations and organizations by themselves as well. When the employees are able to show that there are mitigating steps for sudden rise in competition and confident that employees have provided certain structure to defend against a similar company that could make inroads into profits or market share very soon, it's an indication that the employees have built a good process and that shows the organization has hired high-performing employees.

Going back to the case solutions to a Six Sigma problem will be found in **Process, product or tool, and people and protocol** and continuing the work without stopping is critical.

4.6 CONCLUSION

There are multiple models, processes and structures that can make supply chains successful. At the end, people play a very important role and leaders and employees working closely make supply chain processes and systems effective. The basic principle of supply chain hasn't changed but various improvements have been incorporated by new methods, creative ideas and technology. Just following the Six Sigma methodology itself will not bring success. Six sigma provides the structure which will need to be tweaked depending on the project. The most critical part of a Six Sigma project is understanding the customer's problem. The biggest mistake that employees do in a Six Sigma project is predetermining the problem statement and finding a solution for that. However good the solution is, it will not be of much value since that is not the real problem that the organization needed to solve.

REFERENCES

Gitlow H, Melnyck R, Levine D; *Guide to Six Sigma and Process Improvement for Practitioners and Students, A: Foundations, DMAIC, Tools, Cases, and Certification*; 2015. Upper Saddle River, NJ: Pearson FT Press.

Ecolab Inc, Electric Power Sector, JEA, Inc.; UE Compression, ZF Energy Development; 2012. Journal of Engineering and Science Research.

The Certified Quality Engineer Handbook, ASQ Quality Press; Asq.org; THE DEFINE, MEASURE, ANALYZE, IMPROVE, CONTROL (DMAIC) PROCESS; 2012. Milwaukee, WI: ASQ Press.

Rasmusson R; *SIPOC Picture Book: A Visual Guide to SIPOC/DMAIC Relationship*; 2006. New York: Oriel Incorporated.

Bowen P; *The Fishbone Diagram Handbook - Everything You Need to Know about Fishbone Diagram*; 2016. Brisbane: Emereo Publishing.

Roderich O, *5 Whys: The Effective Root Cause Analysis (five whys improve your mind)*; 2020. Independent.

Yang K, *Voice of the Customer: Capture and Analysis (Six SIGMA Operational Methods)*; 1st edition; 2007. New York: McGraw-Hill.

Packowski J, *LEAN Supply Chain Planning: The New Supply Chain Management Paradigm for Process Industries to Master Today's VUCA World*; 1st edition; 2013. Boca Raton, FL: CRC Press.

4.6 CONCLUSION

REFERENCES

Section 2

The Supply Chain Leader

5

Roles, Responsibilities, and an Industry Overview

Sudip Das

Ishaan Das

CONTENTS

5.1 Objective 92
    5.1.1 For a Supply Chain Management Student or an
            Entry-Level Supply Chain Manager 93
    5.1.2 For a Mid-Level Supply Chain Manager.....................................93
    5.1.3 For a Senior or a Supply Chain Management Executive 93
5.2 An Overview of a Company's Organization Structure 94
    5.2.1 Indirect and Direct Procurement 95
    5.2.2 Indirect Procurement under the CFO 95
    5.2.3 Indirect Procurement under the COO 96
    5.2.4 Direct Procurement under CMO or COO 97
    5.2.5 Shadow Procurement under CTO or GM 97
    5.2.6 Heat Map of Key External Influencers and Decision-Makers 98
5.3 Organizational Structure of an Indirect Procurement Team............100
    5.3.1 Strategic Sourcing Team 100
    5.3.2 Procurement to Pay (P2P) Team...102
5.4 Organizational Structure of an Direct Procurement Team................103
    5.4.1 Global Supply Chain Management Team....................................103
    5.4.2 Supply Chain Planning Team...105
    5.4.3 Global Commodity Management Team 106
    5.4.4 Global Supplier/Vendor Management Team...............................108
    5.4.5 Supply Chain Program Management Team.................................108

DOI: 10.4324/9780429273155-7

5.5 Supply Chain Roles...109

    5.5.1 Indirect Procurement Manager/Indirect Strategic
Sourcing Manager..110

    5.5.2 Direct Procurement Manager/Sourcing Manager/
Strategic Sourcing Manager 112

    5.5.3 Vendor Managers...113

    5.5.4 Commodity Manager..114

    5.5.5 P2P Manager...115

    5.5.6 Supply Chain Program Manager.......................................116

5.6 Preparation 117

    5.6.1 Resume 118

    5.6.2 Interview 118

    5.6.3 Brand – Annual Goals..119

5.7 Concluding Remarks... 120

References.. 120

5.1 OBJECTIVE

Before attempting to bring meaningful changes in your organization, a successful supply chain professional needs to learn their organization well. Understanding the organization implies knowing the people as after all it's all about the people and how they perceive you. First, you need to align yourself with the responsibilities of the role that you have been hired to do and execute per expectations. After that, you need to raise the bar by executing at a level above and beyond the given expectations. You need to learn the processes and align yourself to the existing methodology before maturing and influencing changes that can improve the current functionalities of your organization. The duration for maturing in each phase may differ between an entry-level supply chain manager to a supply chain leader; however, they are necessary part of growing as a supply chain professional. Even if you have a lot of good ideas, being an inexperienced supply chain manager can lead to making crucial mistakes on the job. On the same note, a supply chain leader who never takes the time to listen and understand the organization may end up leading a dysfunctional organization and cause negative disruptions in the long run

5.1.1 For a Supply Chain Management Student or an Entry-Level Supply Chain Manager

There are many types of roles in a supply chain organization and this chapter will help you understand both the organization structure and the types of supply chain management roles and their scopes, as well as the impacts each role has to the organization and their respective career growth paths. For a student, this chapter will also provide guidance on courses that could become more relevant based on the role you want. By comparing job descriptions with the roles and organizations presented in this chapter, you can assess the organizational structure and the skills needed for the job function. This will help you to come up with your own set probing questions for the interviewer. It will also help you to better align with the interviewer's expectations.

5.1.2 For a Mid-Level Supply Chain Manager

If you are transitioning from a non-supply chain management role to a mid-level supply chain manager role within or outside your organization, especially after an MBA, this chapter will arm you with the right set of tools to assess your organization and its functional teams. It will present the necessary information for comparing and analyzing the scope of each individual role and management's expectations. This chapter will also help with navigating a large supply chain organization and how to form quick alliances. For a supply chain manager who is looking for their next move, this chapter will also shed light on skills and experiences sought by hiring managers and recruiters for more experienced roles.

5.1.3 For a Senior or a Supply Chain Management Executive

This chapter will help you with evaluating your upstream organizational structure and calibrate your downstream reporting structure. By comparing between the scope and deliverables of various types of roles presented here and how that relates to your team, you can identify gaps in your organization's structure or overlaps in accountability. This chapter is designed to work out the kinks in your organization on a much wider level and will help you prove your value as a leader. Self-evaluation always gives clarity in vision and this chapter provides that framework.

5.2 AN OVERVIEW OF A COMPANY'S ORGANIZATION STRUCTURE

In the overall organization of a company, where does the supply chain department report to?

Like any other suborganization inside a company, a supply chain department is a part of a larger organization. Supply chain is typically considered to be a strategic department as it is accountable for managing a significantly large amount of the entire company's monetary spend and is also responsible for keeping the business running. Therefore, as a member of the supply chain department you should analyze your organizational tree and understand two things specific to your company. The first is to understand who are the key decision-makers and the second is to understand the priorities of the company and the strategic expectations for the supply chain department that meet those priorities.

With this in mind, these analytical questions will help you understand the organizational structure.

1. Who is the C-Suite officer accountable for the supply chain suborganization in your company?
   a. Does the supply chain department report to the financial chain of your company and does that mean that ultimately the chief financial officer (CFO) is accountable?
   b. Or is it part of the manufacturing organizational structure and ultimately the chief of manufacturing or chief of operations is accountable for supply chain?
   c. Or is it reporting to a different organization like chief technology officer (CTO) or general manager (GM) of a business unit (BU)?
2. Are there more than one independent supply chain departments within your organization that report to two or more separate C-Suite executive officers?
3. Another important detail to consider is that reorganizations are a normal occurrence in any company, and so it is important to know the history of the supply chain department such as whether it has been reorganized and moved around from one pillar to another?

5.2.1 Indirect and Direct Procurement

Let us consider some broad categories of companies.

a. Companies that make software or provide services and need supplier management teams to procure goods or services for internal consumption only and not for using to create and sell a product.

b. Companies that make products and require raw materials or components and machineries to build their products and thus need supply chains to procure goods and services for external consumption only.

c. Companies who need both, i.e. a large company who could be both into manufacturing as well as selling software or services.

Option A is also known as indirect supply chain, wherein a company has a dedicated supply chain or corporate procurement team to manage the procurement of all goods and services needed to run the daily business. Option B is called direct procurement where the company is continuously procuring goods to build products for reselling. Option C is a case where a company has both indirect and direct procurement which is becoming more and more common.

5.2.2 Indirect Procurement under the CFO

Quite often, indirect procurement is called strategic sourcing or corporate sourcing and the supply chain department sits under the financial pillar. If you are new or have transitioned to this role from the direct procurement side, you will soon realize that the prioritization of indirect procurement is quite different from direct procurement. In this role the supply chain organization's peer organizations are legal (contracting), financial planning and analysis, revenue management, accounts payable and other financial functions including finance IT (also sometimes referred to as business applications which manage large scale financial applications and reporting). For an indirect procurement team, customers/people who order from the procurement department can come from all parts of the company. It is crucial to understand who your peer organizations are and how they perceive procurement in terms of value addition to the company. Primarily they may expect three things

(1) Ensuring not to be in an out of contract situation. This translates to on-time closure of procurement contracts with vendors by negotiating commercial and legal terms and conditions. This avoids any disruption in the goods and services needed for the company to run its daily business. (2) Maintain compliance with organizations such as audit, SOX and ISO and other requirements in terms of the processes. (3) Cost negotiation and spend optimization so that the company procures all the goods and services it needs at the best possible price.

Such procurement teams will find its peer organizations speaking in dollars and cents and thinking in terms of fiscal quarters. Being part of the CFO's organization, it will be important to understand the financial impacts, capital expenditure (CAPEX) budget and operational expenditure (OPEX) budget. Being the arm which controls the largest amount of monetary spend in the company, this team can reduce the spend below the budget allocated and impact the bottom line. Expect to be reviewed more as a financial organization with financial metrics (cost savings, avoidance, volume of business, spend growth, etc.) as that is how your peer organizations will be reviewed and compared to each other (Dovgalenko 2020).

5.2.3 Indirect Procurement under the COO

Indirect procurement may also report to the chief operating officer's (COO) organization instead of reporting to the CFO. In that case your peer organizations could be various central function organizations such as workplace resource management, IT and HR. It is important to understand how the organizational expectation changes when run under the COO instead of the CFO. The priorities can potentially shift from working as a tight procurement group under the finance organization to procurement individuals embedded as dotted lines between specific BUs while reporting to the COO. While this may allow you to be hinged to two different cultures, i.e. the parent organization (COO's organization) and that of the customers' organization (BU), this often helps you to better understand the BU's needs and challenges and communicate back to the parent organization as you become the liaison between the two. As a result the team's performance review could be more like an operational review against the business goals of the BU (operational metrics such as service level agreements (SLA) and escalations) as opposed to looking at primarily financial metrics under the CFO.

5.2.4 Direct Procurement under CMO or COO

This is perhaps the most prevalent supply chain role. A company is making a product and the product has a bill of material (BOM) and the marketing team has already created a business forecast. A procurement team is procuring the underlying raw materials or components in bulk according to the forecast to meet the production demand. The clarity in the role creates clear charter and hand off between supply chain and its peer organization who are all ultimately reporting to the chief manufacturing officer or COO. Cost, quality and lead times are perhaps the three key tenets the peer organizations will be expecting this organization to manage. The supply chain organization will be negotiating purchase price of items and leveraging the volume over time to ensure the lowest possible BOM price. The factory is expecting the team to manage the lead time for delivery so that there is assurance of constant supply to build the product, and of course the parts are per specifications and of the highest expected quality. For such organizations, ops reviews tend to be focused on certain key areas such as cost savings and production interruptions, and supplier performance measurement.

5.2.5 Shadow Procurement under CTO or GM

For large organizations, it is not uncommon to create a secondary direct procurement or supply chain team under a different officer in parallel to an existing indirect/direct procurement team. In a company involved in manufacturing this is a normal practice. An indirect team under finance may be working toward managing all the spend for all daily functions of internal employees such as procuring laptops and other software and hardware. At the same time a direct procurement team is working separately under the head of operations to support manufacturing efforts. The charters for each of these organizations are different and they coexist separately. There is minimum opportunity of cross leverage and the terms and conditions of business are also quite different.

However there are situations where the business needs of a particular department or group change significantly from the rest of the company. Strategically such situations demand for a dedicated procurement or supply chain team. An example would be a GM of a new BU attempting to develop a new technology and requiring a new set of suppliers with different risk profiles, hence setting up a second direct supply chain. Another example would be the CTO office running technology operations for data centers and managing an enormous amount of spend which may be more than 50% of total corporate spend and thereby requiring a dedicated technology supply chain manager to manage a secondary indirect supply chain. Supply chain managers from the CFO or COO's office may not meet the challenging business expectations of these BUs primarily for two reasons.

a. Inability to execute per the timeline demanded by business due to overweight corporate processes.
b. Limitation of resources as the central procurement teams may staff up with more generalists who can handle various types of vendors and not having supply chain professionals who are specific technology experts and stay dedicated to their technology areas.

In other words, one solution doesn't fit every problem and neither does a single supply chain department, especially as a company becomes larger and more complicated.

Depending on the clarity at the executive level and their shared commitment to support strategic needs, a parallel supply chain department gets a clear charter to operate with defined roles and responsibilities. However, without clarity, a parallel supply chain department continues as a shadow supply chain team and has overlapping responsibilities which introduces a certain amount of dysfunction. If you are part of a shadow or secondary supply chain team, your performance will be reviewed in light of the strategic purpose of the team's formation. The criteria could be agility and on-time onboarding of vendors, predictable contract closure SLA with vendors for uninterrupted business continuity, technology alignment in proto-development and investment by vendor partners, budget savings, etc. This organization's success may also depend on limited collaboration with peer organizations and the core supply chain organization. Being able to leverage the second organization without being consumed by the core organization's processes is crucial, otherwise the purpose of the formation of the shadow organization gets defeated.

5.2.6 Heat Map of Key External Influencers and Decision-Makers

To be able to influence higher level executives and company decisions is an essential requirement for a supply chain manager. A large transactional decision for a company often requires driving consensus and seeking alignment between multiple stakeholders. A supply chain manager needs to be able to identify and build exec relationships with his customers to drive them to a decision point. In a large and complex organization, identifying who is the ultimate decision-maker on an issue and being able to influence them with the right business advice is a skill that every successful supply chain manager is proficient in.

For example, if you are part of an indirect procurement team and are doing technology procurement you need to develop relationships with the VP of engineering. The functional managers or principal engineers may not feel comfortable relinquishing the purchasing decision or may have a different viewpoint from the procurement manager. However, their choice could be driven based on engineering alone whereas the VP can be influenced by a supply chain manager coming from outside of their own organization. The VP has to view the procurement decision based on both engineering and business judgment. Similarly, in a direct procurement situation, a supplier with better quality products but higher prices may not be favored over the incumbent, with better price and preexisting design advantage, by engineering. If you or your chain of command has the necessary relationships, then you can broker a conversation between the design engineering leader, engineering leader and supplier executive toward an adoption path of this new supplier while working out a more favorable price. You can create a win-win situation where engineering gets better quality, the supplier beats incumbents price to get new business opportunity and supply chain gets better cost and also a second supplier. This may not have been possible by discussing with the stakeholders (engineering and supplier) in isolation.

Creating a heat map of the decision-maker s and their hierarchy is very useful especially when you start a new role as a supply chain manager (SCM). When you are an individual contributor, you should have key design decision-makers and managers or directors as the main constituents of your heat map. You may work with their reports on a regular basis but when it comes to drive decisions you will need to be able to garner sponsorship from these execs. If you are a supply chain exec, your heat map should consist of directors and VPs of your customer organization as well as your own chain of command. Having an understanding of your supporting ecosystem is of equal importance. For example, negotiating legal contracts is a significant task for a SCM. Therefore developing strong relationships with members of the legal team is very helpful both from learning the legal issues surrounding a contract and in order to get faster turnaround of reviews for your contracts under negotiation.

5.3 ORGANIZATIONAL STRUCTURE OF AN INDIRECT PROCUREMENT TEAM

In the previous section, we looked at a supply chain team from an external point of view. We also discussed how to determine the interdepartmental expectations from each type of supply chain team. In this section we will look at a supply chain team and its organization structure internally. We will look at the various types of supply chain departments that may exist and their solid line and dotted line relationships with other departments outside the organization.

5.3.1 Strategic Sourcing Team

This is perhaps the most common form of sourcing team that exists in almost every company. Every company needs to buy various goods and services to run its IT, such as laptops for its employees, networking equipment to run the office WIFI network, HR applications to manage its employees records, salaries, and finance applications to manage the companies accounts and track sales. Each company also needs travel services, workplace resource services, healthcare services, food and beverage services, etc. As the number of employees in the company grows, a company creates a dedicated strategic sourcing team to get more favorable prices for these services through economy of scale. This team works with internal customers and external vendors and manages the business relationships (Monczka et al. 2020).

Typically, depending on the volume and diversity of businesses, a strategic sourcing team could be divided into multiple technology-based or category-based teams. For example, there could be a team which is dedicated to supporting technical operations, such as handling technical contracts and purchases of technical items such as software, internal networks, computer and storage needs. Similarly there could be a group which is working with Airlines, Hotels, Credit card companies, event management companies and similar providers of services. An individual contributing member of this type of team will typically handle a set of vendors and the associated contracts, whereas a manager of such a team would be responsible for the entire category. The senior most leader will be responsible for managing the entire spend of the company in these categories and performance of all the vendors.

The main expectations from a strategic sourcing team are minimizing expenses, negotiating favorable terms and conditions in the contracts with vendors, and handling escalations with vendors. This team is the face of the company to its supplier ecosystem as well as the liaison between internal leaders and execs on the vendor side. The team is also responsible for business continuity, such as the renewal of existing support arguments or the onboarding of a new vendor for a new type of purchasing requirement. Risk management is also another key responsibility. There are various forms of risk, from a vendor failing to provide a service which is essential for running the business, purchased hardware that having unacceptable quality which may have to be returned, indemnifying from vendor's liabilities, competitive and IP protection, etc. All departments who need to engage with external suppliers, require contracts to be executed, prices to be negotiated, new vendors to be identified, etc., are customers of this department (Payne and Dorn 2012).

About 4%–6% of annual revenue is the typical percentage that is spent on indirect procurement for a fortune 500 company. So, for example, a company generating one billion dollars in revenue annually may have a demand in the range of forty to sixty million dollar worth of goods and services in order to run the company for a year. However, this spend might span across a very large number of vendors and therefore often it is not possible for a small strategic supply chain group to manage and optimize every little purchase. Quite often this spend is very skewed. A handful of suppliers may be consuming more than 50% or 60% of the total spend, and then there is a long tail of suppliers for the rest of the spend. A small strategic sourcing team often focuses on the top spends only due to the limited bandwidth. However, in order to manage the rest of the spend, it also establishes self-service procurement policies and preferred vendor lists, catalogues, etc. which it then enforces for the rest of the company to adhere to.

An indirect procurement team under finance also gets involved in managing budget and cash flow of the company. Being the arm of the company through which large spends are executed, a sense of fiscal responsibility and compliance also comes with the territory.

A strategic sourcing department depends on a number of teams to be successful. Two of which are the legal team and the tools and business applications IT team. The former is needed to finalize any legal document and the latter is needed for data analytics, business intelligence and business monitoring to drive procurement decisions.

5.3.2 Procurement to Pay (P2P) Team

A procurement to pay or a P2P team goes hand in hand with a strategic sourcing team. It is responsible for the operational aspect of sourcing which includes creating purchase requests (PRs) and purchase approvals as well as receiving invoices and ensuring timely payment to the vendors. Typically, a hierarchical organization will put spending limits on its officers. For example, a manager can approve only half a million dollars of spend whereas a director may have a limit of one million dollars. Any spend above the limit will need to be approved by a VP or the CFO. Companies set policies like this to control authorization limits to put proper scrutiny before approving a purchase.

A P2P team is responsible for creating all the necessary documentation for a PR. Why is this important? There are three main reasons: (A) execs are signing their names as their authorization for the purchase and are therefore liable, (B) details about the purchase including description and price are documented and put into company history, and (C) approval will result in a purchase order (PO) being issued and subsequent workflow.

Reason A is very important from a trust point of view. It is not possible for an exec to know the minute details of every single contract. So the first signatory puts his authorization on the PR and the next signatory applies his judgment between his knowledge about the purchase, content provided in the PR and trust that his previous signatory has done his due diligence and approves the PR. As the PR goes through the approval levels, the knowledge of specific details decreases and trust on the content and previous approvers increases.

Reason B is very important as the PR becomes the source of truth when all the relevant information has been aggregated. The underlying legal contract, the final quote provided by the vendor, the detailed description of the goods or services to be received, its unit cost, total cost, taxes, delivery cost, etc. all need to be documented and set in stone. The cost and description of the goods or services are replicated from PR to PO. The archived PR and POs are often used to research past purchases.

Finally reason C is important as the approval of a PR results in a workflow. The accounts payable team issues a PO which is then transmitted to the supplier. This now becomes a contract with the supplier. Both internally and externally a set of tasks are done as per the documentation ending with matching the invoice and payment. For a large organization where teams such as internal customers, suppliers, strategic procurement and accounts payable are all located in different places and handle thousands of transactions per week, the content introduced by the P2P team to start a PR is the glue that holds everything together.

Having both qualitative and quantitative data and being able to create easily searchable metadata from the PR/PO is every organization's goal. Many organizations develop specific forms and formats and even coded acronyms to maintain category-based information in the PR. Along with this content capturing for PR, the P2P team is often responsible for the logistics and delivery to the end customer from the supplier and in parallel ensuring the delivery of PO for on-time invoice payment to the supplier.

A strategic sourcing team and a P2P team combined together forms the most prevalent corporate procurement team focused primarily on procurement of goods and services needed to support the workforce and their day-to-day needs.

5.4 ORGANIZATIONAL STRUCTURE OF AN DIRECT PROCUREMENT TEAM

5.4.1 Global Supply Chain Management Team

A direct supply chain is established to procure the goods and services necessary for revenue generation. A global supply chain management team is responsible for optimizing end-to-end procurement of all goods and services to support the production. The scope of this team spans from working with internal teams to understand the requirements and demands, assessing the global supplier landscape to identify suitable suppliers for sourcing the right item to support demand, negotiating contracts and prices for the sourced items and ensuring proper logistics for on-time delivery of sourced goods from point of origin to point of assembly across the world. The three main accountabilities of this team are managing cost, quality and lead time for delivery.

In order to maximize the economy of scale, modern organizations are hyper focused on horizontal integration. Companies are focused on doing only a few things but are focused at being the best in class in those areas and scaling up sales volume and gaining market share quarter after quarter. In order to support this, a global supply chain team is focused on its direct suppliers and tries to create as much competition as possible so that the suppliers offer best cost and services to win the business in exchange for an increase in business volume quarter after quarter. Let's take the computer industry as an example to illustrate this approach.

Previously, computer manufacturers used to design the computer, buy the needed components, assemble them and then manage the sales and distribution in order to sell the computers to an end customer. The change in this industry is a great example of switching from vertical integration to horizontal integration. Now the computer industry is divided into three separate tiers: (1) manufacturers of cost-sensitive personal computers – to lower the overall cost of the hardware, a tier of original design manufacturer (ODM) companies have emerged, many of which are located in Asia and are now primarily focused on manufacturing these computers in very high volume; (2) a second tier of original equipment manufacturer (OEM) companies are sitting in the middle which are developing the design and getting the servers manufactured by ODM companies in high volume but reselling under their own brand to small, medium and large customers who buy in relatively large quantities and need a lot of after sales support. Finally, (3) a third group of companies who are buying various types of computers in relatively large quantities to build data centers and offering computational power as a service. Each layer is restricting themselves to servicing their fixed category of customers to drive volume and economy of scale. Thus, the computer industry has transitioned from vertical to horizontal integration. There have been acquisition and consolidation within each band but not necessarily across the bands.

The finished products of each layer in the above example are the raw materials for the next layer and therefore are part of the scope for the supply chain team of the higher layer. For example, a supply chain management team that belongs to a company which is in the business of data centers is managing a host of OEMs that are responsible for manufacturing the servers, network and storage hardware that will then be used in the data center. Each downstream OEM supply chain team is in turn focused only on the server or the network or the storage depending on their area of expertise. For instance, a server supply chain team is managing a group of server ODMs and buying fully assembled servers, whereas a network supply chain team is procuring from network ODMs. Further downstream, an ODM is managing its specific set of electronic and mechanical items, such as a server ODM supply chain team, which is dealing with the electronic components vendors who supply things such as the processor, memory, hard drive, motherboard, power supply, mechanical assemblies, etc.

Due to diversity in categories, a global supply chain organization is often divided into smaller teams, each of which is focused on a specific category of raw material or subassembly. The specificity could be based on technology, prices, geographic location, etc. Going back to the computer example, a data center team can have two separate supply chain teams: one handling computer procurement and the other handling network procurement. Further downstream a server ODM team can have six global supply chain teams for processor, memory, hard drive, motherboard, power supply and mechanicals.

A category-specific global supply chain team is expected to be knowledgeable about the supplier landscape, business trends and technology transition in its respective area. As a supply chain individual progresses in a direct supply chain organization, their spend volume and domains of influence increase. An individual contributor may start by managing a subcategory such as a very specific type of computer server procurement. Then they may grow to lead the end-to-end supply chain for an entire category such as all server and storage hardware lifecycle management. At this point everything from procurement to logistics of demand management to negotiating warranty cost for maintenance becomes part of their scope. Continuing with the example, the next higher level becomes accountable for all technology procurement categories, which includes servers, storage, networks, etc.

5.4.2 Supply Chain Planning Team

Similar to the P2P team in an indirect supply chain, the global supply chain organization works with the supply chain planning organization, sometimes referred to as the buyers team. This department is the operational pillar of the supply chain as it ensures on-time purchase of the raw materials.

A supply chain planning organization sits in between the global supply chain organization, which is strategic in nature, and the core manufacturing organization, which is completely operational. Being at the middle makes it the most operationally efficient. For example, while a global supply chain organization is divided by technology categories, a supply chain planning organization could be divided into smaller teams to support specific business units (BUs) on all their needs regardless of category. In other words, while the supply chain teams are divided vertically, the supply chain planning team is divided horizontally.

A supply chain planning team is typically responsible for procuring large numbers of diverse items from a large portfolio of suppliers in order to support the product portfolio of one or more BUs, or a specific geo-located factory. As items have different manufacturing lead times, the goal of the planner is to ensure all pieces arrive at the same time in order to have continuous production. The core responsibilities of this team are to communicate forecasted demand to the suppliers for their own planning purposes, ensure timely triggering of POs, delivery logistics and maintain buffer stocks in case of emergencies or issues. Depending on the complexity of the final product, the supply chain planning team can have very complex responsibilities. These teams leverage many types of manufacturing resource planning tools and have an ecosystem of partners providing various kinds of services such as transportation, warehouses and insurance.

An entry-level supply planner may have a few BOMs to manage. As they progress through their career, their responsibility may scale up to handling a larger portfolio of products and/or increasing span of control over the supporting infrastructure, also known as the network design. The basic expectation from the BU, which is the customer of this organization, is to ensure that the production line can never go down due to unavailability of supply. From a view from above, the company would expect this organization to ensure continuous supply in the most cost- and time-efficient manner while minimizing both overstock and shortage risk. The execution success of this team depends on efficient planning and agile communication of the plan internally and externally.

5.4.3 Global Commodity Management Team

Supply chain is an area which is constantly being optimized for improving revenue margins. Looking at the suppliers' supply base is one of the ways to achieve cost efficiency. If the cost of suppliers' raw materials could be reduced, then the savings could be passed through by reducing the cost of the procured items.

Let's take denim jeans as an example. A denim jeans brand may be working with five jeans manufacturers globally that are all making the jeans per the brand's specifications and meeting a particular price target. Assume a pair of jeans has four items in its BOM – denim cloth, metal buttons, zippers and thread. Instead of allowing the manufacturers to procure their own supplies, the brand ranks the amount of money spent on each of the four items and the number of suppliers in each category. Most likely, the denim and zipper industries have a limited number of suppliers while the button and thread industries have many manufacturers. Another way of looking at this is from a cost point of view. Purely from the point of view of material cost, let's say that the denim is 80% of the material cost, the zipper is 10%, the buttons are 8% and 2% of the cost is for thread. From a total material cost point of view, a 10% cost savings on denim will yield an 8% savings while a 10% savings on buttons will only be 0.8% in total savings. The brand can set up a supply chain team which develops direct relationships with a limited number of denim manufacturers and in exchange for assurance of business from the brand, the denim manufacturers give both assurance of supply and lower cost for the specific amount purchased by the manufacturer on behalf of the brand. The brand restricts the manufacturer to source the denim from the specific denim manufacturer and gives the manufacturer freedom to source the zip, button and thread, which lowers the cost for the manufacturer and thus lowering the cost for the brand as a whole.

This is an example of where a special supply chain team is created to do skip level supply chain management. This team manages a selected list of items that are needed along the manufacturing process which are critical. The importance can come from an assurance of supply point of view where shortages due to market condition change is a common occurrence. Having direct relationships and restricting the sources can ensure assurance of supply for the manufacturers. The importance of this team can also be from reducing a percentage of the total material cost which leads to impactful cost savings such as in the example above.

This team of specialist supply chain managers drives business decisions based on market condition changes and ensures a continuity in the suppliers' supply chain. This team also influences both internal and external organizations to choose from the limited list of commodity suppliers so that multiple primary suppliers aggregate to the selected commodity suppliers. Going back to the denim jeans example, the commodity supply chain management team in the brand company will influence internal designers and the manufacturers to source the denim supply from the chosen denim manufacturer or chosen zip manufacturer in order to reduce costs.

This team is very influential for cost control and ensuring supply during shortages. Some companies may have contractual agreements with the critical commodity suppliers but sometimes the team works without any contract but rather based on relationships. Such relationships bank on the accuracy of forecasts provided by the commodity managers to the commodity suppliers and the market share controlled by the commodity managers. In certain industries, the market conditions change extremely rapidly and a commodity team is expected to always have its pulse on the market.

5.4.4 Global Supplier/Vendor Management Team

For large, multi-billion dollar supply chains, there is a point where it becomes difficult for the same team to manage all aspects of the strategic supply chain's internal and external relationships. In such cases, the supply chain team is divided into a sourcing team and a vendor management team. The sourcing team stays operationally focused, liaising between internal stakeholders and suppliers and ensuring supply chain continuity. The global supplier management or global vendor management team curves out the relationship management portion of the supply chain.

The primary deliverables for vendor managers are monitoring and maintaining vendor performance, negotiating master purchase or overarching contracts with the vendors and being accountable for the overall commercial relationships with vendors. Having a dedicated vendor management team is a sign of supply chain maturity where regular business reviews of the suppliers are done on a quarterly or at least an annual basis. While the sourcing managers may negotiate a price for an item, vendor managers ensure total portfolio-based cost reduction. For instance a supplier may be supplying multiple items and a vendor management team maintains the overall portfolio outlook of that vendor and controls its growth or decline based on its performance. Based on vendor's performance in the areas of providing cost reduction, delivering goods on time, maintaining quality by keeping defects low, meeting expectations in providing support and creating new products to support the customer's roadmap of new products, a vendor management team either allows a supplier to grow or decline in revenue by reducing total business engagement. In summary this team manages the business-to-business relationship between the company and the supplier. This team is also expected to monitor suppliers health, merger and acquisitions in supply base, and maintain a dynamic list of upgraded or downgraded suppliers, also known as a preferred supplier list (PSL).

5.4.5 Supply Chain Program Management Team

Along with running the day-to-day business, a supply chain organization, like any other organization has to take internal initiatives to improve its processes and gain more operating efficiencies. These initiatives or programs are often diverse in nature, such as implementing a new manufacturing resource planning tool, identifying a new vendor to create a supply base, launching cost reduction and value engineering efforts to improve product margin, introducing a new network and meeting new compliance regulations. These efforts are cross-functional and often also have external impacts, touching the vendors as well. These programs or initiatives are typically managed by a separate supply chain program management organization.

Some of these teams within the supply chain program management department have specific charters and are focused on creating efficiencies in particular aspects of supply chain. For example, let us consider a network router which needs to be produced for 7 years due to customer demand. In order to support this there could be a supply chain program management team focused on assessing the risks around keeping the production running throughout the product's 7 years life. In order to continually produce the item without any interruption, the constituent subcomponents should be also procurable through the same period of time. Often that's not the case as the subcomponents needed also have their own life cycles. It is quite common that some of the subcomponents available at the beginning of the production will become obsolete and may not be available any more after a few months or a year into production. This is known as end of life (EOL) of the subcomponent. The price of the subcomponent can also increase to an unfavorable level. However, in order to continue production, alternates for those EOL or unfavorable parts have to be identified. This team measures risk of production stoppage and takes measures like finding additional spare material or alternate subcomponents ahead of time. Another example of a specific charter for a supply chain program management team is compliance. These are teams which look at the vendor landscape from an ethics and compliance point of view and ensure the suppliers manufacture their offerings in an ethical and compliant manner based on company values and guidelines. Similarly, there are other teams in this suborganization which look at processes and implement new business processes and do change management.

These teams are all a part of the broader supply chain umbrella and depending on the specific charters, they will work on certain facets of the supply chain. Their purposes are often diverse but typically the objectives of these teams are either gaining efficiency or compliance. Gaining any kind of efficiency in the supply chain ultimately translates into cost savings which is why it is so important. Compliance also translates into avoiding risk and therefore, cost avoidance.

5.5 SUPPLY CHAIN ROLES

Let's recap what you've learned so far in this chapter. First, we took a broad view and reviewed where a supply chain team fits in within a corporate organizational structure. Then we zoomed in to explore the overall supply chain team as an organization and what that looks like with various teams inside it. From here, we viewed the teams specifically and now we will focus on job roles/functions and the responsibilities of an individual within those specific teams. Supply chain organizations throughout the world require individuals with various skill sets and qualifications. The responsibilities for each of these roles get even more nuanced by specific industries. The list below is by no means an exhaustive list and neither are the responsibilities and qualifications described; however, it is a good starting place.

The descriptions below could be well supplemented by searching job descriptions in popular professional social media platforms and job search platforms on the internet. By searching a job title, one can easily find the current job openings and do a deeper dive on the skill sets and qualifications that are being sought after for that specific position. Similarly, these platforms also offer profiles of people who have similar job titles or leadership titles in the same functional area that can then be used as a gauge. By studying their profiles and career path, you can develop an understanding of the list of responsibilities and a map of desired skill sets and experiences to be successful in your professional journey.

Also depending on the size of the supply chain organization and its evolution as a department, the distinction between the roles described below can become blurry and may have overlaps. For example, in a smaller organization, the same person can wear multiple hats whereas in a very large and structured supply chain organization, the roles and responsibilities could be more granularly designed. However, reading all the descriptions given below will help you to develop a comprehensive view point.

5.5.1 Indirect Procurement Manager/Indirect Strategic Sourcing Manager

The fundamental expectation from a strategic sourcing manager or procurement manager in an indirect procurement organization is to be able to establish and nurture relationships with internal and external business stakeholders. The role requires extensive interaction with both technical and nontechnical business groups such as R&D, marketing, HR, professional services and also creating purchasing strategies to meet their business needs in a timely manner while minimizing cost and business risks. These strategies are typically long term focused, clear and measurable. The execution of this strategy involves a range of tasks, starting from gathering the business requirements from the stakeholders, supplier selection through a request for quote (RFQ) process, onboarding the supplier through completion of legal contracts, up till the final purchase of the goods and services. As this role involves extensive interaction with legal, compliance, supplier risk, information security, tax, etc. soft skills such as relationship building, negotiation, communication, coupled with diplomacy and professionalism are essential.

Typically as an individual takes up the responsibility of a procurement manager, there will be existing strategies already in place or deals in flight which need to be taken to the finish line. Various projects will be in different phases of their life cycles as well. However, during the course of managing procurement, there will be opportunities to work on projects from inception, such as creating new strategies or sourcing for a new supplier, etc. Working on a new or a renewal contract or doing a RFP (request for proposal) with vendors is a typical task for a procurement manager. Also, if the size of the procurement team is small, a procurement manager may have to manage too many suppliers and do a diverse set of deals in parallel. This may not provide the satisfaction of acquiring deep product knowledge or knowledge about market dynamics or build strategic relationships with strategic vendors to develop products jointly. Instead there will be constant urgency around finishing the negotiation and closing the contract to ensure business continuity without any interruption. The focus will be on minimizing risk and optimizing cost and ensuring the entire process is completed on time which meets the timeline of the business stakeholders. In other words, this job could be challenging unless you approach the job as a category manager.

For large organizations, typically a practice called category management is utilized to develop subject matter expertise and consistency. A category manager typically manages a specific category of business and thus gets the opportunity to work on the same area with the same set of business stakeholders and suppliers. This gives a manager the opportunity to invest more in relationship management with business partners and supplies, and to analyze and gain knowledge in a specific market condition and create strategies based on this.

Making a business decision to award business to a supplier is a key deliverable. However, this decision comes through influencing, communicating to high-level executives and ultimately leading cross-functional teams into making an object decision. This is an art which a procurement manager needs to master. To be able to gather, analyze and present data also plays an important part in developing this mastery. Stakeholders often have different ways of ingesting information. For example, some stakeholders may believe only in data-driven decisions and may ask for a lot of data to back any advice while some stakeholders may be the exact opposite. A successful procurement manager is mindful of the intended audience and adjusts the content based on the audience. Knowledge of compliance with procurement processes and procedures is also very important.

5.5.2 Direct Procurement Manager/Sourcing Manager/Strategic Sourcing Manager

The role of a direct procurement manager is very similar to that of an indirect procurement manager; however, this role may be more narrowly focused with deeper involvement. A direct procurement manager is usually a specialist in a particular category with a fixed set of vendors specific to that category. This role requires developing a deeper relationship with both the sales and product side of the supplier while at the same time requires knowledge of underlying product technology, benchmarked results against competition and industry pricing trends. The ability to negotiate price and get quarterly or monthly price reduction is expected. Developing subject matter expertise to understand market trends and creating global strategies is also an expected responsibility from a strategic sourcing manager. This role requires the ability to develop new supplier relationships, sustain these relationships by doing yearly or may be more frequent business reviews, and large-scale performance management.

As a sourcing manager, you need to negotiate contracts with suppliers which will optimize your own company's position. Creating new contracts such as a master supplier agreement or a master purchase agreement are pivotal and define the overall business between the customer and the supplier and is quite often a very long and arduous task. Navigating these sometimes takes months but they are essential for doing business. Amendments or contracts defining a statement of work are also quite common contracts. Executing these contracts in a timely manner often depends on proper planning, support from legal and choosing the right time when the supplier is hungry for business. Quite often the internal stakeholders such as engineering or manufacturing teams are too inwardly focused, and for a large organization where multiple teams may be doing product development with overlapping requirements, a good direct procurement manager provides both market intelligence and cross pollinates internal information.

The final measures of success for a procurement manager are of course cost saving and timely delivery of contracts with effective management of lead time (O'Brien 2019).

5.5.3 Vendor Managers

When a supply chain operation starts managing a large number of suppliers which interact with multiple BUs of a corporation, it becomes unmanageable for a single supply chain manager to manage the end-to-end vertical processes for all the parallel engagements. Such organizations tend to break down the vertical job flow into parallel work streams. One example of such a work stream is a sourcing manager and vendor manager. The person who is inward facing and interacting with internal product managers, engineering and liaising between them and the product side of the supplier becomes the sourcing manager. Meanwhile the person who is outward facing and primarily handling the sales and manufacturing side of the supplier becomes the vendor manager.

The day-to-day job involves contract negotiation and compliance, vendor health, KPI (key performance indicator) data reporting and performance monitoring, doing business review, quarterly or monthly cost negotiation of the overall portfolio of spend, escalation mitigation, program managing corrective actions, auditing suppliers, etc. Depending on whether a supplier is a privately held or publicly traded company, information of the supplier may be only available through special sources or the internet, respectively. A vendor manager maintains an external and internal profile of the supply base which should have essential information such as internal spend amounts by categories as well as supplier revenue and profitability.

M&A and competitive landscape also provides market intelligence which is often used to assess risk of doing business with a supplier and is a strategic component for business viability and continuity. Along with risk, it is imperative for a vendor manager to work internally with a cross-functional team on a quarterly or at least on an annual basis to assess vendor performance on factors such as technology, quality, responsiveness, delivery and cost requirements (TQRDC). Depending on organization, these dimensions are represented by numerical scores or more qualitatively with red, yellow and green lights with a vendor manager owning the responsibility of making the vendor accountable for TQRDC. Finally the key ingredients to success as a vendor manager are business acumen, communication and influencing skills by having all the data mentioned above at your disposal.

5.5.4 Commodity Manager

In order to manage cost and ensure on-time delivery of parts it is a common practice to go beyond the first level of supply chain. For example, a data center supply chain who buys servers will go beyond the OEM server manufacturers and manage the supply chain of electronic components such as the processor, memory and hard drives. Similarly a denim jeans retailer will go beyond the apparel manufacturers and negotiate the raw materials such as specific types of denim cloth, zippers and threads. A commodity manager manages the secondary items for various reasons. First of all it helps in optimizing cost as the end customer now manages the cost of supply to their supplier who can no longer get away with arbitrary profit margins. Since the commodity or the raw material cost for making the part is known, the commodity manager negotiates price with the parts supplier based only on the value addition such as labor and SG&A (selling, general & administration) cost as well as a negotiated amount of profit margin. This way, the commodity manager does not allow the parts manufacturer to apply any arbitrary markup on the commodities. By directly negotiating with the commodity vendor, a commodity manager also opens direct communication with the commodity vendor. This allows the commodity item manufacturer to get a forecast of the quantity needed by the end customer which allows it to plan its own manufacturing pipeline more efficiently which is passed up as lower cost and assurance of supply.

A commodity manager often has to maintain a roadmap of the commodities and their underlying technology. Commodities often have latencies. In other words a commodity that is negotiated today will go through multiple manufacturers and assembled into stages of subcomponents before it becomes a final product. For example, a camera lens gets assembled into a subassembly consisting of the lens and the electronic circuitry around it and then gets assembled inside a phone to make the final product. Optical lens being the commodity here is the heart and soul of the end product which could be a cell phone with best in class camera with highest possible resolution. By definition, commodities are produced in much larger quantities and keeping the general market in mind, typically have market-driven pricing and margin pressure. A commodity manager helps in choosing multiple vendors for the same commodity to create a competitive scenario and to get the best cost while at the same time due to low margin nature of the commodity category, limits the number of vendors to ensure the vendor does not lose interest in doing business. As commodities are constantly evolving, a commodity manager has tremendous influence on the product design, as the end products depend on the features of the commodity. Specially due to the latency, a commodity manager plays a pivotal role in influencing the commodity supplier to align them with future feature requirements of their own products. A commodity manager also offers cost reduction or value engineering opportunities to the product manager when a better or cheaper alternate part comes up in the market during the product manufacturing life cycle. This long term aspect is very strategic for both the commodity manufacturer and its consumer. Engineering or deep product knowledge of both commodities and the end product is thus essential to be successful.

When a commodity is both cost sensitive and has cyclical seasonality of supply, the price can soar and soften in rapid succession. In such cases, knowing when to wait and when to buy in is critical. A commodity manager has to track market conditions and future trends in supply along with the seasonality of demands very carefully as based on their recommendation, a company will procure material and manage the costs which ultimately impacts the bottom line.

5.5.5 P2P Manager

Unlike the previous set of roles where you may be executing according to a process, in this role, you are the business process owner (BPO). The P2P manager is a critical role as you get to define policies and controls, as well as processes and metrics for end to end of the supply chain. You are expected to be the subject matter expert for industry-level best practices as well as internal compliance policies. In this role you are expected to drive consensus among a large group of cross-functional teams such as business units, accounts payable, finance, IS&T, legal, logistics and tax to create, review and improve processes and infrastructures for greater operational efficiency and accuracy.

In this role a good understanding of enterprise resource planning (ERP) tools is also critical. As procurement BPO, you have to provide design requirements and procurement process flows for new implementations or modifications of existing systems and tools to realize maximum user friendliness, efficiency and automation while also meeting approval workflow and audit requirements.

Operationally, you may have to handle a team of buyers who is ultimately doing the purchasing at a final negotiated price. This is where the rubber meets the road. As part of the procurement there are multiple steps such as selection of the supplier, negotiating the price and finally agreeing to award the business to the supplier. However, finally the transaction has to happen and this final step culminates in the supply chain issuing a PO to the supplier. The PO is a contract which communicates the quantity of the item that is needed and the price the customer is agreeing to pay. This authorizes the supplier to ship the product to the customer and then it sends the invoice for the goods shipped or service rendered. To make this entire process smooth and efficient, companies often use various metrics to measure time taken to execute various steps, issues that arose and resolution times, cost savings, etc. The metrics range from the durations between making a PR to initiate the procurement process to finally issuance of the PO, time taken by supplier to fulfill the demand from the time of receipt of PO, adherence or compliance with all the checks and balances put in, interruptions and escalations for resolution, etc.

A key deliverable for this role could be creating a data management system which can assimilate data from diverse sources and make the data accessible through various reports and on demand data queries. It should be able to produce reports which are meaningful and actionable. It should be able to answer any data-related queries ranging from finding the health of the supply chain to forecasting and resource planning (Price et al. 2014).

5.5.6 Supply Chain Program Manager

A supply chain program manager is the program manager who implements and runs cross-functional supply chain projects which can potentially involve all parts of the supply chain organization and beyond. This role enables architecting solutions which can lead to better operations such as automation, reporting, decision-making, cost savings, integration with tools or migration to new sets of tools and processes. A key challenge for these tasks is to understand the various domains involved and draw the stakeholders out in order to propose and agree on the value proposition.

The supply chain programs get initiated in many ways but often, the typical driving factors are ERP system transitions, an exec recommendation, or a large and visible execution failure. When an ERP tool is being migrated, typically a supply chain program manager is needed to understand the existing process flows and this is a great opportunity to reevaluate the existing processes along with the data flow. A supply chain PM who has a background in IT and understands the back end master data flow along with domain knowledge of forecasting, ordering, logistics, etc. will be very successful.

For programs arising from executive edict it is very important to be able to create vision, mission statements and execution goals with close collaboration with the sponsoring executive and key stakeholders. A PM plays a pivotal role in order to strike a balance between audacious goals and executable steps from stakeholders while maintaining the project's momentum and not getting lost in executive impatience. For the third category of programs sometimes the goals are much more nebulous. At some organizations, external consultants are brought in as a result of a postmortem from a failure. Typically stakeholders throw each other under the proverbial bus when such postmortems are being conducted. A supply chain PM often plays the role of an evangelist who helps the entire organization to look forward and pivot around the changes.

Besides these large programs, there are also some continuous process improvement programs which are well defined and operationally focused on a supply chain program manager. For example, doing regular supply chain risk assessment of product lines requires understanding the availability and shelf life of all the subcomponents so that the manufacturing never comes to a halt. Additional examples are doing a "should cost modeling" of a product (benchmarking cost paid versus sum of all the subcomponents cost) and benchmarking progress against competition, while dash boarding and reporting KPI for overall internal supply chain health and monitoring (Venkataraman and Pinto 2020).

5.6 PREPARATION

At every step of one's career there is a need for preparation to jump to the next step. From a student who does not have any experience and is looking for an internship or entry-level supply chain job, to an experienced supply chain manager looking for a higher leadership role, everyone needs to either get through an interview or promoted organically within an organization. Either way it is a decision-making process where someone or a group has to choose a capable candidate from a pool of candidates. The decision is often made based on the set of information provided and the perception created. The information is provided in two ways, i.e. via the resume and during the course of the interview. For a hiring candidate, the main impression gets formed during the interview process. For a promotion discussion, the impression gets formed based on the brand created by the supply chain manager. The impression that a candidate creates has to inspire confidence in the decision-maker's mind that their choice of candidate is the best out of the pool of candidates, poses the least amount of risk for the job and offers the greatest amount of potential.

This chapter does not offer any silver bullet for success but tries to connect the three major points discussed in this chapter: the organization, the team and the role with the two aspects of preparations – information and impression.

5.6.1 Resume

When applying for a job, it would make a lot of sense to read the job description carefully to see if it fits any of the profiles presented above from all three meta perspectives. Is the supply chain organization a direct supply chain or an indirect one? Does it sound like a cross-functional role where one needs to work closely with finance or product development? It would make sense to read this description carefully and attempt to imagine the specific job function in light of the team and organization. It may also impress the hiring manager if a candidate seeks validation by inquiring about the organizational structure, team function, etc.

For an entry-level role where prior experience is not relevant, it is important to understand the analytical and soft skills expected in the job. For example, how does one convey their communication skills or ability to build relationships with executives. Highlight experiences which demonstrate desired soft skills in the resume that are relevant to the role.

For a senior supply chain manager, depth of knowledge in product, technology, supply base and conveying knowledge of market intelligence are key aspects. Coming across as a subject matter expert who can articulate a challenge in a crisp and concise manner is often sought after. For an executive level, along with breadth of knowledge in the supply chain and clarity in articulation, additional experience in areas outside of the supply chain draws attention. For example, knowledge of IT systems, engineering, legal or finance are often needed to be successful in the next level and can set one apart.

5.6.2 Interview

If your resume can be compared to facts or data, your interview should be compared to a PowerPoint presentation, where the right amount of data bring home the point and wrong amount of data can lead to confusion. Ultimately a supply chain manager is entrusted with the business and expected to make large monetary decisions and solve supply chain problems. Inspiring confidence is key. One way to inspire confidence is to have clear examples of the skills described in the specific roles as articulated in the roles and responsibilities section of the job description. Art of storytelling is important when giving an interview and it is fine to even describe a past failure as a learning experience and show growth. These days interviews are typically a long 4–5 hour process where different individuals conduct a one-on-one interview with the candidate. Interviewers will test various problem-solving skills and the best way to answer is to present a situation from experience. It is very important to have in-depth knowledge about that situation as the interviewers will probe for details and would try to determine what was your specific contribution as opposed to the team's contribution. Candidates often make the mistake of starting a story with blaming a person or a team as the source of the problem and that can create a negative impression about the candidate themselves. It is also important not to draw down from the same example to different interviewers. Very simply put, though the interviews are one-on-one but the interviewers do debrief together. Recycling the same anecdote or referring to the same experience could be interpreted as shallow experience level.

5.6.3 Brand – Annual Goals

A supply chain leader demonstrates craftsmanship and leadership through their execution. This creates and grows their brand within the company. Being decisive and being able to consistently perform and deliver creates a supply chain manager's brand as dependable and trustworthy. Communication and ability to drive a team of people to a procurement decision is hard but upon being successful, it is looked at as leadership skills.

A method for building your brand is by taking initiative. Based on the job roles described above, a supplier manager may discover gaps and risks. Mitigating those gaps by taking initiative is often the smartest way to grow one's career. Creating annual goals based on identifying those gaps and attempting to mitigate them is a good way to demonstrate leadership and build one's brand.

A key thing to remember is we can not promote ourselves and even our managers do not promote us. Promotion comes from the organization and is approved by a committee. Employees often believe getting promoted is a well-defined process in their organization and feel if they have been performing similar to a person who is at the next level they should be promoted too. Promotion is not a logical decision but an emotional decision made by a committee where the promoting manager makes an argument based on the candidates major contributions and the rest of the committee agrees or disagrees based on their awareness of the candidate's direct contribution and the candidates personal brand. Often the latter plays a bigger role than the former.

Not getting a promotion is also a disheartening experience and can lead to disenchantment and ultimately resignation. Before even developing an expectation to be promoted, it is important to understand if the current job is even scoped for the next level. For example, a commodity management role may not be scoped beyond the category of a senior commodity manager. So developing an expectation when you are in such a role which is not scoped for the next level may not work out. So, what do you do? You have three options: diversify by laterally moving to a different role within the company to get yourself more well-rounded, change from an individual contributor role to a people manager role within the company and look for roles with more responsibility outside your organization.

5.7 CONCLUDING REMARKS

In this chapter, various types of supply chain roles have been discussed. This chapter will help both entry-level supply chain managers and supply chain managers working in a particular silo to understand the scope of supply chain roles in other parts of supply chain. Supply chain teams or the type of supply chain roles often get set in stone during the early growth period of a company and rarely change. For an outsider the organization structure may look confusing or dissimilar and this chapter has attempted to demystify it and establish the basic similarities. The main purpose of this chapter is however to help a supply chain professional to move into various parts of the supply chain organization without difficulty. Finally how to bring it all together in an interview and articulate anecdotes from experience and succeed in securing a new role is also highlighted.

REFERENCES

Dovgalenko, S. 2020. *The Technology Procurement Handbook: A Practical Guide to Digital Buying.* London: Kogan Page.
Monczka, R., Handfield, R., Giunipero, L. et al. 2020. *Purchasing and Supply Chain Management.* Boston, MA: Cengage Learning.

O'Brien, J. 2019. *Category Management in Purchasing: A Strategic Approach to Maximize Business Profitability.* London: Kogan Page.

Payne, J., and Dorn, R. 2012. *Managing Indirect Spend: Enhancing Profitability Through Strategic Sourcing.* Hoboken, NJ: John Wiley & Sons.

Price, P., Henrie, M., Jeffries, F. et al. 2014. *Fundamentals of Purchasing and Supply Management.* Access Education.

Venkataraman, R., and Pinto, J. 2020. *Operations Management: Managing Global Supply Chains.* Thousand Oaks, CA: Sage.

6

The Six Pillars of Supply Chain

Sudip Das

Ishaan Das

CONTENTS

6.1 Objective.. 124
    6.1.1 For an Entry-Level SCM 124
    6.1.2 For a Mid-Level SCM 125
    6.1.3 For a Senior or an SCM Executive.. 125
6.2 Six Pillars for an Autonomous Supply Chain Leader 126
6.3 Pillar 1: Internal Product Knowledge 127
    6.3.1 For an Entry-Level SCM 127
    6.3.2 For a Mid-Level SCM 130
    6.3.3 For a Senior or an SCM Executive..131
6.4 Pillar 2: Internal Demand and Development Process 132
    6.4.1 For an Entry-Level SCM 132
    6.4.2 For a Mid-Level SCM 134
    6.4.3 For a Senior or an SCM Executive..135
6.5 Pillar 3: Supplier Landscape 136
    6.5.1 Technology...136
    6.5.2 Competitive Landscape..138
    6.5.3 Financial Health of the Supply Base.....................................139
    6.5.4 Spend, Dependencies, and Risk 141
    6.5.5 Business Review of Strategic and Operational
           Performance History and Metrics...143
6.6 Pillar 4: Overall Business Knowledge and Value Creation 144
    6.6.1 Return on Investment Analysis 144
    6.6.2 Value Engineering...145
    6.6.3 Cost of Make versus Buy...146
    6.6.4 Non-Recurring Expense Analysis 146

DOI: 10.4324/9780429273155-8

6.7 Pillar 5: Relationship Management 147
6.8 Pillar 6: Execution 149
6.9 Concluding Remarks 151
References 152

6.1 OBJECTIVE

The ability to make informed decisions is perhaps the most important skill to learn and master for any supply chain professional. Independence and self-governance are both extremely important to success regardless of career level and also in proving that you are capable of handling larger responsibilities. A supply chain leader is expected to be autonomous, especially the further you go in your career.

After mastering the fundamentals (Sollish and Semanik 2012) and developing the ability to make decisions on your own and be self-directed, the next foundational skill is the ability to influence a larger audience. Learning how to influence your organization and be entrusted with making multi-million dollar decisions is crucial for growth. Earning your organization's collective trust based on your business acumen and execution capability is the best way to earn these responsibilities and flourish. This chapter will provide a framework for establishing and integrating these skills into becoming integral parts of your personal brand.

6.1.1 For an Entry-Level SCM

In Chapter 4 we discussed various supply chain management (SCM) roles. As an entry-level SCM growing into your role, this chapter will focus on how to boost your own assets and accelerate your career forward. Growth is fueled by one's appetite for knowledge and execution capability. Depending on the job, the daily role can range from procuring products and negotiating a portfolio of contracts to maintaining a set of vendor relationships. Irrespective of the role, two groups of tasks are quite common:

- Interacting with people both internal and external to one's organization, such as product teams and vendor partners.
- Being asked to analyze and leverage all the information and data collected from all the above interactions.

This chapter introduces and presents a framework of six categories. Developing a mastery in these six fundamental categories or pillars will empower you as an entry-level SCM to reach career maturity by excelling at execution in these two groups of tasks.

6.1.2 For a Mid-Level SCM

With maturity comes a higher level of trust within your organization, and as a mid-level SCM this chapter focuses on how to advance and become a leader. At this level, you should already have a certain level of understanding and mastery of these six fundamental pillars, and you should have developed the skills to be an expert in the specific supply chain role with respect to your industry. Whether it is managing a specific technology, establishing a portfolio of vendors, or owning a line of products, you should be able to analyze your own role and skills and also be able to communicate your leadership through these six lenses. Using and more importantly showing your knowledge of this framework will continue to inspire and add to your organization's confidence in your capabilities which translates to leadership opportunities and trust to lead large-scale decisions.

6.1.3 For a Senior or an SCM Executive

As an SCM executive, this framework should translate into managing the overall health of your organization's business. If you can adopt this framework throughout your organization and make it become a part of your team's DNA and core principles, these pillars will evolve from a general set of skills to a dashboard of your overall supply chain strategy. This chapter will establish and answer questions for a larger scale supply chain network and answer fundamental questions for an organization as a whole from the point of view as a leader. How do you build a robust process where a supply chain organization is able to respond to the needs of their business in an agile but consistent manner? How do you benchmark your own supply chain organization's performance? What information do you present to your chain of command when a new leader comes in who does not understand supply chain and they question the value proposition? How can you avoid fire fights and data churns from disparate sources when making business decisions? A rigor of documentation and processes based on this framework as well as an organization-wide regular review will ensure that your entire organization can solve problems and show your leadership value on a larger scale.

6.2 SIX PILLARS FOR AN AUTONOMOUS SUPPLY CHAIN LEADER

In the construction project of a house, different teams of experts work in a certain order based on the stages of construction. First, architects must build the plans and design of a house. Then a cement crew comes in with specific tools and experience to lay the foundation. Next, the framers come in to raise the frame. They are followed by the plumbers and electricians. Each team moves in, completes their task as per the master design while following building codes and then hands off the project to the next team. The one constant is the general manager of the project, who comes in every so often and measures the progress, inspecting the workmanship and adjusting wherever they find deviations from either the design or building code.

In the professional world, jobs are often designed in such a way that pockets of expertise get developed by certain groups and teams. There are designers or engineers who are building the products, as well as financial experts who are managing the cash flow, and sales teams building relationships with customers and servicing them, amongst others. The experts in each domain are often clustered in verticals. We need somebody to be the glue. Experienced leaders from each of these verticals become the glue. They do that by developing a more comprehensive understanding beyond their own verticals. For example, in comparison to the deep expertise of members of an electrical, masonry, or plumbing team, the general manager has a much wider breadth of knowledge rather than depth in one specific area. He can read the blueprint and compare with the actual project, can look for code violations, and can assess progress against deadlines. The general manager's knowledge is striped across multiple areas, and this idea is central to becoming an autonomous supply chain leader.

For a SCM, breadth of knowledge is a key requirement from day one. Quickly learning and becoming a well-rounded member of your company as well as projecting your comprehensive knowledge is the key to be successful as a SCM. The verticals you need to know about are multifaceted as well as time sensitive and encompass the entire company as the supply chain or procurement support is needed all across an organization. Referring back to the analogy of a construction project, business practices, such as building codes, are always changing and so it is extremely important to always stay updated. Business conditions are dynamic and are constantly impacting the supply chain, and therefore constantly learning and staying current on all verticals is a critical skill for being an effective supply chain manager. In other words, remember you are always thinking like a general manager.

At the early stages of your professional journey, as you are an entry-level SCM, being autonomous means being able to develop knowledge of the internal products and their sources of demands as well as external supplier health and market conditions. These are the critical information for you to always stay up to date with, in order to be an effective contributing member of a supply chain team. Similar to reading the blueprints and measuring progress, you have to execute while delivering value and managing relationships.

As you move up the leadership chain and become a mid-level SCM, the meaning of being autonomous reflects the broader and more managerial role, meaning you must continue to develop an even wider breadth of knowledge and handle critical relationships while executing flawlessly. You become the general inspector, except now you are no longer just inspecting houses but also various other types of buildings like schools, hospitals, and offices, and you need to know the specific details and differences between all the projects you are managing.

As a senior SCM executive, you will become responsible for a team of SCMs. You are accountable for all these autonomous general managers and be accountable for all of their actions as well as the different categories of construction projects going on. You are now responsible for the overall performance of your organization.

As you transition from being an individual contributor to a team manager, the categories of products and suppliers keep increasing, but the core set of information per category that you need to manage stays unchanged. In the following section we will discuss the core set of information as a framework. We call this framework as six pillars of knowledge that every supply chain professional needs to develop and project to grow professionally and to project leadership qualities in the supply chain.

6.3 PILLAR 1: INTERNAL PRODUCT KNOWLEDGE

6.3.1 For an Entry-Level SCM

When managing indirect or direct procurement, if you do not know what is the end use of the product that you are procuring, then all you are doing is simply a transaction. Moreover, if your knowledge is only superficial, you will not be able to hold a meaningful discussion with either the subject matter experts on whose behalf the items are being procured or the suppliers from whom you are sourcing the product. As mentioned before, a supply chain manager connects an entire business together and needs to have knowledge of all verticals and information, especially internal to your company. If you find yourself in a situation where all you are doing is acting as a facilitator between the two parties, then you have not created any major value except maybe working with legal, working with purchasing, closing the contract on time, and issuing the purchase order. This approach is not strategic, and it will become harder to deliver results desired by your stakeholders when you are tasked with larger projects and managing others.

In order to become a strategic player and an asset to your company, you have to be able to hold your ground in a subject matter conversation. For example, if you are procuring a semiconductor chip from a vendor, you need to have an understanding of the key high-level features and specifications, manufacturing technology, how does the chip line up with the internal specification of the end product your company is delivering, and what features this specific semiconductor chip delivers in terms of that end product. Most likely the semiconductor chip will have some features which will align with the end product requirements and some features which may not align, and it is important to know these for negotiation and how to prove your value in choosing vendors (Khan 2019).

Now let's dive deeper in understanding why you need to know all of this. You may be asking shouldn't the supply chain be all about just pricing? In all honesty, it can be depending on what you see your role as. If you want to see growth and your scope of opportunities to increase, you need to view your role much more holistically. In case of a complex procurement role, you need to view yourself as an equal stake partner. For example, there will be a marketing team who will be accountable for creating a product that the customer ultimately wants. There will be a designing team who will be accountable for building the product. As a supply chain person, you will be accountable for getting the product built at the right price and right quantity to satisfy the customer's demand. The price you obtain from your suppliers will maximize profitability and ensure business continuity. With a sense of ownership comes clarity on how to influence other stakeholders. Establishing authority is achieved by having product knowledge and being in a position of power. One has to develop the former and then the latter is given as the supply chain person is ultimately managing the price.

So, here's a list of a few things you should spend time to understand in order to have better internal product knowledge:

- Product portfolio, where the procured item will be used in the bill of material (BOM):
- Quite often a company will create multiple variants of the same product family to address different market segments. The variants may have different design requirements and/or cost pressure.

For example, a WIFI router manufacturer may design different broadband WIFI routers for the home and enterprise market. Some homeowner may want to get WIFI from their cable companies and some may get it from phone companies, whereas the enterprise customers may have a network (Ethernet) connection already. So there will be three possible type of inlets (cable, phone, and Ethernet). WIFI usually works on two bands (2.4 GHz radio frequency and 5.0 GHz radio frequency). The router needs to have one WIFI radio for each band. The manufacturer can make two variants of routers: one with only one radio and the other with two radios for more flexibility. If you are the SCM for the radio card you have to be able to explain how many radios of each band you will need in a year. The radio manufacturers will ask you for guidance on whether they should make more 2.4 GHz radios or 5 GHz radios. You have to articulate the split in the total available market (TAM) between these two bands to negotiate the best price (Table 6.1).

This requires knowledge of the underlying wireless technology and product mix as you will be negotiating with the manufacturer and wearing both the marketing and design engineer's hats.

A sourcing manager should be able to articulate the internal TAM in terms of quantity, cost, and the ramp-up timeline. In other words what quantity of

TABLE 6.1

Total Available Market (TAM) for WIFI Router Variants with Different Interfaces and Wireless Radios

Product Variants	Interface			Wireless Standard		Radio 1+2: 2.4 and 5 GHz	%TAM
	Phone	Cable	Ethernet	Radio 1: 2.4 GHz	Radio 2: 5 GHz		
Enterprise			X			X	50%
Home, phone	X					X	10%
Home, cable		X				X	25%
Home, low cost	X			X			15%

radio cards will be needed in every quarter and what is the price target. Let's say the forecast for the total shipment volume of routers during its 3 year life cycle will be 1 million units. Since 85% of that volume will be dual band that means over a 3 year period you have to buy 1,000,000 2.4 GHz radios and 850,000 5 GHz radios. Let's say you have put a target unit cost for the radio as $10 for 2.4 GHz and $12 for 5 GHz. Typically the broadband router, like every product, will have a ramp-up time for adoption. Therefore the underlying card demand will also grow with time. Say the quarterly requirement during launch will be 60,000 units of routers and it will eventually reach up to 100,000 units per quarter in five quarters. Around 85% of that router volume will require both radios (85% of 60,000 is 51,000 and 85% of 100,000 is 85,000). So we would need 60,000 2.4 GHz radios and 51,000 5 GHz radios initially and eventually will require 100,000 2.4 GHz and 85,000 5 GHz radios. That is the ramp-up forecast. The total TAM value is $1.212M (1,000,000×$10 and 850,000×$12). This information is important because it is a tool in leveraging and obtaining a quarterly cost reduction from the wireless radio supplier.

6.3.2 For a Mid-Level SCM

The product offerings of any company are always evolving. In order to stay competitive, companies are always adding new features to delight their customers. Based on market research and customer feedback a marketing manager works with the development team to release these features in a phased manner. The phase-by-phase release of these new features are put in a plan which is commonly referred to as the product roadmap. A sourcing manager's key responsibility is to ensure one's downstream supply base is aligned with this roadmap and the downstream supply chain designs, develops, builds, and delivers the constituent parts on time to ensure timely release of the targeted features as per the roadmap. In other words the roadmap of a customer becomes the input for the roadmap of its suppliers' and the sourcing manager is the conduit for this information flow.

Example: the battery development business unit (BU) of an electric vehicle (EV) manufacturer is developing batteries to support the company roadmap. The manufacturer's 3 year roadmap has a plan to increase the range of the electric cars from 80 to 120 miles in the following year and 160 after that. As a sourcing manager for batteries, you need to understand how this roadmap impacts the overall BU's roadmap. Keeping the overall EV size and weight similar between generations, the batteries have to pack in more capacity to support the increase in mileage. There the lithium ion batteries for each generation have to have more energy density to produce more ampere hours to get more range. As a sourcing manager it is your responsibility to understand and convince the Li ion battery manufacturers to invest and develop more efficient batteries to support the extended range. The manufacturer has to invest in materials technology research to develop lighter material with higher energy capacity.

Similarly if one is an indirect strategic sourcing manager, understanding the internal technology roadmap of one's information technology (IT) team is equally important. IT teams are responsible for providing IT solutions for human resource (HR) management, financial resource management, etc. Consider a situation where one is responsible for supporting the HR-IT team for all its HR management solutions. For example, the HR-IT team may have a plan to manage its payroll administration software from on premise solution to cloud, or move away from its current salary administration solution from one vendor to a new one, or identify and introduce a new software tool. As the strategic sourcing manager managing the HR-IT roadmap you need to understand various HR management solutions at a high level as part of your portfolio. Usually implementation of any IT software solutions involves not only buying the software but also procuring warranty support, professional services, and training. As a sourcing manager you have to understand not only the implementation timeline but also the scope of various functional modules of software that you have to negotiate.

6.3.3 For a Senior or an SCM Executive

As an SCM you are responsible for a team of professionals who has invested time in gaining product knowledge, otherwise the team or individual team members will not gain trust from their internal stakeholders. If the stakeholders do not feel confident that the supply chain team has their back and if they do not believe that one can accurately represent their needs in front of the supplier, the stakeholders will engage and negotiate with suppliers themselves without waiting for the supplier manager. If that happens the supply chain team will also lose the ability to influence the decision-making process. As a supplier manager for a category, one will be working with the same group of internal customers and external suppliers over a long period of time doing multiple projects. Especially if one is managing a technical supply chain category, without authoritatively understanding the product portfolio and roadmap, one will not be respected or sometimes not even be included in the procurement decisions by the stakeholders.

However, on the flip side if you as a leader have built a team that has complete knowledge of the products, the team will hold its ground when debating with stakeholders and gain trust and build a strong relationship. Whether it is R&D cost to develop a new battery which will enable longer range cars or implementation cost of adopting a HR-IT solution, your team will be able to exert its influence. For example, the battery manufacturer may have to increase the next generation battery unit cost to recover a portion of their R&D investment, in that case how does it impact the profit margin of the car? The return on investment (ROI) will only become better if the EV manufacturer can reach a certain production volume or if the battery manufacturer innovates a new efficient way to lower the cost of production. Or say in the time period the company is planning to launch the next generation car, the manufacturer may still have kinks in the battery which cannot be ironed out and hence there could be quality issues and it should be called out as a risk. All these situations impact the cost, production lead time, and quality of the end product. A supplier manager can present the optimization options and help one's internal product team to choose the right battery.

In the case of the HR-IT solution, the cost of the implementation may have been touted as just the cost of the software and support. However, based on past contracts with the previous vendor the strategic sourcing manager realizes that there will be additional cost to port the existing data from the old software to the new software or for integration with existing sources of data. This may change the math on ROI and one may question the viability of the migration and offer improved feature development with the existing vendor at a lower cost. This is when the sourcing team is viewed as a source of category leadership.

6.4 PILLAR 2: INTERNAL DEMAND AND DEVELOPMENT PROCESS

6.4.1 For an Entry-Level SCM

All supply chain activities ultimately converge to a purchase transaction. The transactions can be described as a certain count of items exchanging hands costing a specified amount of money over a specified period of time. As a supply chain manager you may have to forecast these transactions. You will be providing guidance on quantities that you will be procuring at what periodicity and the value of the transactions. These information will translate into production quantity, delivery time, and ultimately revenue for your suppliers. The accuracy of your forecast of internal demand and subsequent follow through with the actual demand-based transaction will be the basis of your success and growth in the profession (Snyder and Shen 2019).

As supply chain managers endeavor to make their organization successfully release products and features every quarter, so do the suppliers under the supply chain manager's guidance. Simply put, as the supply chain manager's parent organization expects to show revenue growth, so does the supply base. And the supply chain manager plays a pivotal role in this cascaded process. The accuracy gives downstream credibility to the suppliers' sales team who takes cues from the supply chain manager and influences internally to ramp-up or ramp down production or do new product development. Therefore, to be able to predict how much does the parent organization needs and when does it need by are the two questions a supply chain manager will be constantly hounded for to answer. Understanding the forecast and taking ownership for its accuracy in terms of actual demand is key as these two translate into risk and revenue, respectively, for the suppliers.

An analogy of this could be drawn from a 4×100 relay race. Let's say a BU is making a hair dryer and the electrical motor is the most expensive component inside the hair dryer. The production of the dryer is the anchor leg of the race with the successful ramp-up as its finish line. However, in the first leg, the supplier manager has to provide the forecast for how many types of motors and when they are needed by.

The plan is to produce a high-end hair dryer for the US market and mid-range hair dryer for Asian market. The BU has 30% and 20% share of the total available US and Asian market, respectively. Expectations are that it will take four quarters to ramp-up to the targeted market shares. However, the BU is trying to launch the products in fall so that it can address the US winter holiday demand which is typically highest in the year. All these factors translate into a forecast profile with monthly demand quantities. There can be a possibility of pulling in and trying to ramp faster in two quarters instead of four if the US market responds favorably at launch.

A supplier manager will translate the volume of the hair dryer revenue projection into monthly motor forecast and explain the possibility of the US market ramping up faster than Asian market buoyed by holiday shopping through scenario-based forecasts to better prepare the suppliers. Along with negotiating price, the supplier manager will provide the quantitative internal forecast to the motor supplier in the first leg.

The sales team from the electrical motor manufacturers immediately starts running with this forecast and gives it to inside sales for supply planning. In the second leg, the motor company develops the prototypes and gets final approval and starts pipelining materials like copper wire, armature, and other subcomponents to ramp-up its production based on the forecast given by the supplier manager. The supply base takes the risk and sends its own forecast and starts buying according to a schedule.

In the third leg, the BU starts production as the motors get delivered and builds up inventory for the launch. Based on the knowledge of the design and development process of the hair dryer and the supplier's lead time for producing the motor, the supplier manager establishes a lead time for delivery of the motor at the hair dryer assembly plant. The purchase order starts being issued from the BU to the motor manufacturer based on this lead time. The BU starts receiving motors according to its actual production schedule and launches the product. Finally, in the anchor leg adjustments are made. Based on actual customer adoption of the hair dryers in the two markets, either the BU needs motors as per forecast and keeps placing demand as per the forecast or there could be a variance. If the variance is positive, more motors need to be supplied, whereas if its negative then production volume needs to be cut down.

6.4.2 For a Mid-Level SCM

Although there are manufacturing resource planning (MRP) tools which send forecasts, purchase orders for demand, digitally across the world in microseconds and have taken business to business communication to new heights, ultimately the decision makers are still human. Trust plays a very important role here. The suppliers may demonstrate appetite to take risk and purchase material on its own dime, thereby reducing the delivery lead time. On the other hand a supplier could be risk averse and does not purchase materials without a contract or purchase order would result in longer lead time. Both kinds of behavior stem from the trust between the supply chain manager and its suppliers' sales account manager. Suppliers make a lot of decisions based on verbal cues from the supplier manager instead of waiting for formal cues like purchase order or an electronically transmitted forecast. This risk is taken based on trust developed over a long-term relationship.

MRPs are lightning fast but the data that are loaded into the MRP from the BU side is done by planners who are at a downstream end of the information flow. So whether to increase the forecast or cut it down, they are last to know.

Similarly at the receiving end, there is a process lag between the team which receives the notice of upside or downside and the team which cuts down further downstream production. This is where a supply chain manager makes the difference. Based on past experience one knows whether the ramp plan created by marketing is too aggressive or not. Quite often a business relationship gets soured between a company and its supplier because of an over optimistic or bullish marketing forecast. Imagine the impact of a supply chain manager's role on one's supplier's revenue. A supplier manager's fair understanding of internal demand and the development process is the guiding light for the supplier. Suppliers are investing both human and cash captial in delivering to this forecast. A supplier manager who takes accountability of that ultimately gains trust of his suppliers. A supplier manager needs to be able to communicate the following to gain trust:

1. Assessment of the viability of forecast in an unbiased manner so that suppliers are not left hanging with excess inventory. A supply chain manager should have confidence in his own forecast based on knowledge of past products ramp and company's investment of its own resources. How important is this product for the BU and how much revenue is riding on this product family; knowledge of these helps to raise confidence in the suppliers.
2. Communicate development hiccups and launch delays immediately. One constituent part can have design issues leading to delaying the production and can put all other suppliers into a holding inventory situation. A supply chain manager's responsibility is to minimize it.
3. Communicate the health of the ramp-up directly and not depend on the MRP process only.
4. Finally, to review the quality of forecast and address internally and externally in postmortems or quarterly business reviews (QBR), if there is chronically high forecast variance.

What happens when one takes accountability for the forecast numbers and associated soft guidance to help both the BU and the suppliers to be successful? One gains trust and the power of influencing.

6.4.3 For a Senior or an SCM Executive

Senior SCM executives have dual accountability. Externally they are peers to the leaders of the suppliers organizations. The senior exec is ultimately accountable for setting the tone of the relationship between the supplier and customer organizations. He will also be the owner of escalations. Escalations are bidirectional which means it can be from customer to supplier or vice versa. For example, if a product starts failing then the customer organization would require support from the supplier to know root cause of the failure and a quick resolution plan and hence they would escalate these to the supply chain leader for ensuring supplier's attention to the problem. On other hand, a supplier may escalate if the forecast does not materialize quarter over quarter. When it comes to forecast and actual demand, the supply chain leader's integrity is the most important thing. If the leader cannot stand behind the forecast coming from the supply chain organization, suppliers will stop believing the forecast and will start judging the forecast. When a supplier executes based on forecast, it takes a leap of faith based on past run rate and customer assurance and exposes itself to substantial financial risk in building products without any payment. If the trust is broken, the supplier will either end up asking for purchase orders prior to pipeline material or deprioritize the customer over other customers and will end up reducing supply and putting the customer at a disadvantage. Overall it will put strain on the relationship and loss of negotiation power on either side.

6.5 PILLAR 3: SUPPLIER LANDSCAPE

Like the knowledge of law is to a lawyer, so is the knowledge of the supply base to a supplier manager. Irrespective of the specific supply chain function one does, in order to execute well and also to be perceived as a leader, this is a muscle one needs to develop. At minimum a supplier manager needs to view one's role as not somebody who just deals with a supplier but somebody who represents both the supplier and its product or technology inside one's own company. The parent organization will look at the supplier and the supply base based on the lens the supply manager puts on in front of the company. Let us dig deeper by understanding the previous sentences from a few facets (Weigel and Ruecker 2017).

6.5.1 Technology

The supply base typically has its own roadmap. Due to continuous improvement and constant innovation, the suppliers are constantly evolving their products. Same supplier is also talking with multiple customers and collecting feedback to improve their product which will allow them to retain and or gain more market share. In this regard a supply chain manager is the conduit for marketing information from one's supply base. Organizations often dislike being bombarded by business developers from suppliers drumming for business. Giving access to suppliers sales and marketing folks to internal product developers and engineers is both unproductive and risky from an intellectual property (IP) protection point of view. Confidential information can accidentally leak out to suppliers In this regard a supplier manager has two key roles: first is to control the access of members of the supply base inside one's organization and second to be able to represent the supply base technically. It's a fine act of liaising. An ideal supplier manager should be knowledgeable enough to represent one's suppliers technically so that one's internal customers are making a decision based on full view of the market. Often the internal customers have their own bias based on past products or past job experience and may have a bias toward a particular supplier or technology. A supplier manager listens to multiple technology presentations from multiple suppliers and presents only relevant suppliers to the product developer. A successful supplier manager understands both technology and can filter in the right supplier for the optimum business and technological alignment. The depth of understanding about the product is a key differentiator between an *entry-level* and a *mid-level* supply chain manager. A *mid-level* supplier manager has enough command over the underlying technology to stand their ground in front of the product manager. For a *senior exec* that depth is desired but not necessarily scalable if the product portfolio is really large. However, a *senior leader* is expected to communicate the technology to their superiors, and a successful senior leader ensures that the supply chain team has that technical depth to challenge, influence, and educate.

Referencing the earlier example of an EV manufacturer, the supply chain manager responsible for battery should know one's battery suppliers technological roadmap and be able to articulate the timeline in which supplier will release products with various energy densities. At the same note he or she should know how technologically the batteries are getting lighter in weight as a high-density but a heavy-weight battery may not work for an EV manufacturer who is trying to create a sports model. However, the same EV manufacturer may have a truck in its design board and to that product team the weight of the battery may play a less important role than the ability to go 300 miles in full charge. So a supplier manager should typically meet with multiple battery suppliers, understand their product roadmaps, and schedule the release of new batteries. This will help him or her articulate a high-level summary of battery technology landscape and suppliers' capability development in light of the technology advancement. Let's say a current battery vendor is not investing in a battery for sports EV; to know that in advance and find an alternate vendor and align them with the product development team during the design phase is a supply chain manager's responsibility.

6.5.2 Competitive Landscape

The world of supply chain is as dynamic and volatile as the stock market. The supply base ecosystem is driven by the same market forces as the company which is procuring goods and services from them. A procurement manager's job is to navigate the disruptions so that there is assurance of supply. The disruptions can come from all possible directions. Some examples could be – M&A – the supplier can get acquired by its competition who changes the roadmap or the number of suppliers in the market reduces creating less competitive situation; disaster: supplier's production get disrupted by natural disaster, fire, and political disturbance; tariff: supplier cost increases due to tariff; shortage: suppliers supply base gets impacted due to raw material shortage or rise of basics such as oil price; cyclical business: suppliers cut down investment and cause market constraint which drives price, etc. Depending on an industries' price sensitivity and profit margin, failing to understand the competitive landscape may lead to margin drop or drop in revenue. On the flip side hedging smartly may give a competitive advantage. For example, if an airline hedges on long-term oil contracts at the right price it can enjoy lower oil cost than its competitions.

A supplier manager is perhaps the only person in the entire organization who has the responsibility to worry about the future of one's suppliers and communicate back. In a way it is a supplier manager's responsibility to be able to read the tea leaves and influence the parent organization to prepare for the future. There are often disruptions which happen without prior notice like a natural disaster or a fire or a geopolitical disruption. The company often reacts to this kind of situation collectively depending on the visibility of the impact. An *entry-level* supplier manager will typically make their organization aware as soon as an incident occurs and identify the products impacted. A *mid-level* person will have the foresight to influence their peer organization for future products and to take longer term measures. A *senior leader* is expected to build a process for business continuity as supply chain disruptions are common. A senior person would understand the trends and bring in resiliency to the supply base (Srinivasan et al. 2014).

For example, consider a computer server OEM reacting to supply chain constraint of memory compared to capacitor shortage. Occasionally, memory vendors get into a bumpy transition from one semiconductor manufacturing process technology to the next and it leads to memory supply shortages in the server industry. Memory being a key component both in terms of importance and cost, supplier managers have no difficulty communicating the change in the competitive landscape. Often supply chain managers and supply chain leaders of the organization fly across the world to meet with suppliers and get assurance of supply or do spot buy and hold inventory or may sign long-term contracts and get creative and involve multiple parts of the organization from manufacturing to legal to finance to ensure business continuity. On the other hand, the recent shortage of MLCC (multilayer ceramic chip) capacitors has caught a lot of companies off guard. Capacitor being the penny part often the assurance of supply is taken for granted. However, the capacitor manufacturing landscape has changed drastically. Lot of suppliers heavily invested their capacity to supply the mobile phone and automotive industry which led the computer industry to suffer from shortages. These kinds of situations are slow burners where a company may not pay attention to this early enough and then react when it is too late. However, the point of this example is both types of situations are equally critical as ultimately they end up disrupting the production of servers.

Understanding the changes and anticipating their impact, raising awareness and taking measures ahead of time is the supplier manager's responsibility.

6.5.3 Financial Health of the Supply Base

All companies have expenses and have to make profit to survive. A supplier manager's responsibility is to track this for one's supply base. The supply base can consist of mature and publicly traded organizations or a small private startup. However, even within a mature organization, the actual product may come from a small division who can be having challenges in making profit and surviving.

As an *entry-level* supply chain manager looks externally and probes into the suppliers financial records, one has a responsibility to keep an eye on the profit margin of the supplier on the item one is procuring from the supplier. At the end of the day a supplier who is struggling to pay his expenses will not be able to invest in designing new products or investing in additional capacity and will ultimately end up failing to supply its products. For publicly traded companies this information is a little easy to gather from public sources. It's important for a supplier manager to be aware if the supplier's year over year or quarter over quarter revenue is dropping or if the supplier has announced a loss. Publicly traded companies also publish 10k forms which are comprehensive financial reports required to be published by the US Securities and Exchange Commission (SEC) and offer valuable information on both the overall financial health and the financial health of specific divisions. For a privately traded company this may be harder to get but still it's possible to get it both directly and indirectly. It's possible to get it from the supplier if it has been made obligatory through contracts put in place with the supplier as part of doing business or by verbally inquiring. There are also companies who maintain financial risk ratings and reports for private companies and sell it as a subscription service. It is a supplier manager's responsibility to track and alert one's organization if the supplier's health exhibits signs of concern.

Negotiating product price and annual cost reduction are two key responsibilities of a supplier manager. Depending on the volume of purchase a supplier manager yields a lot of negotiating power over one's supply base including taking away business from one supplier to another. However, with power comes responsibility to ensure the supply base is doing a financially viable business. If the supplier is solely doing a product for the customer or more than 8%–10% of the supplier's revenue is coming from a single customer then the risk increases further. A drastic windfall in business caused by the supplier manager by suddenly taking away business from that particular supplier could be disastrous for the supplier's survival. This is neither ethical nor smart business practice. If there is a supplier performance-related issue or there is a valid reason for the move, the supplier manager needs to communicate this. This could also potentially creates a bad reputation for the customer. Also when negotiating a supplier manager usually develops a good understanding of the cost breakdown. For example, in a lot of businesses the supplier will be mandated to show its bill of material (BOM), labor, engineering, SG&A cost, and profit. A *mid-level* supplier manager needs to know what is the right profit margin based on past history or 10k reports from the supplier or its publicly traded competitors.

As a supplier manager one should probe into all financial issues and have discussion with the supplier to fully understand the cause and ramification of the financial health change. A *supply chain leader* usually has a good understanding of financial weakness of the supply base. Creating a formal process to collect the health metrics and documents for individual suppliers is the leaders responsibility. Communicating and navigating their customers away or around financially weak or unstable companies is a critical responsibility of the supply chain exec.

6.5.4 Spend, Dependencies, and Risk

Perhaps the simplest way for a supplier manager to demonstrate their knowledge of the supply base is to be able to paint the spend picture. The spend picture typically has to be drillable. The overall spend by the parent organization and its quarterly increase or decrease tells whether the business is growing between the supplier or trending down or staying flat. The underlying volume of items is also a key piece of information as it tells whether the parent organization is buying more of the same products from the supplier or incorporating the supplier's parts in a multitude of its own product lines. Therefore the next level of drillable nuances a supplier manager needs to be able to articulate is how the spend is spread across different product families. The spend trends of the parts and impacted products summing up to the overall spend picture is something supplier managers should always have at their finger tips.

Often organizations will do a balance of trade where they both buy and sell to its supply base. For example, consider a software company selling an enterprise resource planning (ERP) tool to a server manufacturer from whom its also buying the servers to run the ERP itself. It is important for a supplier manager to build relationships with the sales person within its own organization to create an overall win-win situation between the two companies as both can offer mutual discount on cost of the servers and the ERP tool to build longer term locked relationships.

It is important to understand if a part is sole sourced or single sourced. Quite often it is very difficult for large organizations to multisource every item it purchases. Specially for indirect procurement quite often the supply base consists of a large number of sole sourced suppliers for various goods and services it procures. This introduces dependencies as often the supplier will enter through competitive bidding and fight to win the business. However after winning the business as the volume of business grows and more products are purchased, the supplier manager may not be able to control all engagements, and the profitability and margin profile of the supplier will change. This often creates a situation down the line where a supplier manager faces an unhappy customer base who now finds it difficult to do business with the supplier due to cost while deeply engaged with the supplier. The dependencies often run deep and is quite hard to get out of and be replaced by an alternate supplier. In the meantime the supplier expects a certain amount of revenue growth and raises price as the customer is now locked in. This often leads to escalations or souring of relationships. An *entry-level* supplier manager needs to be able to develop a deeper understanding of the dependencies and the pockets of products and understand the utilization. The first step towards portfolio optimization is to be able to understand which products are no longer being used or unnecessary and stop them. This reduces the spend in a legitimate way, and the supplier starts going back to a competitive mode to win back business. The second step is to communicate the dependencies to the customer base and optimize it. A *mid-level supply chain manager* usually has a deeper understanding and has more authority to influence both customers and suppliers to come to a reasonable agreement. In addition, a *leader* usually has a good understanding of the legalities when suppliers are not performing to the agreed upon contracts and can manage separations. They also have abilities to influence the closure of key deals to close in a timely manner balancing between the need of the customer and the revenue target of the supplier.

Products have a life cycle, which means products go through a launch, ramp-up, ramp down, and stoppage of production. Some of the common terms used are launch, early access (EA), alpha and beta customers, first customership (FCS), and general availability (GA). On the other end of the spectrum some of the common terms are end of sale, end of life, and end of support. Each end product is a collection of multiple subcomponents which can be a hardware and or a piece of software and like the end product itself the subcomponent also has a life cycle. Since the end product is dependent on the entire BOM that means all the subcomponents need to be in production so that it can be produced. In reality the subcomponents' life cycles are not synchronous which means it is supplier managers responsibility to communicate the life cycle of the finished product downstream to the supply base to ensure the suppliers continue production to sustain the finished product. On the flip side a supplier manager also communicates upstream, i.e. internally, which products are being put on end of life well ahead of time for the internal team to qualify alternate parts. This communication is often described as risk assessment. It assesses the risk around business continuity so that the production line does not stop unexpectedly due to the end of life of parts.

As part of this exercise a supplier manager checks if a part can be multisourced so there is a qualified alternative from the get go. For single sourced parts, one has to articulate the risk and also check if the single sourced part manufacturer has multiple manufacturing sites so that there is a plan for disaster recovery.

6.5.5 Business Review of Strategic and Operational Performance History and Metrics

The final step in the supplier landscape assessment is how a supplier performs specifically with its customer. A supplier can be doing great in all the previous four areas and can still be disliked by the customer or vice versa. A supplier manager keeps a pulse on the day-to-day operational transaction between the supplier and one's customer. This is where the rubber meets the road. A supplier manager is also responsible for doing operational review of the business on a regular basis to iron out all kinks and manage the supplier interactions with one's organization.

Typically there are four areas where a supplier manager ensures alignment and they are development, cost, delivery, and roadmap. (1) Development implies whether the supplier worked closely with the design team during the product development phase which involves putting resources and engagement with team members, prototype delivery, and designing as per specification. This also shows the new engagements that the supplier is getting, which will translate into future revenue for the supplier. (2) Cost performance implies whether the supplier is meeting both individual price targets for products and overall cost quarterly cost reduction. This also indicates whether the existing products are ramping up and supplier business is growing. (3) Delivery represents the operational performance of the supplier. This is an indicator of the supplier successfully delivering to maintain business continuity in terms of on-time delivery of the right quantity with no quality issue. (4). Roadmap indicates investment made by suppliers to meet future requirements of the customer. Simply put it is a cyclical process where a supplier engages to develop products at the right price point and delivers them as per plan and then reinvests some of the profit to develop the next generation of products.

A supplier manager typically surveys the stakeholders and takes feedback in these four areas either formally through survey tools on a quarterly basis or informally. One's key responsibilities are both to communicate the feedback and show improvement. A supplier manager and the supplier's account manager work are joined at the hips to ensure communication between the two companies and demonstrate change as requested by the stakeholders. An entry-level supplier manager usually collects and compiles feedback for QBR and owns the process. A *mid-level supplier manager* should be capable of driving the corrective actions coming out of QBR and ensures the suppliers mitigate the issues in a timely manner. A *exec* typically ensures that the QBRs are happening as per schedule and as per process and tracking the metrics across their supply base. A team leader is responsible for ensuring that the necessary supply base is being reviewed continually and problems are getting fixed.

6.6 PILLAR 4: OVERALL BUSINESS KNOWLEDGE AND VALUE CREATION

Irrespective of whether one is doing direct or indirect sourcing, or any sub-function within a supply chain org, one's job function exists to create certain values. Some of the values are part of the charter as we discussed in Chapter 4, for example, price negotiation, contract negotiation, etc. These are areas where a supplier manager would be expected to deliver and ensure overall organizational success. However, business environments are dynamic and therefore its wants and needs evolve constantly and thereby creating a lot of other opportunities to create additional values. The difference between an *entry-level* or *mid-level* supplier manager is essentially in the depth of knowledge. For a *leader* it will be expected that their team is capable of producing these business values in a systematic way (Trent 2016).

6.6.1 Return on Investment Analysis

A supply chain manager controls costs which impact the bottom line. Similar to a business development manager or a product manager who opens up a new line of revenue and has to present analysis to get buy in for investment, a supply chain manager is often asked to do ROI analysis between procurement alternatives. For an indirect procurement this may mean projecting a long-term outlook and compare between a multiyear contract and single-year contract. For a direct supply chain it may mean cost of choosing various architectures and offer a ROI comparison.

Knowledge of financial capital expense (Capex), operational expenses (Opex), and depreciation is very helpful to articulate ROI to finance as the goal of these exercises is to project the overall cost reduction. An example will be an indirect procurement manager asked to do the ROI analysis before a major investment, such as putting a new contract for data center expansion or a new factory build out. Usually these models involve more branches of the organization as higher the investment amount is. Finance and legal are two important partners whom the supply chain manager collaborates with. The supply chain partner helps in building the ROI and also directly impacts the ROI calculation by negotiating both cost and key terms.

6.6.2 Value Engineering

Value engineering are opportunities for a supplier manager to introduce a mid-life kicker for a product and boost profit margin. Typically products become cheaper or more discounts are given as it matures and thereby reduces the profit margin for the product which ultimately leads to rolling out the next generation. A supplier manager often finds out new alternatives such as a new supplier who has lower cost or the same supplier offering a slightly different part which has lower cost due to higher volume. A supplier manager works with the stakeholders by introducing the new opportunity which may lead to a design change to replace the more expensive part with the new cost-reduced part. Value engineering is a typically opportunity to create value by reducing cost.

The other way to do value engineering is to discover new avenues to save cost. For example, consider a customer buying different goods where the cost of transportation is embedded in the cost. Now different suppliers are using different logistics provider and one finds that cost is not being managed. A supplier manager can work in self-managing the logistics and aggregate to one or two specific logistics provider and negotiate direct cost to optimize the overall spend of the transportation which ends up reducing the cost of the supplies.

An indirect procurement manager involved in renewal contract for software identifies usage pattern of one's customer bases and certain categories of licenses. If one finds that the customer base is using more of one type of license and not using another type, then they can study and project the usage after discussing with customers and create a plan for buying more of the high usage type and getting buy in for decommissioning the low usage type. In other words a supplier manager looks for patterns and different areas to find innovative ways to optimize and reduce the overall cost of procurement.

6.6.3 Cost of Make versus Buy

This is a cost negotiation method similar to ROI analysis. This involves estimating the cost of building the product themselves by directly procuring the underlying subcomponent versus buying it as a finished product from the supplier. In certain cases there could be significant barrier of entry but barring those cases, it is possible for the end customer to skip the first level of suppliers and invest in design and build the product on its own. The cost of the material and labor are the two key terms to be estimated. This is supplemented with the cost of sustaining the products and the dependencies and risks introduced. On one hand the customer will reduce dependencies on the supplier and that has multiple values like IP protection, no delivery delay, and reduced cost but it may incur higher cost by having to hire people to develop and sustain the subcomponent's product life cycle. This is also sometimes known as cost of vertical integration, for example, a mobile phone supplier deciding to build its own semiconductor processor instead of buying it. Similarly it could be to control a runaway enterprise software cost and replace it with a home brew software.

The analysis is often used to do a comparable evaluation of goods and services to negotiate down cost with a vendor. However, sometimes these decisions are strategic when the parent organizations attempt to reduce cost, risk, and dependence on a supplier.

6.6.4 Non-Recurring Expense Analysis

Often enterprise procurement involves customization. The customization may mean additional features in a product or embedded software development. For an indirect procurement of an enterprise software development it may imply additional resources needed to develop interfaces or integration with existing sources of data. Sometimes this could be as significant as a full product development based on customer's product specification.

It is a procurement manager's responsibility to manage this cost. The suppliers often break the cost into two parts: a product cost and a one-time non-recurring expense (NRE). Negotiating NRE requires both close coordination with the stakeholders to understand their requirements and figuring out how to price it. They may require understanding how the supplier is building its resource cost as part of the NRE. If the supplier requires 10 developers for 6 months then is the NRE cost in line with the salaries or resource cost of 60 man months at suppliers location-specific market? Quite often NRE cost involves labor cost, prototype development cost, hard tool versus soft tool cost, cost of support at the service level required by customer, etc.

This area is often overlooked and not negotiated to the fullest extent. Also this area goes through scope creek and changes midway which if not managed in a timely manner increases the upfront cost for the customer. A supplier manager also provides continuity through one's experience in the specific category based on past experience and provides guidance to one's customer on various types of NRE expenses and its negotiation.

6.7 PILLAR 5: RELATIONSHIP MANAGEMENT

The enterprise world runs on maintaining laser sharp focus on its customer base and a supply chain manager is the face of the customer. The supplier gets to hear its successes and failures in the form of customer satisfaction and dissatisfaction through the supply chain manager. The supplier manager also plays a pivotal role to draw attention to resolve escalations and send appreciation to drive better organizational alignment between the parent organization and the supplier (O'Brien 2014).

Although there are many software tools to communicate up to date demand or inventory levels or invoices and other data and metrics of the business, at the end of the day it is humans who are running the business on both sides. A supplier manager is expected to build relationships both with internal leaders and externally with the leaders in the supplier organization. Internally, if a leader of the BU (customer organization) prefers to do business discussion directly with the supplier and without the supplier manager, it could be considered as a failure of the supplier manager. The supplier manager may not have been able to demonstrate their value to be involved in the conversation. It could happen because the customer may have not seen examples of urgency from the supplier manager (how quickly the supplier manager has solved price, lead time or quality-related issues) but most of the time it happens when the supplier manager has not developed relationship with the customer. Meanwhile the supplier has built a stronger relationship than the supplier manager. Although we have discussed the importance of building relationships with the internal stakeholders, this section will focus on the key folks in the supplier's organization.

A supplier manager has to ensure that one has a relationship with both sides, and when both sides, namely internal customers and external vendors, view the supplier manager as the natural path of least resistance then a supplier manager has done their job well. In some organizations this is forced but that often leads to bureaucracies which are not always appreciated. However, when this happens naturally then the supplier manager has been able to build the right synergies both internally and externally. Escalations, expedites, and cost negotiations are some of the typical situations when a supplier manager needs to exert one's influence over the supplier on behalf of one's customers. Similarly revenue growth with new design wins; faster communication of accurate internal demand, introduction to executives to present new products, or getting a purchase order before the end of the supplier's fiscal quarter are some of the typical reasons when a supplier manager needs to exert influence over his customers on behalf of one's suppliers. A supplier manager manages the relationship with fairness but is able to maintain a healthy level of tension so that supply base does not become complacent.

Relationships are often managed at a peer-to-peer level based on title. For example, an *entry- or a mid-level* supply chain manager who owns a supplier relationship is usually considered as a peer of the sales manager of the suppliers org. A *senior manager* or director of supply chain usually interfaces with a director or VP in the suppliers organizations. The supply chain *exec* is at the peer level of the VP of the supplier organization or even CEO depending on the size of the customer's spend with respect to suppliers total revenue. A similar hierarchy is also managed internally between the supply chain and customers' organization. A successful entry-level and mid-level supplier manager can influence both internally and externally. This means an entry-level or a mid-level supply chain manager should be able to provide information and business context to leaders of the organizations on both sides.

A supplier manager should have a direct relationship with the account manager, account director, and at least a sales or product vice president. Quite often an escalation meeting is measured by the level of the attendees. If a VP level person attends from one's parent organization, a supplier level VP or SVP is expected to be present. As a supplier manager it is one's responsibility to make sure the right level of attendees attend from the supplier side as it exhibits the ability to influence. Quite often escalations happen due to unforeseen events and status meetings turn into daily calls. If the internal audience is up to VP level then ensure that there is visibility in the supplier organization up to VP level so that there is the same level of accountability on both sides.

Developing this kind of relationship takes time and give and take in an ethical manner. Suppliers will often spend efforts in building relationships with supplier managers. However, every organization has guidelines on what is appropriate. Although there may not be anybody checking for compliance or violation of ethical code of conduct, there is a side effect of being over accepting and that is losing one's ability to influence. Although going for dinner or to a game in a supplier's box may give opportunity to meet and greet with the VP level executive of the supplier organization, it has an ethical cost where a supplier manager may get more biased toward a supplier. Although relationships are formed over dinner occasionally, a better way of building relationships is doing a fair give and take with the supplier.

A successful supplier manager develops relationships which follows him through one's career. A successful supplier manager makes one's sales account manager successful to improve the sales manager's visibility within the supplier organization and to extend one's influence within the supplier organization. As a supplier manager prefers to be the path of least resistance so does the account manager. So one way of building relationships is to create a win-win relationship with the account manager. A typical way of doing this is by giving credit to the supplier manager where credit is due but in front of the supplier account manager's execs or offer some non-cost related gives which does not impact or does not put one's organization business interest in jeopardy. This empowers the account manager who is now more motivated to represent one's account inside one's own organization. The supplier manager builds relationships with higher executives through him, and during escalations nurtures those relationships to make one's parent organization more successful.

6.8 PILLAR 6: EXECUTION

To avoid repetition, this section is not going to restate how a supply chain manager can obtain best cost without compromising quality and delivery lead time. The roles and responsibilities of a supply chain professional has been covered in Chapter 4. Instead this section will focus on the capability to put it all together and communicate as a subject matter expert. An expert who is capable of measuring and articulating financial metrics, KPI, strategy-driven targets and maturity milestones, and get into solving complex problems (DeSmet 2018).

Assuming one has developed subject matter expertise in the three areas mentioned earlier in this chapter, such as internal product knowledge and its sources of demand as well as knowledge of supplier landscape, one has to still speak up, communicate, and share the information. Modern organizations are meeting-driven organizations, where subject matter experts gather and give status and take consensus-driven decisions for ongoing projects. As a supplier manager, one should understand one's role in such forums. It is important to be able to paint a picture covering the three areas. The details often depend on the level and appetite for information of the audience but it is important for the supplier manager to connect the dots and paint a futuristic picture. Employees often spend more time internally focused than externally and that is where a supplier manager brings in value to make the team collectively make some choices today which helps them to secure better business continuity in the future.

Often during escalations or during value creation opportunities where ROI or similar analysis are done, a supplier manager gets an opportunity to showcase one's prowess in the first three sections. In a very simplistic way, the knowledge base is described in the first three sections, the showcasing opportunities are described in the next two sections, and doing so is referred to as execution in this section.

From an *entry-level* supply chain professional to a supply chain *leader*, the journey often involves growth in portfolio of categories of products to manage to growth in portfolios of operational activities to manage. Categories and spend dollars are a good measure of growth. For example, a supply chain manager handling batteries for EVs can grow into a supply chain leader who manages the entire electronic category which includes batteries, electronic motherboards, sensors, etc. The next step of progression could be managing the entire supply chain which includes from sheet metal, mechanical parts, engines, to all things electronics. Operationally the role can evolve from managing supply chain to managing the entire spend portfolio, which may include supply chain, logistics, supply chain IT, etc.

Depending on the career aspirations, a supply chain professional can also measure one's career progress by moving laterally and learning diverse categories of products with a new set of supplier ecosystems. It's also equally important to rotate and learn other aspects of the supply chain by rotating through various roles. Along with doing the day job of negotiating, sending requests for quotes or negotiating contracts like a master purchase agreement, a supplier manager can also get oneself well rounded by learning various kinds of roles and also developing understanding of various legal, compliance, and financial aspects of the supply chain. The key to execution is to be able to project knowledge of the three areas during the execution of these roles.

Depending on one's level, the communication is done by either sharing various supply chain metrics in a team setting or doing full blown operational reviews. For an individual contributor supply chain professional, collaterals like one page supplier profile or a white paper of supply chain landscape for a key category are often the right vehicle to project subject matter expertise. Although a lot of experienced supply chain managers are excellent story tellers, one of the effective ways is to be able to document product knowledge or supplier landscape in a simplistic but exec level PowerPoint. An organized supply chain professional keeps this ready. A supply chain leader should always have a PowerPoint to articulate category strategy and spend summary. Knowledge of the previous QBR always help as in a meeting forum this becomes a pivoting point. A data-driven and analytical supply chain professional, depending on the audience level, can reduce or increase data content in one's storytelling to influence effectively.

6.9 CONCLUDING REMARKS

This concept of six pillars is essential to develop a 360-degree view for any supply chain professional. Supply chain teams in a large organization often develop tunnel visions due to the operational nature of the jobs. This can lead to over indexing one particular pillar. For example, a supply chain commodity manager may have a deep knowledge of the pillar 3 (supplier landscape) and pillar 5 (relationship management with supplier executives). Similarly a supplier manager may be hyper focused in pillar 2 and pillar 6 where they know about the demands and chase shortages and mitigates risk but may fail to catch product transitions due to technology changes (pillar 1). Supply chain is a unique field which spans all across a company from product development, sales, finance to operations. Supply chain is also about creating options for each of these areas, for example, finding a new supplier and lower cost or higher quality or reduced lead time to make sales successful. It is also about improving the bottom line by reducing operating cost by using shipping instead of air freighting or having better payment terms with vendors. It is also about finding out sourcing optimum vendors to partner with who are ready to invest in technology and resources to improve the product development process. The six pillars will help a supply chain person to develop that well-rounded view and identify new solutions and to grow the organization and themselves. The six knowledge pillars present a powerful tool to develop leadership and decision-making skills.

REFERENCES

DeSmet, B. 2018. *Supply Chain Strategy and Financial Metrics: The Supply Chain Triangle Of Service, Cost And Cash*. London: Kogan Page.

Khan, O. 2019. *Product Design and the Supply Chain: Competing Through Design*. London: Kogan Page.

O'Brien, J. 2014. *Supplier Relationship Management: Unlocking the Hidden Value in Your Supply Base*. London: Kogan Page.

Snyder, L., and Shen, Z. 2019. *Fundamentals of Supply Chain Theory*. Hoboken, NJ: John Wiley & Sons.

Sollish, F., and Semanik, J. 2012. *The Procurement and Supply Manager's Desk Reference*. Hoboken, NJ: John Wiley & Sons.

Srinivasan, M., Stank, T., Dornier, P. et al. 2014. *Global Supply Chains: Evaluating Regions on an EPIC Framework – Economy, Politics, Infrastructure, and Competence*. New York: McGraw Hill Education.

Trent, R. 2016. *Supply Chain Financial Management: Best Practices, Tools, and Applications for Improved Performance*. Plantation, FL: J. Ross.

Weigel, U., and Ruecker, M. 2017. *The Strategic Procurement Practice Guide: Know-how, Tools and Techniques for Global Buyers*. Heidelberg: Springer.

7

Making Your Mark as a Leader

Sudip Das

Ishaan Das

CONTENTS

7.1 Objective..154
    7.1.1 For an Impatient Entry-Level SCM...155
    7.1.2 For an Impatient Mid-Level SCM..155
    7.1.3 For an Impatient Senior or an SCM Executive..........................155
7.2 Maturity and Rigidity in a Supply Chain Organization 156
7.3 Reading Guidance..157
7.4 Opportunities..159
    7.4.1 Dashboard, White Paper, and Spend Summary........................159
    7.4.2 Price – the Lowest Hanging Fruit...161
    7.4.3 Approval Flow 162
    7.4.4 Reactive Escalation Management 163
    7.4.5 Communicating Supply Chain Strategy.....................................165
    7.4.6 Escalation Response..165
    7.4.7 Data and Presentation 167
7.5 Complex Opportunities 168
    7.5.1 Vendor Change and Consolidation 168
    7.5.2 Vendor Managed Inventory...169
    7.5.3 Multilevel Supply Chain..169
    7.5.4 Logistics...170
    7.5.5 E&O Management and Disposal 171
7.6 Worries of a Leader...171
    7.6.1 Automation 172
    7.6.2 Decision Support and Data Analytics...172
    7.6.3 Lost Opportunities 173
    7.6.4 A Whole Lot of Issues..174
7.7 Concluding Remarks...175
References...176

DOI: 10.4324/9780429273155-9

7.1 OBJECTIVE

A mature supply chain organization is typically quite rigid or set in its way of doing business. Actions involving switching an existing vendor with a new one; increasing buffer stock quantity for risk mitigation; switching logistics partners and routes; changing terms and conditions with vendors for ease of doing business; implementing new automation or enterprise resource planning (ERP) tools for faster communication with suppliers or any such new initiatives typically takes a lot of time to get buy-in and to implement. As practices become old, workflows get deep seated, relationships with existing supply base forged, an organization forms its habits. These patterns become standards and take deep roots in a supply chain ecosystem. Subsequently a mature supply chain may have a tendency to become a slow moving, process-oriented, rigid juggernaut mired with contracts, compliance issues and processes.

As a supply chain manager (SCM), you may often hear complaints from your customers that the supply chain is so complex and bureaucratic. They may also complain that only if they know the right people or escalate then only they can get things done fast enough with the supply chain department. However, if a supplier manager or a leader tries to solve the customer's problems, one may face organizational resistance from the supply chain organization itself. The resistance can come from various quarters. For example, the end customer may ask for expediting a transaction with a specific supplier but it may not be possible due to process compliance issues where the supplier manager is required to show evidence of multiparty competitive bidding which will drag out the transaction. Or every transaction needs a contract which makes it slow due to lack of availability of legal resources. Dotting the i's and crossing the t's become more important than meeting the customer's urgent schedule. Similarly there could be challenges where a customer wants to engage with a new supplier who has not been on boarded or is not part of the preferred supplier list. Here also one may have to either influence the customer not to engage with the new vendor or go through a long drawn justification, approval and contracting process with the new vendor before being allowed to engage. Being very limited and trying to manage customers through a narrow scope of rules and processes in the name of risk avoidance and ensuring legal protection, where business leaders have less say than the legal counsels, is a telltale sign of a rigid supply chain organization. However, this is the reality.

So, how does a supply chain leader create an impact in such an organization which is quite set in its way? Depending on the seniority of one's role, this chapter will present some options on how to move the needle in spite of the inherent rigidity. Instead of viewing their own organization as rigid and inflexible, how could one view it as an optimization problem to solve for? Through leveraging knowledge, negotiation skills and relationships, one can identify and sell the opportunities that they have discovered.

7.1.1 For an Impatient Entry-Level SCM

Assuming you have read Chapter 6, you are now in full command of your portfolio. That means you can now articulate the spend, the products, the health of your business and your suppliers. Along with that you have built the relationships, have executed successfully and built credibility. In other words you have gained trust of your customers. Where do you go from here? This chapter will present you with a list of paths to explore. You have to be also able to assess the changes you have brought in or improvements you have introduced and be able to articulate how those have moved the needle and made impacts along with the lessons learnt from them.

7.1.2 For an Impatient Mid-Level SCM

At your level negotiating the right price, executing proper supply chain related escalations, delivering cost savings, mitigating risks, etc. are all expected table stakes. Having great influence over your internal customers and suppliers, and being viewed as somebody who unblocks any kind of business continuity challenges are now expected behaviors from you. These have become part of running the business activities for you. Where do you go from here? While contemplating to seek a new role internally or externally could be an option but do not feel frustrated with the rigidity. Your answer could be moving the needle by finding new efficiencies. This chapter will help you with that.

7.1.3 For an Impatient Senior or an SCM Executive

Your organization is running like a well-oiled machine delivering great efficiencies. Your teams have great performance track records with well archived results and yet you are not feeling challenged enough. New opportunities for learning are flat lining for you. This can happen, especially when your company matures and revenue growth slows down. Till the next growth spurt comes, you have now entered from *growth* to a *sustain* phase. In this phase, the spend growth slows down significantly and pressure for cost reduction will increase. However, how do you maintain your own personal growth momentum, that is the question under discussion here. This is the phase where you have to constantly find new ways of improving the company's financial bottom line. The book will present you with some options to explore, which may lead to creating more value for the shareholders.

7.2 MATURITY AND RIGIDITY IN A SUPPLY CHAIN ORGANIZATION

Assessing the state of maturity of the supply chain organization is the first step, before you want to improve its efficiency. An organization may collectively feel the strategic need for starting a supply chain group at any given point during its life cycle. Exploring the reasons, such as when was it established and why, could give an idea about the current maturity of the procurement department. For example, a direct supply chain team may get formed early for a manufacturing company to build products. In contrast to that, a company getting their product made through a third party may feel the need to vertically integrate and start its own supply chain team once it reaches a certain sales volume. Another use case could be where a procure-to-pay (P2P) team established to run indirect procurement may decide to start a strategic sourcing team to do better contract and cost negotiation.

Like any growth organization, it is easier to move the needle when rules have not been set and everything is done for the first time. Opportunities are endless in such an organization. You can move the needle in multiple ways, such as defining procurement processes that everyone needs to follow, establishing supply chain strategies like designing a network of suppliers, consolidating many suppliers to a few, delivering significant savings by starting to negotiate instead of just paying the invoices, etc. If every day feels like you are firefighting, you may be in a growth organization and there are a lot of unrealized efficiencies. If the entire organization is young, it is usually more open, creative and less resistive to try different approaches. Fear of failure is low as the number of decision makers is less.

However, as maturity comes within the supply chain department or outside within the larger organization, risk aversion starts setting in.

Complexity also increases with maturity as the number of decision makers increases. The environment starts becoming more rigid. A rigid organization is typically a hierarchical organization with multiple layers of managers with smaller spans of control and functional silos which operate independently. Having a centralized procurement team to manage all indirect or direct procurement activities through a command control style of leadership is also an example of hierarchical structure and rigidity. A typical example is often large corporations trying to route all its diverse procurement needs through a centralized procurement department. A single procurement effort goes through multiple handoffs between various subdepartments such as buyers, reviewers, legal, compliance, finance, etc. where each of these subdepartments are independently inundated with their own respective workloads. Consistency and accountability becomes a challenge.

However, there is a silver lining: a procurement organization while it could be rigid but it has all the opportunities of not to remain as an organization in silo. A SCM's job is both cross-functional as well as dynamic. It has a very large circle of influence ranging from various types of internal partners such as finance, product development, legal and external suppliers. Changes in business conditions also enforce flexibility. So, in spite of belonging to a hierarchical organization a supply chain team member does not have a tunnel vision. Instead, because of the plethora of data sources and constant exposure to fresh new ideas from these sources, there is an opportunity to develop a 360-degree view.

However, the challenge comes when the SCM attempts to do problem-solving in a mature and rigid environment as opposed to a growth environment. For example, if a supplier is not performing well, replacing the supplier can be a quick decision in a fast growth environment but it can take months or even years in a mature organization. A fast growth organization may be quick to cut its losses whereas in a mature organization opposition to the change can come from multiple sides due to various reasons such as supplier is designed in, supplier has other business relationships which are working well, supplier is preferred by the customer and deeper executive relationship.

7.3 READING GUIDANCE

In subsequent sections of this chapter we will present different ways to deliver efficiencies in both high growth and mature or rigid environments.

In some cases the opportunities are feasible for individual contributors and some are in scope for senior leaders and people managers. The opportunities are presented with the perspective of value creation by an individual employee. The scopes of the solutions discussed though not restricted, are bucketed based on the level of seniority of a supply chain professional. The scope of solution will differ significantly between what is feasible for a SCM who is an individual contributor and what is achievable by a member of the supply chain leadership team who can leverage and deploy larger amounts of resources both in terms of budget and headcount. Also there are certain opportunities which are simpler and can be achieved even if there is rigidity. And then there are more complex opportunities which could be easy to implement in a growth organization but a lot harder in a mature environment as it will demand significant management efforts.

As the scope will scale based on complexity and resources, the opportunities presented here are by no means exhaustive or prescriptive. Our recommendation is to treat the opportunities presented here as kernels of a few thought-provoking examples and collectively as an approach. As one faces different types of challenges from their customers and suppliers, some of them could be solved by drawing ideas from these examples.

The use cases or opportunities are divided into two sections. There are some quick wins which are relatively easy to implement. An individual contributor will be able to deliver most of them as individual best practice even in a very rigid supply chain without raising too many eye brows or needing permission from one's superiors. Similarly a manager can also implement most of these as part of operational excellence without putting the burden of additional manual processes on his reports. These low hanging fruits have a high return on investment. The impacts in the form of value addition are very easily communicable.

In the second section more complex use cases are presented. These are longer term changes which often need sponsorship from the leadership team. While a senior executive of a supply chain organization may find the second set of examples more relevant, however, a relatively younger member should not be ignoring this section assuming that these scenarios are beyond their scope. Complex solutions often emerge from the person who is closest to the problem and not necessarily from executives. It's often the individual contributors who act like owners and drive high-impact initiatives.

7.4 OPPORTUNITIES

7.4.1 Dashboard, White Paper, and Spend Summary

How much is our spend with a specific strategic vendor and how it is growing or declining year over year? It was $20 million in 2018 and it has been increasing by 5% year over year. Apparently a simple answer to a simple question. However, in reality, a lot of Fortune 500 companies will struggle to answer this question in a consistent and reliable way for all its vendors, without putting major tools and processes in place. Imagine your VP is going to meet with the CEO of one of your suppliers. The day before the meeting, your VP asks you for an overview of the business. You are tasked to include some basic information, such as total spend by various business units with this vendor and the overall spend as a percentage of the total in the specific category (Pandit and Marmanis 2008). The VP's objective is simple, that is to get more favorable terms in lieu of more future business. Basically you have to come up with a white paper which gives an overview of the health of the relationship. However, be mindful that organizational leaders have usually trained themselves to doubt any data at their face value and will challenge and try to punch holes and find whether you have really taken all spends into account or not. This is their way of doing validation before owning up to the data and underlying messaging. And if they find any reason to doubt the number that they have been provided, your credibility may sink. The first impression you create is often the only impression which sticks about you in your executive's mind.

It is actually quite surprising once we realize how a simple spend dashboard could be an easy win. Unless there is business critical information which is changing dynamically and frequently, no one may pay close attention towards a dashboard. Maintaining a dashboard also takes resources in terms of time investment and integration efforts with disconnected sources. Then why is dashboard important? A dashboard of a set of frequently needed, comprehensive and critical information often becomes pivotal for negotiation. Last moment requests for a set of information by an executive is very common like what do we buy, how much do we buy and from whom do we buy from. A lot of organizations will burn midnight oil reactively and scramble to provide the information, not leaving enough time to authenticate. The success in the negotiation on the following day often becomes a function of which party has done its homework well.

If you are an individual contributor, based on frequency of ask, identify a key set of information. For example, a simple set of information will be total spend and how that spend is divided among top ten or twenty suppliers and for the rest of the long tail list of suppliers.

- Total spend in 2018 – $100M
- Spend distribution by vendor – Vendor 1 ($22M), Vendor 2 ($15M), …, Vendor 10 ($1.5M), remaining 30 vendors ($13M).
- Spend distribution by category – Category A ($30M), Category B ($15M), …, Misc. ($8M)

Repeat the above set of information for 2 more years. Based on the data, now you can project a year over year trend by both vendor and category.

Spend with Vendor A has been declining for the last 3 years by about 10%. Vendor B's spend has been increasing, etc. Similarly many different pivots can be created, for example, spend distribution by geography, by business units, by product line and by technology.

Depending on the business demand, building a comprehensive dashboard with a few but critical sets of information is very important. There are organizations that put systems in place to provide this information. However, quite often the right pivot is never available and human intervention is required to massage the data to go along with the messaging. As an individual contributor you can always move the needle by (1) building and evangelizing a dashboard with a key set of information and (2) maintaining your own set of information and making it available reliably and consistently. As a senior leader you can move the needle by creating a format for communicating key set quantitative metrics or qualitative descriptors as a standard. Depending on the appetite for complexity, the metrics can come from complex financial or supply chain tools or it can come from manually stored information. However, it's important to ensure that the data is either self-managed or very easy to access and maintain and not something whose maintenance itself causes unnecessary stress in an organization.

The pros of creating dashboards, white papers or spend summary reports is that it becomes the source of information and hence will be able to weave a story based on facts and project trends. In other words, it empowers a SCM to influence the chain of command. The cons on the other hand is sustainability. To be consistent in terms of publishing based on a quarter or two or even more frequently, the content cannot be stale and therefore needs constant attention.

7.4.2 Price – the Lowest Hanging Fruit

The way an organization decides what price to pay for goods and services it procures is quite diverse. Also what an organization agrees to pay at the beginning of a procurement cycle and what it ends up paying during the life cycle of the product could be quite different. As price is the most important part of procurement, therefore whatever your role is in the supply chain hierarchy is, empower yourself to dig deeper at any given aspects and identify opportunities for cost savings and it will always be appreciated (Templar 2019).

Different organizations set prices differently. For an item which is continuously purchased, for example, a component which gets integrated into the product, a supplier manager may negotiate price quarterly or monthly. On other hand there could be a highly expensive piece of software where the price negotiation may end with a handshake between the VP of your organization and the VP of the supplier's organization. However, if we dig deeper and compare with the final invoices and do spot checking throughout the procurement cycle you will notice that the price has diverged from the original amount. The divergence may happen due to various reasons, such as the original agreed upon component is modified, additional cost is incurred due to expedited manufacturing or shipping request, additional tariff is imposed and order volume threshold is not met.

Bottom line is if you start digging deeper and try to understand what the average cost of goods paid and compare with original cost goals, a different picture will start to emerge. In case of software purchase you may find the actual number of licenses have far exceeded the committed limit of licenses and therefore the cost has gone up. As an individual SCM you can trace the final invoiced or the total cost of ownership and may find out opportunities to reduce cost. You may discover situations where due to lack of forecasts, purchase requests are always coming at the last moment causing expedited air shipping. Better forecasting and planning can reduce the logistics cost by switching to alternate freight arrangement.

You may also find the initial assumptions made during cost negotiations have changed, for example, the volume has increased and yet the cost has not been renegotiated. Sometimes a simple price reduction is possible just by asking the vendor for a price reduction. Suppliers are dealing with multiple customers and they will often charge different amounts for the same item for different customers. Often the pricing difference could be because of the volume of business. However let's say a component is being manufactured for a long period of time, suppliers will do cost reduction and will also discount more as the product becomes old. These discounts will stay unrealized if nobody asks for it.

For a senior supply chain person it's important to do regular cost review and review the total cost of ownership of the portfolio of items. This is an important task which needs to be resourced and reviewed to keep control on price. Establishing a cadence and calendaring a schedule to ensure regular price review is important. It builds a price focused rigor in the organization.

7.4.3 Approval Flow

Procurement process is a cross-functional process which involves participation of various teams. For example, there is a product development team which controls the engineering development or design of the product. There is a finance team which controls the purchase order to payment for the raw material and other goods and services required to manufacture the product. There is a manufacturing and fulfillment team which controls the production and delivery to the customer. In each of these cases there are multiple organizations which have various roles to play. An organization may initiate a request for approval, another may have to approve in order to complete a task. One of the most common process is the purchase order request and approval process which culminates in the issuance of the purchase orders. For example, a team wants to buy a set of widgets and they raise a purchase request via email or through a ticketing system. It goes to finance who ensures that there is enough budget and whether funds have been allocated against this request. Then it goes to executives who would ensure that this is aligned with an approved business objective (Cohen and Roussel 2013).

The key point behind this topic is to improve efficiency. As these approvals are ultimately done by humans, the end-to-end duration for an approval process takes a lot of time. Valuable time could be saved by streamlining approval flow and automating steps where applicable. Approving flows are often hard coded within an organization. Due to financial compliance requirements, approval authorization limits are tied to levels or positions. For example, an organization may have a spend limit of ten thousand dollars on a manager. In other words, a manager has been empowered to approve or authorize up to a ten thousand dollar purchase. In the same way a director may have a hundred thousand authorization limit and a vice president may have a million dollar limit. Anything above may have to go to a senior vice president or a C-suite executive. Often these approvals take a lot of time and create periods of inactivity and churn. These delays become part of the practice and are often left un-addressed for years till one day business gets impacted and draws executive attention.

As a supply chain organization touches all these functional teams such as product development, finance, leadership team and manufacturing, it gets visibility to these approval chains and durations. There is an opportunity to create efficiency. It could be as simple as moving one more approver from serial to parallel flow. For example, there could be multiple executives who need to approve after financial approval who could be put in parallel and can save time. Also often approvers could get added more for the purpose of keeping them informed than actually seeking approval. They could be moved as watchers instead of approvers. Another way to gain efficiency is by putting an aging restriction or wait period limit. Beyond a certain time period, the approval request will automatically move to the manager of the approver if it ages beyond a certain amount of time.

Changing approval flow may be quite complex as it may require compliance approval, approval and cooperation of IT for implementation, and other challenges such as organizational inertia. It could be difficult for a SCM and easier for a supply chain executive. However, an informal heads-up notification to the approvers goes a long way and is easily achievable. Quite often approvers and particularly executives have hundreds of approvals to go through above and beyond their busy schedule. Understanding the executives personal criteria is the key here. Sending the upcoming approvals together as a list and providing the executives with a write-up with certain key information in a few lines can develop a bond of trust between the narrative creator and the approver. As humans are creatures of habit and everybody appreciates efficiency, any effort to reduce the information search by the approver will produce efficiency in the approval flow and reduce the approval time significantly. This will also create a trusting relationship.

7.4.4 Reactive Escalation Management

Events such as shortage, production or performance disruption, contract violation, damaged delivery and large-scale field failure are realities. Anyone of these hits every supply chain every now and then. Quite often organizations forget how to react to such incidents when they occur.

As a result, often the efforts to mitigate them starts in a dysfunctional way at the beginning. A mature organization may eventually develop a standard process with boilerplates for communication and predefined designated owners so that it does not redefine the reactive process every time. This is an opportunity for a supply chain organization to put in place (Manners-Bell 2018).

The most typical opportunity is handling a shortage situation. Often different parts of the organization get to know about the situation in an asynchronous manner. In addition to that downstream organization may often react and panic without fully understanding the mitigating actions taken and the get well timeline. So step one for any organization is to create a template with a minimum of three part communique. It should articulate the following: first, the item of shortage and its time horizon; second, the mitigating actions taken and their expected time line; third, the impacted products and their expected recovery timelines. Having the rigor to send this communique daily or weekly helps the broader organization to react much more calmly and focuses them away from dealing with the uncertainty of shortage to the timeline of getting healthy.

Along with influencing the organization to internally mitigate a disruption in a methodical manner, the second opportunity for supply chain organization is to influence the same organization on how to externally mitigate the same disruption. A supply chain leader can also program an organization on how to do external escalation with suppliers. Let's take an example of field failure. A product has all of a sudden started to have an increased number of customer returns and it has been root caused to a specific component. The component manufacturer needs to take responsibility in further investigating and providing a solution which could be to replace with an improved part or applying a new software patch. The supplier manager has to engage the supplier to create a solution which requires the suppliers understanding and replicating the problem in their own lab, do a deeper dive to come up with a solution and take accountability which may mean reimbursement, exchange, monetary compensation, etc. How does a SCM escalate within the supplier organization and draw the supplier's executives' attention? If the SCM fails to get the required attention, how they communicate internally and seek help from their leaders is critical. To be able to do so in a time-sensitive manner requires templated communication and sense of urgency, so that the escalation can move up quickly. This way the supply chain hierarchy can demand attention with the suppliers organizational hierarchy leading to quicker resolution.

7.4.5 Communicating Supply Chain Strategy

Supply chain is a dynamic environment – technology changes, business environment changes, supplier performance changes and so on. The supply chain strategies often have to adapt and react immediately. Quite often the broader organization may not have clear ideas about these strategies and forgets the reasons and benefits of engagements with strategic partners. In other words while being immersed in day-to-day operations, the broader organization forgets the strategic intent behind engagements with partners and may often try to fragment the supply chain by trying to engage with many partners as they desire. A SCM's role is to keep the supply base consolidated and not to create fragmented point solutions with many suppliers, but creating a consolidated supply base which enables the company to get the most attention from its suppliers and get the best cost, quality and lead time. In a large organization with many product teams, they often do not see the big picture and tend to focus on efficiency from their perspectives only. It's the SCMs responsibility to strike the right balance.

A SCM should be always armed with a 30,000 feet view of their commodities and suppliers. Expecting the internal customers to be knowledgeable or having a view of this high-level perspective is a mistake. A SCM should always be able to articulate the total spend for the commodity, total spend with key suppliers in the commodity, the benefits of choosing strategic supplier partners, future trends in market and future growth of the suppliers and opportunity for the customers.

Often this kind of strategy with key performance indices and spend metrics are asked to be provided by executives when they are about to meet their counterparts. Not having the up to date overarching strategies creates a bad habit inside the organization. Having the strategies reviewed and communicated helps refresh the memory and also allows the organization to revisit the strategies at a regular cadence. A supply chain leader should always endeavor to have strategy reviews and updates in a quarterly or whatever cadence is appropriate.

7.4.6 Escalation Response

Supply chain is a dynamic world where operational problems will happen. After mitigating the issues, management may do an inquiry to identify the root cause. Some organizations may even have process teams who have formal processes to interview the stakeholders, audit the issue, document the lessons learnt and put preventive measures in place. However, this section is not about the post-event analysis but more on how to manage the problem during the event.

Problems can happen in different shapes and forms such as shortage of parts, logistics disruption causing lead time increase, quality impact leading to disqualification of parts and product build being stopped. Step one is to see the problem as a business owner. It's important to view this as a vendor and vendor product-related issue and hence take ownership. This perception should be independent of the fact that other parts of the organization are also involved in solving it. For example, a logistics disruption could be caused by the lack of freight carriers, and logistics teams are involved to solve it. Or in case of a quality problem, the product engineering team is involved. A supplier manager should take an overarching business owner's view and take accountability of owning the problem till normalcy returns.

The three most critical aspects of escalation handling are speed, communication and driving engagement. Recognizing the problem and rushing in to fill the vacuum avoids the organization from wasting valuable time in determining ownership during these nebulous periods. Shortening that demonstrates leadership. The next step is taking ownership of the cross-functional communication with clarity. Defining the problem, articulating steps already taken and expected next steps shows the situation is being handled and is moving toward a possible solution. Communication should not be only about the problem and its impact, as it makes the communication open ended which to an executive will sound like a doomsday message and cause alarm. The message should have the complete picture including measures put in place and the supply leader should communicate a sense of control over the situation. A leader should be reassuring and calming but also preparing the organization for reality. Also a leader avoids the organization from churning and chasing paper tigers. When a problem first shows up, due to lack of information, it may look more complicated than it actually is. As pieces fall into place with time, it becomes more defined and appears to be manageable. Communication should have at least the following three parts: description, status with action items for the audience, and next steps or milestones.

The final step is engagement. As the program manager and owner of the situation, a SCM has an opportunity here to drive the engagement. It could be to directly engage with the logistics provider on the supplier side in case of logistics-related challenge; or with the product engineering support team from supplier side for quality-related issues; or it could be engaging with the executives on the supplier side to ensure resource allocation and executive visibility. The main purpose is to match the urgency of all the parties involved. This may require setting up meetings at an unusual cadence such as daily conference calls or whatever the situation demands. It is a win for a SCM, if they have jumped into the fire fight before everybody else and have taken ownership. If their leaders find that the escalation process is already in place, stakeholders are being communicated with action items and next steps and all parties are fully engaged, then the SCM has done the job right.

7.4.7 Data and Presentation

Presenting data is a science and knowing how much data to present is an art. There are many books on presentation and data visualization. This segment will, however, offer a different perspective on presenting data. When a SCM does a presentation, most likely the topic of discussion will be one of the following three categories. It could be something old like a presentation on operational status as part of an existing practice where the audience is habituated to the practice; it could be something new like a proposal to gain a new efficiency and change and could be a surprise to the audience, or it could be about something unexpected like response to a disruption where audience is expecting the way forward. Now apply a lens of *risk* to all three categories.

Before doing the presentation a supplier manager should consider the element of risk that the content will introduce to the audience. In case of a status meeting, following a preexisting format that the audience is used to is important. The past has already trained the audience what to look for. Deviation from the status quo is a risk. Putting additional content or data does not always result in a favorable response, especially for a large audience as it forces them to process unexpected sets of information. In something new, there is a risk of decision-making. Therefore, the audience's appetite for taking the risk and having the appropriate authority to take it needs to be weighed in. Often in a large meeting there would be a key decision maker and a lot of critiques without authority. So it's important to assess whether the key decision maker has been given the right amount of time to process the risk and make an informed decision, instead of getting distracted by the critiques. Being accountable is also a risk for the decision maker. Disruption-related meetings, on the other hand, are all about articulating the risks and its mitigations.

7.5 COMPLEX OPPORTUNITIES

The opportunities outlined in the previous section are achievable by all levels within a supply chain organization. These are efforts which an individual contributor can do and even a supply chain leader can ask from their teams. There are often lean six sigma black belts who find opportunities to improve the supply chain via lean methodologies (Myerson 2012). The items discussed in this section could be a bit more complex as they may require sponsorship from a manager or a director or even higher level of authority to move forward. These opportunities require more collaboration and sponsorship from stakeholders (Hugos 2018). These could be due to cost of implementation, duration of implementation and or could be simply due to reluctance to change.

7.5.1 Vendor Change and Consolidation

As an organization engages with a vendor, the relationship starts taking roots. The roots grow in terms of familiarity with the vendor's products and ease of integration through generations of the products; partnership grows between executives on both sides (supplier and customer) and develops into a trusting relationship. All of these roots also create inertia and prevent the organizations from looking further and exploring new partners. Sometimes things happen naturally like supplier's performance drops to an unacceptable level or a supplier gets acquired by a competitor but overall the organizations tend to maintain status quo with its supply base to ensure predictability in procurement price, subcomponent performance and its delivery.

Shaking the organization and bringing in change is hard. A supply chain leader can always create an impact by bringing in new efficiencies through consolidating vendors and getting better cost, introducing new efficiencies like better warehousing capabilities of the vendor and new technologies from more advanced vendors to build better products. At the same note as any change can also introduce risk, it needs the power of influencing and credibility with the stakeholder organizations. Most importantly, it requires patience to bring in vendor changes and wait for the strategic intent of the change to materialize. Creating a win-win reason with stakeholders helps to bring this type of changes; however, it also requires the change manager to put safeguards to mitigate risk and have fall back options.

7.5.2 Vendor Managed Inventory

There are many books written on vendor managed inventory (VMI) and how to manage it (Sabri et al. 2007). However, it's a common perception that VMIs are possible when an organization reaches scale or has a huge purchasing power, a supplier will put material next to the customer's factory at its own cost. Taking a step away from that perception, what this section offers is a deconstructed VMI concept. Typical VMIs of course have a warehousing and logistics aspect. A vendor manages the transportation and puts a certain quantity of material at a warehouse which is not part of its own distribution or channel. However, if this aspect is set aside for a moment, the essence of VMI is the supplier setting aside a portion of its inventory ahead of time in anticipation that the customer will need it within a short period, like within a quarter. Narrowing down this definition further will imply a vendor is setting aside a portion of inventory for a specific customer.

If we ignore the warehousing and transportation aspect and do not care where the material is stored and assume that delivery of the product to any point in the globe is only a plane ride away then it really does not matter where the material is warehoused. By this definition the key information for a vendor to implement VMI is (1) understanding the quantity which it needs to set aside and (2) with a high predictability the customer will pick it up within the stipulated amount of time. This now makes it a possibility for having a VMI within every supply chain leaders' grasp by providing a steady forecast and standing behind it. If the vendor develops a high degree of assurance, this actually works to the vendors' benefit. It will now have a steady customer who takes a predictable amount every quarter. A supply chain leader can look at their forecast and take some bets and create a baseline forecast which they can stand behind. Suppliers would typically set aside a quantity with predictable delivery lead time to the customer's factory. In exchange the customer gets an assurance of supply which helps with its business continuity plans.

7.5.3 Multilevel Supply Chain

A supply chain organization often comprises SCMs who are dealing with direct suppliers. As efficiencies are maximized in terms of cost, the next level of cost optimization could be obtained from the secondary suppliers. It is important to understand the role of a primary supplier in this regard (Ravikanti 2019). For example, let us say a supplier is providing a 100 dollar part for a telephone. The part is the overall camera assembly and has various subparts such as camera, microphone and audio video processing semiconductor parts. Let's assume all these parts cost the supplier 40 dollars as it goes to the camera vendor, microphone vendor and semiconductor vendor and buy those subcomponents. The supplier gets the order, builds the full subassembly, puts the part in VMI and waits for the telephone manufacturer to pick it up. Then the supplier sends an invoice to the telephone manufacturer and gets paid. However, the supplier has to already pay the camera, microphone, semiconductor manufacturer 40 dollars before it has collected the 100 dollar itself. So in a way the supplier acted like a bank where it had invested its own 40 dollars to fund the purchasing of components which it took a 100 days to get paid back.

Typically the supplier may take a short term loan to fund the 40 dollar purchase. In that case assume it has to pay 2 dollars as interest to the bank for 100 days. To cover this cost, the supplier typically will put a 2 dollar markup over the 40 dollars as its cost, and to maintain a 60 dollar profit it should increase the cost to 102 dollars.

This situation provides two opportunities for a supply chain leader to create cost efficiency. One is to go directly with the camera, microphone, semiconductor supplier and directly negotiate and reduce the total cost to below 40 dollars. The second opportunity is to purchase the secondary components and eliminate the markup. Controlling the secondary supply chain also ensures better supply chain continuity and reduces shortage risk and better quality. However, this also comes at a cost as the supply chain leader will require more headcount. The cost benefit needs to be articulated to the operations leader to get headcount funding.

7.5.4 Logistics

Following the line of argument in the previous section where the supplier acts as a bank by funding the component cost, similarly the supplier also funds the transportation and logistics cost. Transportation and logistics cost impacts all movements including subcomponents, finished goods, expedite delivery, etc. The shipping lanes cost changes based on seasonality, disruptions and geography. This cost is often spread across and hard to separate. However, at a scale, it is significant and can be managed by directly negotiating with logistics providers and making the suppliers use preferred logistics providers.

A common analogy would be travel services as most companies will restrict all its employees' business travel through a preferred airline or a preferred hotel chain. This ultimately gets the company preferred rates with the airlines and hotels and even assures a particular class of service. Similarly for logistics and transportation of goods, establishing preferred logistics partners can yield efficiencies in terms of transportation cost, insurance cost, ease of going through export and import regulations. There is also a risk mitigating aspect as it can buy preferred service and priority during disruptions.

7.5.5 E&O Management and Disposal

Excess and obsolescence (E&O) management often has a stigma of wastage and is shunned away from mainstream supply chain activities. The pragmatic way of thinking is every supply chain has E&O exposure. What makes it often unpalatable is lack of monitoring and the consequent element of surprise. Any executive will react in an unfavorable manner when an approval for scrapping a large dollar amount of material is placed for their signatures. This happens particularly if the buildup is not monitored and communicated with regular cadence. This also happens when new sources of E&O are discovered. It is one thing to have excess at finished goods level for lack of sales, but when E&O is caused at a factory due to unused raw material or subcomponent then usually category or commodity managers are put to an inquest. So it is equally important to understand E&O risk and build up at all possible sources and levels (Levitt 2017).

In addition the path often overlooked is recovery from disposal. Since scrapping is considered money down the drain, organizations may fail to invest in a disposal manager who can get better value of the scrapped items from the after sales market. Lack or inaccuracy of data on the material to be scrapped also diminishes the recovery value from the aftermarket disposal vendors. This often boils down to similar rigor as monitoring E&O as it requires the rigor to maintain proper inventory, parts information and knowledge of aftermarket demand. Disposal vendors are often engaged by operations to pick up a pile of stuff and lack proper audit between what is disposed in the financial books and what exactly is thrown out in reality. Proper asset management and disposal may increase disposal revenue by 3 times with a headcount investment.

7.6 WORRIES OF A LEADER

A supply chain leader faces a lot of challenges which they have to navigate every day. The challenges could be caused by disruptions or they could be caused by volatility in internal demand or due to rapid change in market conditions. Some of the sources of worry could be internal, such as human error, lack of automation or delay in technology adoption. Some of the sources of worries are discussed below, which are not exhaustive by any means but will help a leader to think and prepare.

7.6.1 Automation

As the world of the Internet and underlying software and hardware technologies are evolving, coping up with the expectation from external customers and internal employees have become extremely challenging. On the flip side, automation costs money and time. As everybody in the world with their smartphones and laptops are managing their personal life at the tip of their finger, similar efficiencies are now being demanded by both the customer and the employees. The investment on the former increases top line revenue and the investment on the latter raises the bottom line.

This has been a constant dilemma for all industries and the supply chain is no exception. Supply chain automation has been traditionally looked at as investment in ERP, manufacturing resource planning (MRP) tools and are a legacy of large-scale consultant-based implementations (Kurbel 2013). While they have produced robust systems which are fault tolerant and span globally but at the same note have been victims of expensive and sluggish upgrades. The technology debt has increased as a result.

Some supply chain organizations are going toward more open data architecture where data have been democratized and made accessible to all internal employees. Teams are creating last mile solutions by accessing this data and pivoting and presenting the data based on their specific needs and use cases.

7.6.2 Decision Support and Data Analytics

As an organization grows the need for more data analytics (Sinha et al 2020) increases to support decision-making. However, there is a cost to obtain continuously improving and more connected and drill-able or richer dataset. A supply chain leader has to be mindful of striking a balance between anaytics software development resources spent for automation which improves workflow as opposed to resources spent to create dashboards which drives decision. When it comes to having a limited amount of software resources, over indexing on dashboard development may lead to less investment in workflow automation which will lead to inefficiency and manual work and employee dissatisfaction. On the other hand a limited amount of dashboard prevents the organization from finding new sources of efficiencies which can help in optimizing the supply chain. It is important to track the actionability of dashboards in terms of hit rate from viewers. Dashboards are often developed based on executive requests or one-time ask for data for a specific purpose. That often leads to waste of bandwidth. Dashboards should be monitored in terms of its daily or weekly usage by its viewers followed by automation of the follow-up actions and dashboards should be decommissioned if certain usage thresholds or the actionable criteria are not met.

On the other hand dashboards need a lot of back-end interface building with different sources of data so that a new set of pivots or data slices can be created. In addition more decision supporting tools involving artificial intelligence and machine learning are now becoming available. Having a clear understanding of end goals for new dashboards needs to be documented and reviewed before investing in complex dashboard creation. Ability to keep the data fresh is also an important aspect which should be measured along with the effectiveness of dashboards.

One low-cost approach to create dashboards is to create a data lake and democratize data. Instead of building dashboards, build a data lake where disparate sources of data are brought in by building the necessary data plumbing in the back end. At the front end instead of hardwiring the dashboards, organizations are providing drag and drop tools to pull slices of data into a visualization tool for display. This allows teams within supply chain organizations to build their own final dashboard without depending on a lot of IT involvement and accelerates the overall process.

7.6.3 Lost Opportunities

A supply chain leader often has to get more creative to get cost savings. However there is a cost to it. For example, a product company gets its product manufactured by an integrator. First step of cost optimization is to negotiate the total cost with the integrator. The next refinement step would be to separate out key material in the bill of material and negotiate the cost of those items directly with the sellers and negotiate the remaining cost with the integrator. The key materials in the bill of material are typically the more expensive parts and may cover 60% of the bill of material cost. This is an iterative process and the next source of cost negotiation could be the next 20% of the bill of material.

For a company with a portfolio of products, category management is the most common approach. A category manager manages the cost of a particular category of raw materials across the portfolio of products. Opportunities to optimize cost for categories which are most prevalent across all products and which covers most of the bill of materials make logical sense. However, to get more cost savings, more cost optimization is needed. For example, apart from material cost optimization, there could be opportunity to optimize logistics and transportation cost. For a mature supply chain organization, there are category managers even for penny parts (i.e. every part in the bill of material including the very low-cost parts), warehousing, secondary suppliers, etc. A supply chain leader has to be able to strike a balance between recovering cost savings and headcounts responsible for category management.

7.6.4 A Whole Lot of Issues

To come up with an exhaustive list of challenges that a supply chain leader faces is beyond the scope of this book. However, a lot of these challenges can be prevented by putting regular audits in place. For example, a typical challenge worth mentioning is supply chain risk mitigation (O'Sullivan 2019). This requires auditing the entire supply base and finding if the supplier has alternate manufacturing locations to switch in case of a disaster. Auditing the vulnerabilities and raising awareness is an essential step toward preparedness. Another common challenge is shortage mitigation due to forecast and demand volatility. The demand disappears and supply dries up and then suddenly demand reappears and then immediately gets into shortage. Putting an appropriate buffering strategy could be a solution. Similar to supply chain risk mitigation another key area of concern for supply chain leaders is ensuring supply chain security and maintaining corporate social responsibility (CSR) in the supply base. The former requires more rigor in terms of maintaining an audible chain of custody. For electronic products, it has become critical to audit regularly to prevent malicious snippets of software codes from being introduced deep into devices. CSR is critical for every company and creates a huge impact in terms of revenue and customer perception. The supply base has to be audited to ensure the supply base is doing its business correctly and following the high standards of CSR as required by the customer.

Compliance (Cook 2013) and legal are two key partners for any supply chain organization. Auditing for supply chain compliance and managing timely closure of legal contracts with suppliers are very important. Legal resources are often scarce within an organization and it is even harder to maintain legal resources dedicated to the supply chain. Meanwhile supply chain often generates a pipeline of master agreements, statements of work, purchasing contracts, etc. which need to be closed in a timely manner and therefore an item of constant attention from leadership.

7.7 CONCLUDING REMARKS

This book is centered around professional growth of supply chain professionals. In light of that vision, the book is essentially divided into two parts. The first part, which concludes with this chapter, is about understanding how to successfully fit in, meet the expectation of the organization and create value and become a leader. The preceding content articulates the different types of supply chain roles and their teams and their stakeholders. The scope of their individual roles and the strategic intent of their teams are articulated in this section of the book. Apart from showing how an individual or their team fits into the grand scheme of things, Sections I and II also articulate who they have to collaborate with and what they care about so that one can create a win-win solution for both.

The second part (Section III) of the book is about developing and demonstrating leadership in supply chain for an individual. To grow professionally, one has to bring in impactful changes to their organization and overall business. Supply chain is a dynamic area which is like the stock market, it gets impacted by every little disruption. The disruptions can come from within the broader organization like a product slip or a customer quality problem or it could come from external sources like a natural disaster, merger and acquisition, and litigation. With challenges and unpredictability come opportunities and they can be handled at every professional level of supply chain. Although there are complex opportunities at a leadership level like implementing an MRP system or changing the forecasting process, there are no dearth of opportunities for other levels too. Even at an individual contributor level opportunities in spend management, reducing lead time improves both revenue and customer experience. An autonomous supply chain leader is a businessman who improves the business in many possible ways.

The challenges and opportunities for supply chain professionals in various verticals are reviewed next. The verticals are looked through the lens of a supply chain professional and reviewed through the framework of six pillars.

REFERENCES

Cohen, S., and Roussel, J. 2013. *Strategic Supply Chain Management: The Five Core Disciplines for Top Performance.* New York: McGraw Hill Education.

Cook, T. 2013. *Managing Global Supply Chains: Compliance, Security, and Dealing with Terrorism.* Boca Raton, FL: Auerbach/Taylor & Francis.

Hugos, M. 2018. *Essentials of Supply Chain Management.* Hoboken, NJ: John Wiley & Sons.

Kurbel, K. 2013. *Enterprise Resource Planning and Supply Chain Management: Functions, Business Processes and Software for Manufacturing Companies.* Heidelberg: Springer.

Levitt, J. 2017. *Surviving the Spare Parts Crisis: Maintenance Storeroom and Inventory Control.* South Norwalk: Industrial Press.

Manners-Bell, J. 2018. *Supply Chain Risk Management: Understanding Emerging Threats to Global Supply Chains.* London: Kogan Page.

Myerson, P. 2012. *Lean Supply Chain and Logistics Management.* London: McGraw Hill.

O'Sullivan, S. 2019. *Supply Chain Disruption: Aligning Business Strategy and Supply Chain Tactics.* London: Kogan Page.

Pandit, K., and Marmanis, H. 2008. *Spend Analysis: The Window into Strategic Sourcing.* Fort Lauderdale, FL: J. Ross.

Ravikanti, S. 2019. Multi-tier Supply Chain Visibility. https://www.cgnglobal.com/blog/multi-tier-supply-chain-visibility (Accessed on Dec 21, 2020).

Sabri, E., Gupta, A., and Beitler, M. 2007. *Purchase Order Management Best Practices: Process, Technology, and Change Management.* Fort Lauderdale, FL: J. Ross.

Sinha, A., Bernardes, E., Calderon, R. et al. 2020. *Digital Supply Networks: Transform Your Supply Chain and Gain Competitive Advantage with Disruptive Technology and Reimagined Processes.* New York: McGraw-Hill.

Templar, S. 2019. *Supply Chain Management Accounting: Managing Profitability, Working Capital and Asset Utilization.* London: Kogan Page.

Section 3

Practices, Perspectives and Leadership in Different Industries and Their Key Nuances

8

Changing Landscape in the Printer Industry

Ashok Murthy

CONTENTS

8.1 Objective..180
8.2 Exploring Business Opportunities for Printers: An Overview 180
    8.2.1 Growth Opportunities 181
    8.2.2 Home and Workplace Printing Practices....................181
8.3 Macro Trends Impacting the Industry Landscape......................182
8.4 Shift in Supply Chain Management 183
    8.4.1 Convert Printing into Service 183
    8.4.2 Intellectual Property Protection of Consumables...................184
    8.4.3 Business Consolidations 185
8.5 Operational Features of Printer Business............................186
    8.5.1 SKU Proliferation..186
    8.5.2 Unit Volumes Are Not Large......................................186
    8.5.3 Products Are Generally Repairable.............................187
    8.5.4 Value-Added Reseller 187
    8.5.5 Selling to VARs and Distributors 188
    8.5.6 Firmware Challenge..188
8.6 Supply Chain Implications in Printer Manufacturing......................188
    8.6.1 Printers 188
    8.6.2 Consumables..189
    8.6.3 Spare Parts 189
    8.6.4 Cost of Goods (COG) 189
    8.6.5 Print Engine...190
    8.6.6 RIP Controller 190
    8.6.7 The Consumables..190
    8.6.8 Spare Parts 191
8.7 Relationship Management..191

DOI: 10.4324/9780429273155-11

8.8 Case Studies 192
   8.8.1 Dead Asset Reduction 192
   8.8.2 Predictive Consumables Ordering 194
8.9 Performance and Leadership 195
8.10 Concluding Remarks 196
About the Contributor 196
References 197

8.1 OBJECTIVE

This chapter focuses on electronic manufacturing, particularly on printers. Printers are categorized as light commercial electronics and are used primarily in offices, home offices and for various types of commercial activities and small-scale publishing. A printer is a good representative of the light commercial electronic industry because it is a very competitive market, and the supply chain plays a significant role in maintaining profitability and viability of a company. Printer manufacturers typically design and own just a few critical components to maintain their uniqueness or competitive edge and leverage the supply chain to procure off the shelf components as often as possible to drive down costs.

8.2 EXPLORING BUSINESS OPPORTUNITIES FOR PRINTERS: AN OVERVIEW

The printing business worldwide is about an 800 billion dollar industry as of 2020 (Smithers.com 2020), including printers, supplies, parts and services. While some parts of the printing business are shrinking, others are growing. Workplace and at-home printing, which is what we normally think of as printing, has been shrinking since the 2008 credit market collapse. In 2008, Lehman Brothers, a global financial service company which provided banking, mortgage and other financial services, declared bankruptcy (O'Toole 2013) and several other financial organizations had to be financially rescued. The business then saw an uptick in 2010 which was mainly to back fill the supply chain pipeline as the stock had depleted previously. Workplace and at-home printing includes desktop and corner printing machines in offices and light commercial spaces as well as printers in homes and home offices. The industry has declined sharply in the last 10 years but it may get further impacted by the pandemic (IDC.com 2020) This is generally considered a secular decline driven by reductions in customer need which in turn is driven by a gradual replacement of printing by digital alternatives.

8.2.1 Growth Opportunities

The category that is growing is industrial and package/label printing, driven by the explosive growth in e-commerce and its downstream effects on the warehousing and shipping and logistics industries. Also growing are mid-volume variable content direct digital printing which includes things such as low volume books on order (the 'long tail' publishing business) and customized printed content for direct mail to the mailbox. Overall, the printing business is growing but not uniformly, with some parts shrinking and others growing. The rest of this narrative concerns itself with the home and workplace printing segment.

8.2.2 Home and Workplace Printing Practices

The dominant players in the printing industry are global brands such as HP, Xerox, Fuji-Xerox, Canon, Konica-Minolta, Ricoh, Okidata and Lexmark. Overall the industry has a shrinking revenue profile and profitability is under stress. The list was a lot longer about a decade ago but has shrunk due to M&A activity, disinvestment and closures (Elliot 2020). In the printers used for home and workplace printing, there are two essential parts: the print engine and the controller. The print engine has a limited set of suppliers because it's a fairly unique electromechanical component. Previously the printer manufacturers would typically engage with print engine suppliers on an exclusive basis or often have the print engine manufacturer as a subsidiary company to maintain exclusivity. However, due to a lot of litigation in this industry and lack of investment, the print engine manufacturers have been sold or divested and as a result the exclusivity has greatly reduced. Some of the print engine suppliers have now been supplying for more than one printer manufacturer as a contingent OEM (original equipment manufacturer) supplier. Managing the relationship between the printer manufacturer and the print engine supplier is a key relationship that a supply chain team is responsible for in order to lower cost and increase a company's competitiveness in the field.

The controllers are, however, more closely guarded by the printer manufacturers. Typically, the controllers and associated software are designed and developed by the printer manufacturers themselves and uses exclusively in their products. This helps them to manage the customer experience and the branding of the product. The customer interacts with the printer through the controller and it is that part of the device which the customer ultimately sees. The components to build the controllers and rest of the printers are typically managed by the supply chain team. This includes plastic and other enclosure material, electronic circuitry, cables, toners, inks, etc. and these materials are all managed by the supply chain organization within a company.

8.3 MACRO TRENDS IMPACTING THE INDUSTRY LANDSCAPE

There are a few key factors which are putting pressure in this market. Printers have been commoditized where the quality of output does not differ that much anymore between brands. A customer can pick up a solution from any of the brands and it would satisfy most of their needs. The secular changes in customer preference are the underpinning for all the industry's woes. It directly affects top line revenues and impacts the profit margins of all the competitors. This is quite evident in a lot of light industrial segments as manufacturers are struggling to maintain brand loyalty. A customer is less interested in a new feature which they will rarely use and instead more inclined to make a purchase decision based on price. The printer manufacturer base is still large and there is an overcapacity in the segment with too many products chasing after a declining customer base which results in an annual ritual of price decline. Due to a lot of competition, manufacturers are unable to charge a high premium by introducing new features. This has put the manufacturers under a unique price pressure as they are caught between rising operating cost and declining or flat retail prices.

This makes the printer industry an extremely cost-sensitive business. The main printer is sold at very low margins and the profitability comes afterward by selling services and replacements parts including toners, ink cartridges, etc. A company's supply chain plays a major role in controlling costs and ensuring risk mitigation. A sudden shortage and slight increase in a component price can easily erode the profit margin of the main printer. This would mean the after-sales support items also have to be extremely cost optimized by the supply chain organization to maximize the revenue and regain profitability.

Antitrust laws that protect consumable suppliers also pose a unique challenge to this segment (Fung 2017). As the printing business falls under a 'razors & blades' business model, the printer is usually sold at near zero (or often below zero) gross margin. The printer manufacturers make their overall margin by selling supplies (and parts) at a higher gross margin. Antitrust laws protecting consumable suppliers basically undercuts this model by allowing pure consumable suppliers to take up the most profitable part of the business model (i.e. the toners) and not invest in the printer design. For example, in a printer, the ink cartridges are consumables. Suppose these ink cartridges are in a special bottle with a microchip attached by the manufacturer so that no other competitors can make ink cartridges or the manufacturer pays retail chains to not sell competitors' ink. These types of activities are considered to be illegal according to antitrust laws as these create monopolization. However, this also creates a set of competitors who may not even try to design or manufacture printers but only manufacture ink.

Overall the printer price has been declining for years. In spite of a new generation of printers coming to the market with better technology and new features every year, the price barely goes up a little bit even when it's very new, and then it gradually starts to come down on an annualized basis due to the competitive market and overcapacity.

8.4 SHIFT IN SUPPLY CHAIN MANAGEMENT

Manufacturers are doing a variety of things to sustain profit levels and grow revenue. The market has been consolidating and the number of printer manufacturers is declining. The consolidated, larger suppliers are now attempting to shift the business from selling printers to selling printing as a service.

The dominant businesses have taken the following approaches to solving the deep problems stated above, with varying degrees of success:

8.4.1 Convert Printing into Service

In this model (Fernandes 2018), first espoused by Xerox Corp. many decades ago, the product is leased under a full service agreement that provides for maintenance, consumables and all related 'care and feeding' of the printer. This model has changed over the years from being a large enterprise level 'supplies & service' maintenance contracts with manually tracking (checking ink levels of toners, paper jams, replacements, etc. for large corner printers) to now mostly managed and delivered (printer communicates toner levels via internet and replacements are dispatched) over the internet. Most manufacturers have a version of this kind of managed print service. But there is also a significant third party presence in this category who offers similar but cheaper alternatives using third-party toners.

From a supply chain point of view, the cost of the service level components needs to be negotiated to a point where the printer company can compete with the third-party offerings. The original equipment can usually afford to charge only a slight premium if they are directly selling to the end customer. If they are selling the parts and inks as part of service, then also the cost needs to be managed well to maintain profitability. Moreover, these replacement items are essentially spare parts which the supply chain logistics needs to manage optimally. Shortage of inventory would lead to failure to meet customer service level agreements (SLAs) while excess inventory will lead to excess and obsolescence (E&O) risk.

As new printers are now communicating back to the manufacturer their toner levels and issues and receiving replacement parts, the execution requires efficient spares management throughout the supply chain network. At the other end of the spectrum, a supplier manager is periodically buying spares in the right quantity which are then delivered in the right depots for efficient distribution. In order to enable this, the request signals from the printers need to be aggregated into a forecast and demand early signals to ensure prepositioning of the material.

8.4.2 Intellectual Property Protection of Consumables

This approach basically means designing a product and consumable that is protected by Patent Law thus prohibiting the intrusion of third-party consumable suppliers. This strategy has not worked, by and large. The latest litigation in 2017 delivered a loss to the defendant Lexmark International (Fung 2017). The courts in this case ruled in favor of Impression Products Inc., the plaintiff, and upheld their rights to continue to provide refilled toner cartridges to customers of Lexmark printers under the so-called 'right to tinker' protections afforded to customers. In other words the print manufacturer could not block the nonbranded companies from servicing the printers, including manufacturing toners.

The key differentiator from a consumer's perspective is value for money. Consumer behavior may change from a residential customer to a corporate customer. A corporate client may look at the total cost of ownership and appreciate the nuances between down time in hours versus days while waiting for things such as toner or replacement parts. A residential customer with a few printers which rarely breaks down an ad hoc time and material (T&M) contract may make sense. A printer as and when breaks, the manufacturer is called, repair work gets done and then invoice gets paid. However, for a large corporation with a global footprint of a large number of printers, a different level of simplicity is needed. It's not possible to do long engagements with follow up and invoice payment every time a printer breaks, instead a single point of contact like a toll free number or email is needed where one communication takes care of business. Similarly, for residential customers, the cost per page is perhaps the only factor and therefore it is a lot easier for them to use third-party ink cartridges if the price of the branded ink cartridges appears to be too high. A supply chain manager on the other hand has to really know the market price of the cartridges and parts to negotiate the right prices of the raw material while ensuring the quality level. The third-party solutions are a major revenue threat and therefore cost per page and quality of the ink are the two key factors which keep the customer loyal to the brand.

8.4.3 Business Consolidations

The market is consolidating gradually. The net effect of this will be to reduce competition among the printer manufacturers which will may ultimately lead to stabilizing pricing and reduce the dependency on supplies sales alone.

From a supply chain perspective it translates to forging new relationships between the acquired printer manufacturer and a different print engine supplier which it has to do business with now after being acquired. It also translates into consolidation of the printer portfolio between two merged print manufacturers and optimizing the overall product SKUs (stock keeping unit) while minimizing manufacturing cost and optimizing supporting sparing inventory across the portfolio of two merged manufacturers. As both companies bring in their supply chain, distribution and logistics partners, these branches also have to get consolidated and optimized. All the new supply chain relationships and cross team collaboration need to be managed by the supply chain and should be ultimately reflected in optimizing the bottom line and increasing profit margins along with top line revenue growth.

8.5 OPERATIONAL FEATURES OF PRINTER BUSINESS

This overarching business industry, loosely called 'Light Commercial Electronics', differs significantly from consumer electronics and from industrial electronics. Here are some of the characteristics:

8.5.1 SKU Proliferation

A proliferation of product SKUs (stock keeping units) or product variants is driven by very fine-grained categorization of customer needs. This is a 'specsmanship' game played by all players causing many similar but not exactly the same products which cause a variety of supply chain issues and inventory overhang problems.

This increases the mix of different products and reduces the volume of each. This high mix low volume situation poses a challenge for the supply chain manager as they have to manage more inventory of more variants which ultimately drives up costs. Trying to minimize and leverage as many common parts would be an important task for a supplier manager. In terms of developing knowledge, understanding the products from a customer's perspective and connecting the desired features to the specific subcomponents would always be advantageous for a supplier manager. This would help them in understanding demand and changing supply/parts accordingly.

8.5.2 Unit Volumes Are Not Large

Each product model type is produced in mid-volume quantities which, coupled with a large number of SKUs, makes for a very complex service and consumable supply chain.

As both office printers and home printers change significantly in every generation and become faster and cheaper and with more paper handling capacity, understanding these product transitions is key for a supplier manager. As unit volumes for each SKU is already small, it gets further complicated if one SKU is transitioning from one generation to the next. The forecast for all the myriad of SKUs across generation changes become more complex. A supplier manager typically has to stay close to both the product management and engineering teams to track the transition timelines to accurately line up inventory against the demands. At any given time a supplier manager will have a few items going through new product introduction

(NPI) or development, a few going through mass production (sustaining) and a few going through end of life (EOL). Understanding specific underlying components demand increase (NPI), or flatter but high-volume demand (sustaining) and declining demand (EOL) is critical for supply chain managers.

8.5.3 Products Are Generally Repairable

Most products in the industry come with a 12-month warranty under which the products are repaired. Printers and other small items (generally less than 40 lbs. weight) may be repaired at a depot, while others are repaired on-site by trained service technicians.

From a supply chain point of view this requires understanding the spares forecast and logistics. Logistics would involve putting supply at various locations which may range from regional depots to storage at a customer's place. The size of the inventory at any location will be dependent on the count of printers and generations of machines. The volume of inventory also depends on the failure rate of the machines. Typically there is a quality department which predicts the failure rate of the parts and comes up with the forecast of the components that need to be stored. Optimization of spares to minimize wastage while meeting SLAs with customers is always a big opportunity for supply chain operations to increase profit margins.

8.5.4 Value-Added Reseller

Printers and other products in this industry are generally sold in the IT VAR (value-added reseller) channel, although the lower end products are increasingly sold on internet marketplaces such as Amazon and through manufacturers' own websites. Still, given that printers are IT products with a significant level of networking complexity and security considerations, a large percentage continues to be sold by resellers (VAR). These VARs are generally stockless sales entities who place demand against the IT distributor network.

From a supply chain perspective it is important to understand the sources of demand. The demand may come from internet-based sales or demand pull-in requests from VARs. It is important to have an understanding of the inventory level at VARs as VARs may return excess inventory which can lead to over inventory for a manufacturer. A supply chain manager needs to understand and take steps if the demand becomes soft or if there is any demand spike. One of the best ways to understand and react to that is to be able to make calls when demand drops or increases as soon as it happens or is expected to happen.

8.5.5 Selling to VARs and Distributors

Manufacturers sell into the large distributors and the VARs pull from the distributor stock. This is because the manufacturers cannot afford to deal directly with the much smaller VAR businesses while the distributors are set up to do so. Also, the distributors provide other value-added services to the VARs such as payment terms, soft loans, marketing programs and sometimes packaged services in a box. Of course, distributors also provide liquidity to the manufacturers.

8.5.6 Firmware Challenge

Most of the products do not have a clear firmware migration pathway. This means that in many cases, the firmware must be re-flashed by a trained technician on-site or in the factory. This is unlike some IOT-type consumer products that were designed from the start to support FOTA (firmware over the air).

8.6 SUPPLY CHAIN IMPLICATIONS IN PRINTER MANUFACTURING

Types of SKUs: There are three major SKU categories:

8.6.1 Printers

As mentioned above, there are a great number of model types in existence. Plus, the manufacturers have a legal obligation to repair any printers in the field for 7 years after manufacturing of the SKU has ceased. Generally, models cease manufacturing after 2 years or sometimes even sooner. This means that disposing of unsold inventory is a near constant issue. As mentioned above, printers are sold through the 2-tier model. Manufacturers sell into distributors and distributors sell out to the VARs generally when the VAR has closed on a sale.

8.6.2 Consumables

Generally, there are 4–8 consumable SKUs per printer (1 for each color, i.e. 4 colors and then 2× for high capacity versions thereof). Consumables are also sold into and through distributors. It is also increasingly common for manufacturers to sell directly through their own websites or through retailers/online retailers. There is strong motivation for manufacturers to get customers into services contracts where the toner is provided as part of the service. These kinds of contracts are administered by the manufacturers and sometimes sold through the channel. Most common versions of such managed print service contracts are with larger customers because a reliable infrastructure must be set up to collect data from the customer premises that drives the billing cycle.

8.6.3 Spare Parts

There are many unique parts in each printer, resulting in thousands of parts that must be tracked and delivered. Service parts are generally not customer replaceable, though some items are classified as customer replaceable units. The vast majority, however, are classified as field replaceable units that must be replaced by a trained technician. Disposition of these spare parts is a huge inventory problem, especially for printers that are too large for a customer to reasonably put into a box and ship back to the depot. In such cases, the service tech must have easy access to spare parts prior to his service call (see example later for more on this)

8.6.4 Cost of Goods (COG)

Printers are comprised of three major subassemblies:

1. The print engine, sometimes also known as the IOT (input output terminal)
2. The controller, sometimes also known as the RIP (raster image processor)
3. Software, usually network based, but now has become increasingly cloud based as well

Most US manufacturers OEM the print engine from one of the Japan-based manufacturers and pair it with their own controller and software. The controller is what drives the customer facing behaviors of the printer and provides API access to the printer functions to any software that may reside outside the printer itself. Since the look and feel and branding is directly affected by the controller functions, the controller is usually (not always – sometimes OEM versions of controllers are sold paired with the engine) insourced.

From a COGs point of view, the print engine and the print controller are both about 45% (each) of the total build cost, and the remaining 10% is the warranty cost accrual that pays out against actual service costs.

The two major components are dealt with in different ways.

8.6.5 Print Engine

This usually comes from a certain preferred partner. This partner is either an external entity or may be developed in-house. In both cases, procurement strategies are in flux constantly. Print engine development is a long cycle time R&D process that requires the procurement management team to stay closely coupled with development engineering that usually leads to selection of print engines to satisfy future needs. This 'pathfinder' role of engineering stays in close contact with the procurement organization so that when the engine selection is approved, the formal contracting process can follow seamlessly.

8.6.6 RIP Controller

This is a special purpose computer. Usually these boards try to stay within a family of CPUs, memory, etc. and upgrades are to the FW/SW. Even so, there is so much specificity of the controller to the engine with which it is paired, that in practice, there is always a new controller to go with each engine.

8.6.7 The Consumables

Usually, consumables are organized at the corporate level as a separate line of business or a management profit and loss. This line of buisness is responsible for the profit & loss (P&L) implications of consumables including:

1. Marketing & sales – driving sales growth, coming up with new ways to capture consumable sales percentage, reducing third-party encroachment on sales, reducing gray market offsets, etc.
2. Mix management including procurement of supplies
3. Inventory management and balance sheet impact

8.6.8 Spare Parts

Demand for parts is driven by service operations. During the warranty period, parts are 'free' to the customer which means that all parts and labor expenses are charged to the internal warranty accrual. After the warranty period, service is a for-profit activity. VARs like to own service and may become authorized service providers or bring in an outside entity as a service delivery partner.

As such, parts disposition is a very complex supply chain issue driven by on-site service warranties which are usually based on next business day service legal agreements. This means that the service tech must have quick and easy access to the right parts. Given that there may be hundreds of printers (over the past 7 years) in the field, each with hundreds of parts, there may be thousands of parts necessary to fulfill next business day expectations. These parts must be available in the following location categories:

- National depot – usually at the end of a runway near a UPS or FedEx hub
- Regional depot
- Local depot
- Trunk stock – in the technician's service vehicle
- On-site – consigned at the customer's location

Special service contracts with larger and more important customers may carry significant SLA violation penalties. Parts disposition becomes much more critical in these contracts.

8.7 RELATIONSHIP MANAGEMENT

The key relationship in the printer supply chain is typically with the engine suppliers. This is one of two most critical parts which is often purchased entirely from a sole sourced supplier. Typically, there is a lot of investment which goes into this relationship and some of the partnerships have been existing for more than 40 years. This is a long-term relationship which develops in a codependent manner in terms of product development and investment. Often the financial goals are planned together. Till 2016–2017, most of the printer manufacturers managed these relationships very closely.

In the last few years, however, some of the long-term relationships have changed due to a huge amount of turbulence in the market created by cross acquisition. One of the large US manufacturers bought a Korean print engine supplier. The Korean supplier was previously a supplier for another US-based manufacturer. The latter had to then source the print engine from an alternate source. This led to litigations and divestiture and effectively some of the major long-term relationships in the industry got broken.

8.8 CASE STUDIES

These are two examples where a supply chain person irrespective of the level can move the needle by delivering tens of millions of dollars of savings. The first example is about developing supply chain understanding: do they really need to make that buy? The cheapest buy you can ever make is the one you don't have to make. This requires understanding the data around the demand side for the supply chain.

8.8.1 Dead Asset Reduction

Maintaining serviceability requires support from the entire supply chain organization. After-sales servicing of products is a critical subsection of the whole printer supply chain. It gets more complicated because manufacturers are legally required to maintain serviceability 7 years after a product is shipped for the last time. So if a manufacturer does direct business with end customers, then there are typically hundreds of products in the field at any given time, and then each product has hundreds of parts. So the printer manufacturer winds up with an enormous number of service parts. In addition most of the service operations have next business day as SLA which requires spares to be on-site or very close to the customer, as mentioned above.

The technician must receive the part and finish the repair or replacement within the next business day. However, different parts are at different places. The parts could be parked at a national depot, regional depot, trunk stocks (service technician's service vans) or even on-site. Typically, the manufacturer winds up with an enormous amount of bad assets sitting in those places and it is very hard to keep physical assets across all the stocks globally constant. The challenge is that a lot of the spares worth a lot of money remain unused and eventually age out. It is very hard to do physical audits and do cycle counts to manage the inventory optimally and thus parts end up becoming dead assets.

One solution that was implemented involved the controllers to be designed in a way that they would publish the diagnostic data to the customers notifying them of incidents. The customers were motivated to make a service call. Typically, they would call an 800 number but the manufacturer pushed them to open a service ticket from the computer and start the process digitally. As they had to use their computer to create a request ID (RID), the ticketing process would pick up the data which were available from the networked printer. The ticket would pick up the incident details from the printer and would send it along as part of the RID ticket. At the back end, once the ticket was received by the warranty service providing arm of the manufacturer, the details would be analyzed to identify the failed part and the appropriate replacement part that would be needed by the service technician. Then the back-end system would send messages to the national depot, which typically sits in a central location, such as Louisville, KY, which can serve a large portion of the entirety of the USA in any direction. They would drop ship a kit of possible spare parts to the customer location and the service tech would then typically meet the part at the customer location. And then on the back end of this, the technician would record which part out of the possible parts was actually used. Then there would be a close loop which makes the corrections to the rule engine that triggered the kitting operation and improves accuracy in providing the right part. As the rule engine improved with time, it enabled a significant reduction in stock because supporting next business day SLA became a feasible option through a national stocking depot as opposed to stocking at various regional levels and the information coming directly from the printer. In summary this was re-architecting of the overall spares management process through accurate information gathering from the networked printers and improving the rule engine. This enabled reduction in the stocking depots and led to overall reduction of dead assets while maintaining next business day SLA and thus improving profit margins.

Logistics details

- Printers published diagnostic data and made available to external software via an API. Local browser-based apps picked up diagnostic data when customer initiated an online service call
- Diagnostic data ran through AI engine in the back end and produced symptom to solution packages including selecting three service parts and accompanying instructions

- The service part package was kitted at the national service depot and drop shipped direct to customer
- Service tech also received the customized instruction package for that printer as well as tracking information for the parts kit. Service contracts may be same day service or 4-hour service
- Service tech used the parts they needed, put the damaged parts and any unused parts back in the box and sent it back to depot
- Depot analyzed the parts usage. Fed back to a team that maintained the rules engine on a 24-hour basis
- Significantly reduced stock at the various stocking locations. Reduced inventory, write-downs, audit problems, physical reconciliation costs, etc.
- Organizations involved: engineering, manufacturing, customer support, IT, sales/marketing (it was positioned as a value add from Xerox), finance and supply chain

8.8.2 Predictive Consumables Ordering

Consumables are high mix and high-volume business. There could be up to eight consumables per printer SKU. There are four colors which are available in two capacities (regular and high cap). So there could be eight colors but then depending on the printers the toner cartridges could be of various shapes, which requires the toners to be platform (series) and model specific. With up to eight consumables per printer model, this creates a huge number of consumable SKUs. The problem gets more complicated as the manufacturers aspire to sell these consumables up to the 7-year limit because that is almost the only thing that makes a serious and reliable profit. However, in the same issue as the previous case study, if the toner runs out and the printers stop working, the customer gets unhappy.

Replacing toners had always been a very manual pull process where customers would react once they run out of toner. This was replaced by a push process through a management service. An app was created on the customer portal and provided as an optional contract service. It would analyze the printers that were placed under the umbrella of this optional service which the customer had the option to close out at any time. The app analyzed all the print usage and volumes and sent it back to the manufacturer. The manufacturer's back-end tools would analyze and predict when the printers were expected to run out of toner. It would automatically send alerts to the customer when the projected date was imminent so the customers could place orders and would receive toners prior to the projected date. This was offered as an unlimited free service and was adopted by the customers because it was predictive and made customers' lives easier. In addition, previously when a customer's toner level went down and needed to be replaced, the customers would search on the internet and look based on pricing. Typically the pricing search would not yield the manufacturer's toner at the top of the search list and would instead show a third-party cartridge on the top of the search result. However now, the customers had a higher attachment to the consumables produced by the manufacturers.

Logistics details

- Browser-based app that communicated with all Xerox (and non-Xerox) printers on the network
- Analyzed toner usage, page print volumes statistics, who used how much, etc. and resource provided to admin or manager as a free value add
- Cloud-based projections of when toner would run out based on page counts and including adjustments for cyclical/seasonal variations (e.g. reduce in July 4 week, increase in December prior to Christmas, etc.)
- Customer-approved preorder of toner with free shipping. Shipping could be cheaper because there was time available
- Free service warranty if toner service was maintained
- Very popular with customers
- Currently sold as a service by major consumable brands

8.9 PERFORMANCE AND LEADERSHIP

For the engine side of the supply chain operation, or at least the procurement operation that deals with engines, the nature of the business has its own nuances. Engines are jointly developed. In addition, as printers are continuously evolving, the joint development continues through product development cycles. Therefore, it is a very different business mindset, as both sides typically have to collaborate to develop engines. So, for example, a printer manufacturer has to start collaborating with the engine suppliers maybe 2 or 3 years ahead of time. The strategic sourcing team manages the relationship between the engineering groups on both sides as the design goes through development to finalization. In parallel the procurement team starts negotiating pricing and contracts with the sales side of the OEM. Typically, it starts from a design lead whose role is to be the path finder. They are hunting through different engine offerings and different technologies, and then identifying targets about 2 or 3 years ahead of time, and then this process starts. Then it is a gradual hand off from the engineering side to the commercial side managed by procurement.

Consequently, the procurement team has to stay pretty close during the design development. Quite often, leaders on the team come from a business operations or engineering design background and have in-depth experience with the process and challenges of the design cycle. In other words, an individual in the supply chain needs to be technically savvy as they have to balance between technical requirements and commercial requirements, translating engineering needs into supply chain impacts and understanding technical roadmaps.

8.10 CONCLUDING REMARKS

Beyond technical and operational knowledge, one important trait for a supply chain leader to exhibit is sensitivity to cultural behaviors. Without cultural sensitivity, they fail, and that failure is actually not uncommon. Japan and the US have a significant supply base for printer engines and other components as well as providing a major customer base. Understanding how to do business with Japan is very important. Understanding how to do business mutually is important for all parties involved. Having the sensitivity to understand Japanese business behavior, culture (Bucknall 2005) and how procurement works along with understanding social interaction in a business setting are key factors to success. All countries have their cultural nuances, but some may be more complex, such as in Japan, where decision-making is hierarchical and factors like seniority, age, etc. need to be understood and appreciated. Typically the tone of collaboration has to be forged at the top and set by the senior-most leader which is then echoed and followed through the rest of the organization.

ABOUT THE CONTRIBUTOR

Ashok Murthy is a technology executive and currently serves as CEO of Devices-Unlimited, Santa Ana, California. He has played multiple leadership roles at Xerox and Fuji Xerox including General Manager, Chief Technology

Officer and Senior VP of Business development. He is an expert in printing technology and holds over fifty patents. He has set up and managed operations across the globe in Asia, US and EMEA. He has a PhD from MIT and is a graduate from the Indian Institute of Technology.

REFERENCES

Bucknall, K. 2005. *Japan: Doing Business in a Unique Culture*. Raleigh, NC: Boson Books.

Elliot, I. 2020. Mergers & acquisitions: A $20 billion growth opportunity, E&S Solutions. https://info.eandssolutions.com/mergers-and-acquisitions-a-twenty-billion-dollar-growth-opportunity (accessed December 21, 2020).

Fernandes, L. 2018. Global print 2025 evolution or revolution? Quicirca. https://www.xerox.com/en-mu/services/managed-print/insights/quocirca-future of print   (accessed December 21, 2020).

Fung, B. 2017. How a Supreme Court ruling on printer cartridges changes what it means to buy almost anything Washington Post. https://www.washingtonpost.com/news/the-switch/wp/2017/05/31/how-a-supreme-court-ruling-on-printer-cartridges-changes-what-it-means-to-buy-almost-anything/ (accessed December 21, 2020).

IDC.com, 2020. IDC forecasts a sharp decline in total page volumes printed on office and home devices in 2020 as COVID-19 affects the document printing market. https://www.idc.com/getdoc.jsp?containerId=prUS46609820 (accessed December 21, 2020).

O'Toole, J. 2013. Five years later, Lehman bankruptcy fees hit $2.2 billion, CNN Business. https://money.cnn.com/2013/09/13/news/companies/lehman-bankruptcy-fees/ (accessed December 21, 2020).

Smithers.com. 2020. Global printing market to top $821 billion by 2022. https://www.smithers.com/resources/2018/mar/global-printing-market-to-top-$821-billion-by-2022 (accessed December 21, 2020).

9

Saving Lives through Medical Devices

Upayan Sengupta

CONTENTS

9.1 Objective.. 199
9.2 An Overview of Medical Devices 200
9.3 An Overview of the Manufacturers 201
9.4 Macro Trends Impacting the Industry Landscape.............................. 202
9.5 Technology Trends... 203
9.6 Regulatory Compliance in Supply Chain.. 204
9.7 Quality Management... 205
9.8 Supply Chain Management 207
    9.8.1 High Mix Low Volume.. 207
    9.8.2 Pricing.. 208
9.9 Organizational Leadership 209
9.10 Individual Performance..210
    9.10.1 Inventory Control 211
    9.10.2 Sustaining..212
    9.10.3 Customer and Product Knowledge 213
9.11 Concluding Remarks...214
About the Contributor..215
References...215

9.1 OBJECTIVE

This chapter will focus on the medical device industry and discuss how its supply chain is managed. One of the most important and unique aspects of this industry is the stringent regulatory requirements which ultimately control its supply chain. The industry also has unique differentiation between its customer and the ultimate or actual end user. To add, the industry is also

DOI: 10.4324/9780429273155-12

differentiated by an amalgamation of a diverse set of technologies and driven by the innovations in each of these respective areas.

9.2 AN OVERVIEW OF MEDICAL DEVICES

At an overly broad level, medical devices can be defined as pretty much anything that is used to treat, diagnose or sustain life. If a device claims to do any one of these three things, i.e. the product either treats, diagnoses or helps in sustaining life, it automatically becomes classified as a medical device. Once a product is classified as a medical device, the product is then regulated by appropriate regulatory organization which controls and oversees the specific market in which the product will be sold. Medical devices can range from large machines which are often referred to as big irons (Boulton 2014) such as MRI (magnetic resonance imaging) machines, CT (computer tomography) machines, nuclear cameras and X-ray machines to smaller patient monitoring machines like ECG (electrocardiogram) machines, bedside monitors, imaging screens and ultrasound machines. Even much simpler resources can be classified as medical devices, such as surgical gowns, masks and visors depending on the specifics of an individual product. While most of the above listed items are classified as personal protective equipment (PPE), which are not under the medical devices umbrella, there are some which have specific properties which then can be, from an externality standpoint, claimed as medical devices. A lot of sterile equipment that does not have to be powered can also be considered as medical devices, and even packaging could be categorized as medical devices depending on whether it is specially sealed or sterilized. For example, inhalers are just in a special type of packaging but it is also categorized as a medical device. Medical devices run the gamut from big to small items, powered devices to low-power devices, devices that get implanted within our bodies like a pacemaker to devices without any power sources. These intricacies increase the complexity of the supply chain that manufactures these devices.

The supply chain, therefore, is categorized by the type of medical device. For example, when you look at big iron like X-ray machines or MRI machines, the supply chain tends to be very complex and deep, because of the complications of the machinery itself as well as the size, expense, etc. One of the biggest challenges that manufacturers have is that these are relatively low-volume products and very high product mix. So, the supply chain has a very long tail of different types of components to manufacture these products because the machines are going into different installations and every installation could possibly be different. And therefore, there is a lot of customization in these types of products, which continues to make the supply chain more complicated.

Even though big irons are very complex items which have a lot of heavy inventory sitting with low turnover, all the parts that form these machines are critical because even a $2 part can prevent a million-dollar scanner from shipping.

On the other hand, if you look at products like stents or intraocular lenses, they are very high volume and get shipped around the world all the time. While the supply chain here is equally as complex, the challenges are completely different. These products do not tend to have a very deep supply chain because there are not a lot of components. But because of the nature of these products, you have got to have more localized manufacturing due to the frequent usage by people from all over the world. So, you have to be able to have a supply base that is more local to the customer set. For example, some of these products have a half-life of one day or even shorter, and so it is impossible to rely on international shipping and centralized manufacturing, especially with heavy usage and inventory turnover. Since these products have to be injected into the body immediately after manufacturing, they have to be manufactured to facilitate that. Either they have to be produced in close proximity to the patient or have the necessary transporting logistics between the source and the patient. Thus, the challenges tend to be different from a supply chain standpoint for larger machinery even though all these supply chains are encompassed under the medical device industry.

9.3 AN OVERVIEW OF THE MANUFACTURERS

Globally the medical devices industry has an annual turnover in the neighborhood of about $500 billion in revenue (The Business Research Company 2020). The industry is also growing rapidly at about 5% as a global aggregate, COVID-19 notwithstanding. The US market grows closer to about 3%–4% annually. Meanwhile, Europe is growing closer to 1%–2%. Big growth is coming from Asian markets as well as parts of African and Latin American markets. China is growing at 10%–15%, India at about 10% and Brazil is growing at about 8%–9%. Certain countries in Africa such as South Africa,

Nigeria, Zimbabwe and Kenya have also been growing at about 12%–15% in the last few years.

The US still makes up the largest revenue base for major companies in the industry. Out of the $220 billion global revenue, the top 10–15 companies generate 80% of that revenue. The remaining 20% is generated by a long tail of thousands of medical device manufacturers spread all around the world. From a manufacturing standpoint, most of these companies are either based out of the US or Europe from a legacy standpoint. Asia certainly has a number of smaller companies that are growing but Japan is one of the few Asian countries that has some well-established medical device companies.

China has seen a growth of several large companies over the last few years but it still has a long way to go when it comes to being a real design base for medical devices. The three large areas right now are Europe, US and Japan. Within Europe, Germany by far is the largest market and the largest manufacturing design base for medical devices followed by the UK, France, Italy, and Spain. The supply chain specifically for medical devices is very widely scattered around the world. Most design-based organizations are based out of Europe and US and while we tend to have final assemblies coming out of these two regions, the secondary supply bases are all in lower production-cost countries. For example, Mexico has a fairly large supply base. China and other parts of Asia like Vietnam and Thailand have also grown their footprint. There is also some manufacturing in Brazil; this is partly due to export regulations which encourages localized products and manufacturing out of Brazil.

9.4 MACRO TRENDS IMPACTING THE INDUSTRY LANDSCAPE

While the top 10–15 manufacturers make up 80% of the total revenue, they have not really changed in the last 20 years. There has been a lot of consolidation with some pretty big mergers and acquisitions, but in general, those companies are still around in one way or another in some shape or form, and there has not been any new major entrant into the marketplace. On the long tail list of manufacturers making the remaining 20%, there has been constant changes. Companies are either getting bought out or going out of business. There is a lot of start-up activity and so, in terms of the large-scale competition, the top 10–15 companies tend to compete mostly with each other, and then the manufacturers in the long tail end tend to compete with each other. Large conglomerates have also entered into healthcare through acquisitions, particularly in China and Japan.

The constituents in the upper echelon of the medical devices industry have not really seen any large-scale newcomers. There is a high barrier of entry for coming into this marketplace partly due to the need for having very high levels of regulatory capabilities. Other industries where there is similar level of regulatory controls, like aerospace, also exhibit similar trends and challenges for newcomers. For example, Boeing and Airbus pretty much makeup the entire aerospace market. With a few smaller companies, the constituents have not really changed for the last 40 years in the aerospace industry. Similarly, with automotive, there is high barrier of entry because of very strict quality requirements all through the supply chain which mirrors the healthcare industry.

The supply chain is the most vulnerable area in the medical devices industry because it is one of the highest areas of risk when it comes to quality management in healthcare. One can imagine having a supply base that is spread all over the world and having to be able to control how design is flowed down to manufacturers, how local manufacturers source components, how they follow the work instructions, how testing is done at those sites, etc. When a medical device manufacturer could potentially have a supply base of 10,000 suppliers, something is likely to go wrong and that tends to be a really a big focus from a risk management standpoint for all the manufacturers.

9.5 TECHNOLOGY TRENDS

Medical devices are going through tremendous amounts of innovation with proliferation of more and more electronics into the devices. As a lot of consumer devices such as cameras, video, and computers have become features in smartphones and smart watches; consumer electronics are also starting to enter into the medical device market. With mobility and social media built in these devices, these devices are becoming a part of our well-being and daily life.

Any technology that we see on the consumer side, be it nanotechnology, 5G, robotics, 3D printing, etc., are all being applied in some way, shape or form within healthcare. So healthcare is a really an amazing blend of electronic technology, physical technology, chemical technology, and biological and clinical science. One of the unique facet of an industry like healthcare is that it takes a long time to shift because the consumer of the product usually tends to be a clinician, not a patient. While the patient ends up being the end person on whom the system or the product is actually used on, the actual user is still for the most part are the clinicians.

Even with the new changes in technology, smartphones or smart watches cannot help us manage our health, and we still need doctors. We are still going to use a clinician who is trained to understand our specific symptoms and has a deep understanding of health sciences. So regardless of what technology is out there, the medical devices industry is still looking for someone who understands health and well-being. One of two reasons for this is, first, the device must deal with humans, and second is that the device has got an extremely large burden of regulatory requirements. The reason for this is because of the way regulations work in healthcare, which are designed to prevent harm to a patient.

So even if an innovation shows promise or shows an opportunity to perhaps improve a patient's life, the regulator has to constantly evaluate whether it is going to hurt more people than it is going to help. This is a very delicate science and therefore, in general, regulators are very risk averse because they have to judge what is the potential for harm if something goes wrong. So while it is possible to build a remote ECG on our wristwatch or build a noninvasive glucose monitoring feature in it, there can be significant issues with actually producing these items. If because of a manufacturing flaw or a user error, which is actually common even in the clinical space, somebody has a false negative or false positive, this can lead to significant issues downstream especially if the result is used by the user to determine whether to seek treatment. The amount of liability and potential harm that could occur is very high, which is why regulatory control is needed, and this makes the difference between consumer electronics and a medical device.

9.6 REGULATORY COMPLIANCE IN SUPPLY CHAIN

Regulatory compliance is one of the most critical aspects of medical devices manufacturing and its supply chain (Amato and Ezzell 2015). The first thing to look while sourcing a supplier is whether the manufacturer is certified to a specific standard. In consumer devices and in general manufacturing, we look for ISO 9001 certification and so on. Medical device brand owners in healthcare look for ISO 13485 certification from their manufacturers. Having that certification and being able to maintain the necessary standards implies that the manufacturer has the right controls in place to be able to manufacture a medical device. So, as design owners are out there looking for either a third-party manufacturing organization or even maybe setting themselves up as a manufacturing shop, they need certification. At some point before launch to be able to distribute the product, the product has to be manufactured under the guidelines of ISO 13485 (Abuhav 2018).

There are other standards that can be needed depending on the device. For example, there are standards of maintaining sterility or a sterile manufacturing environment depending on what is being made. But in general, for qualifying a manufacturer for its capability and quality control, the first thing that one needs is to review the manufacturer's ISO 13485 certification. ISO 13485 standards covers the overall management of quality. Review against this standard will indicate what is the overall process of quality management and the system put in place by the manufacturer to ensure quality and will also reflect how the manufacturer performs within that environment.

So, this includes how documents are managed in the supply chain organization. Manufacturers will demonstrate the way they have put control over key documentation which will help oversee the design, production, quality management, and distribution of end products. Then they look at whether the right internal organization is in place to manage quality which includes everything from the right leadership on the manufacturing floor to the right leadership at the plant and at organizational levels to ensure that there is escalation management within the organization. Then they look at the process and controls that have been put into creating very clear work instructions. Work instructions are created at the engineering level, then they are validated by manufacturing. After that they are forwarded to manufacturing for mass production. This critical transfer of instructions and knowledge is referred to as design transfer.

9.7 QUALITY MANAGEMENT

After the engineers have created a design or a prototype, the underlying manufacturing information such as the design drawings, specifications, and bill of material of the constituent parts needs to be communicated to the mass manufacturer. This critical step is known as design transfer.

This process covers various aspects of design which includes verification and validation of software for performance, process of first article inspection (to make sure they meet expected quality) of the incoming subcomponents or parts produced by the suppliers, inspection of pilot production units from the assembly line, etc. It also ensures that the right tests which should be conducted both at the subcomponent level and also at the final assembled system level are being defined and run to deliver the expected performance. It also articulates what is the right amount of spare parts that need to be in production to provide warranty support for the products.

Overall, there are multiple stages of development between producing the first prototype to a stage when a product is ready for mass production and product launch. The step described in the previous paragraph is stage one. Eventually the product gets to a stage where the prototypes are perfected to the desired performance including appearance, packaging, etc. The completion of these stages are referred to as milestones. Milestone one is basically the completion of pilot production. Milestone two is getting to a prelaunch production state. Milestone three is when all the supporting activities are complete, and the product is made available to its customer for the first time. Milestone four is when the product is in full production mode. These are the different steps and the evidence of exit and entry into each stage needs to be shown as part of the manufacturer's quality management system. It tracks the development process and mitigation of defects or issues found during the development process and continues to do so during the mass production phase. In addition, an important part of the quality management system is how a manufacturer manages customer complaints, quality issues in the field, and quality issues in the production plant or issues in its supply base. It needs to track its return material authorization which is how many units are returned by customers and for what reasons they had to be replaced. One of the possible reasons could be defects in a component. Then the inventory of that specific defective part at the manufacturer's production sites needs to be traced and purged and alternate replacements need to be qualified.

As part of the quality management system, a manufacturer needs to demonstrate that it has installed the effective mechanisms in place that evaluate all the different quality metrics to ensure that it is not shipping bad products out into the field. It also needs to be able to show its ability to recall the product, to trace the product, and to ensure that if it needs to destroy any products that are out in the field, it has that capability. So that is the overall architecture of the quality management system (White 2018, Manz 2019).

9.8 SUPPLY CHAIN MANAGEMENT

9.8.1 High Mix Low Volume

Suppliers are managed based on three metrics. First is lead time, which is the duration for the manufacturer to deliver the product. Second is the cost of the product, which has to meet a certain target. The third metric is quality. So the supply chain managers build service level agreements and manage the performance of the suppliers around these three tenets. So, if you want higher quality, the cost is going to go up, and your return will probably go down. If you want to drop your lead time, either your cost of the product is going to go higher or your cost of inventory is going to go up. A supplier manager is constantly managing these different elements (O'Connor and Periman 2018).

Companies manage lead time and cost in multiple ways. They also try to maximize the amount of inventory in the pipeline with least amount of cost. For example, they may negotiate favorable payment terms like 90 day payment terms with their suppliers to maximize liquidity. For example, manufacturers may opt for ocean freight over air freight to save cost but that may, however, extend the delivery lead time from the subcomponent manufacturers. Manufacturers (buyers) may also negotiate for favorable international commercial terms for delivery. For example, a seller (subcomponent manufacturer) may invoice a buyer (final product manufacturer) once the freight is on board the truck from the seller's dock and the buyer has to pay within 90 days from that day. Or it could be the other way, where a seller can send the invoice only after the buyer has received the goods at its receiving dock. The latter adds additional time as the supplier cannot invoice the buyer until goods arrive on the buyer's dock. However, longer lead time also implies more supply is needed to fill the extended duration (invoice terms + transportation time). For example, if the lead time is 90 days (13 weeks) and supply is needed every 7 days to support continuous production, 13 weeks of supply has to exist in the pipeline. So at any given point 12× of weekly supply will be held in the inbound pipeline. The buyer may not be invoiced for the entire amount but is however liable for the entire 12× amount of supply as the supplier has shipped this amount based on buyer's purchase orders. So overall the high level of component level inventory in the supply pipeline poses a risk which the supply chain manager needs to manage. Also this risk cascades down because the pull from the final buyer will create similar pull in the second, third, and fourth tier supply base.

Overall this creates a higher than usual E&O (excess and obsolescence) risk. For example, all of a sudden, a design issue is discovered at one of the component levels. That will require scrubbing the entire inventory both in the on-hand inventory and the inventory in the inbound pipeline as well as any possible buffer inventory kept aside which needs to be scrubbed too. The entire batch of components need to be either updated, modified, or replaced with alternate components. If the component has to be replaced, it can so happen that almost a year's worth of inventory needs to be scrapped. The cost ultimately funnels up to the design owner. So as a supply manager, one has to really look into these aspects regularly and weigh in the performance of suppliers through these three different elements (cost, quality, and lead time) and build a framework with their suppliers. Based on performance, manufacturers often start separating and categorizing the better performing subcomponent suppliers and consider them as the core suppliers.

Usually core suppliers enjoy more friendly terms and conditions, such as better or less than 90 day payment terms in exchange for improved lead times, and deeper partnerships with more involvement in design development. There will also been a set of suppliers which are not core suppliers but at the next closest level who can be managed as niche suppliers. The supplier manager needs to evaluate the criticality of the components coming from this noncore supply base as usually there is less leverage and the risk needs to be mitigated by having more than one source.

Mitigating risk is very critical because typical medical devices are high mix and low-volume parts as opposed to low mix high-volume parts. In a high mix there are a lot of variants which overall leads to a higher number of critical parts and less number of suppliers, whereas in a low mix and high-volume situation, switching suppliers is easier as a lot of suppliers produce similar parts in high volume. For example, an MRI machine is sold in much smaller volume than a mass-produced consumer electronics device. There could also be more variants which require managing more parts and suppliers. In comparison, there are high-volume low mix items such as surgical equipment, sterile gloves, and blood pressure cups. Those are produced in very high volume like a million a week with few parts.

9.8.2 Pricing

In the medical devices industry, the market has a relatively high barrier of entry. This has limited the number of suppliers and has also limited price fluctuations. That being said, there is definitely strong price competition.

One of the unique things about healthcare is that the patient that the product is being used on may not be the consumer or even user of the device. Often, it tends to be the clinician. And so, clinician preference weighs heavily into the demand. For example, if a patient needs surgery to take care of an ablation of the coronary arteries, or needs a prosthetic to be installed, replacement of a hip, etc., in most of these situations, the patient is not the consumer who dictates and asks for a specific cobalt hip replacement by a specific manufacturer, etc. The patients may try to influence the clinicians, but the clinicians are the experts here and make knowledgeable decisions. The final choices of medical devices will obviously be based on the clinicians' experience and other considerations which they feel will be good for the patient.

So, the pricing depends on the economics of healthcare (Medicare Payment Advisory Commission 2017). The clinician preference plays a big part into who negotiates the price and where the demand is coming from. Another facet that heavily influences demand in healthcare is what is actually getting reimbursed. For example, in the US, healthcare is driven by reimbursement. A practitioner who conducts a procedure or treats a patient gets reimbursed by the CMS which is central for Medicare and Medicaid Services. Due to differences in reimbursement amounts based on healthcare plans, one may choose a specific procedure or device that gets more reimbursement versus something which gets less. This also influences the demand. Therefore, manufacturers aim to get their products coded such that they have higher reimbursement rates, which does not directly affect their cost, but it indirectly impacts the demand.

9.9 ORGANIZATIONAL LEADERSHIP

The medical device supply chain leade has one of the toughest jobs in the company. Even a very inexpensive few dollar part can prevent the shipment of a million-dollar piece of equipment and impact the manufacturer's revenue. Small things such as incorrect labeling on the devices can lead to inventory being held up in a warehouse. This can prevent shipment of supply from a supplier to a distributor or whoever the ultimate customer is, such as a hospital or clinic. The breadth of accountability makes the supply chain management a very challenging position for the chief procurement officer. Besides managing supply, a supply chain leader may also manage factories, warehouses, fulfillment centers, or own the logistics of delivery.

Besides ensuring continuity of supply for uninterrupted production, a supply chain leader is also responsible for managing the right amount of raw materials and spare components at all levels. This is needed to build additional amounts of products if there is a surge in demand. The organization should also have a strong sense of cost control. Managing the right amount of excess material is essential. However, there is a fine balance as one can go wrong very quickly in either direction. If there is an upturn in demand and not enough material available, that would lead to a revenue miss. On the other hand if there is excess material and not enough orders, then that would lead to cost overruns. So, an organization needs to be very cost conscious and strong in controlling cost. It also needs a stable and easily accessible IT infrastructure to drive decisions. Finally it needs to have a supply base which is well managed and have a good quality management system. These are some of the key dimensions that a supply chain leader will look for in his organization or aspire to institute.

9.10 INDIVIDUAL PERFORMANCE

Among the individual contributors, there are four critical roles in the medical devices supply chain organization.

The first is obviously those that are bringing in the raw materials, the buyers, and the material planners. Second, there is a team of supply chain managers who are monitoring and managing the factories and similar manufacturing organizations that are building the products as per the design transferred to them.

Third is the forecasting and demand planning side of medical devices. The individuals in this role are responsible for forecasting future demand. They work very closely with the commercial teams to get a really good sense of where demand is trending. Therefore, in most cases, they are embedded within the commercial teams where they report to the supply chain. In many situations, the CFO of the organization will ask for two sets of roll ups: one from the commercial organization and one from the supply chain organization to triangulate where the business is actually headed. Commercial teams usually tend to be fairly aggressive and optimistic whereas the supply chain organization tends to be a little bit more realistic, or maybe a little bit more on the pessimistic side. And so, things do tend to kind of even out. So, this is a critical part of a healthcare supply chain because high mix and low volumes creates a risk and therefore, requires a higher level of forecast accuracy (Karuppan et al. 2016).

The fourth individual role in the supply chain, which is also very critical, is the quality management role. There are supplier quality managers and material quality managers who are looking at all elements of quality along the supply chain at all the time. These four roles really tend to be the bedrock of a strong operating supply chain in the medical device industry.

Besides these functions there is also a strong need for functional roles such as process management and process improvement. The scope of these roles spans both within the buying organization and for auditing and improving supplier organizations too. Quite often, an organization will have a few experts in processes and a few junior-level team members working on improving and owning processes. There are also specific functions like logistics and transportation which are relatively more ubiquitous roles which exist in most supply organizations. However, specific to healthcare, it is often quite difficult to take someone who understands let's say consumer electronics and put that into the perspective of a medical device. This is because there is a relatively long learning curve. Not that it can't be done, as it has been done by many in the industry, but it is nevertheless a daunting task to ramp-up coming into the medical devices supply chain. One of the fundamental challenges in ramping up is the regulatory compliance and managing quality aspect that is unique to the industry. All companies had learning and development tools and courses to help new employees to get up to speed faster, and for the most part, these courses and tools are very, very helpful. However, to get somebody fully up to speed and fully productive, it tends to take somewhere around 6–9 months to be really productive depending on the role. That is why employee turnovers are difficult to handle. A supply chain organization strives to keep turnover low because it takes a long time to groom somebody into a strong supply chain manager with expertise in quality and the regulatory aspect who also has the ability to understand and navigate the complexities of the medical devices supply base (Min 2014).

9.10.1 Inventory Control

Like any other industry, medical device manufacturers also want to turn their inventory into products as fast as possible and then turn it into revenue. However, in healthcare, due to the high mix with high complexity, there is always a higher amount of inventory all along the supply chain. A successful supply chain manager is one who is really smart about tracking these pockets of inventory and understanding if they are current and usable. Version control is an extremely important aspect of healthcare because a product may call out a subcomponent which the quality management system has dictated to be a specific version. In other words successful supply managers not only understand their inventory but track their inventory to specific version level or design revision level to understand what is usable and what is not and needs to be purged. They also ensure that the components' relevant information and documents are also accurate. The risk of not doing so is too high and not only does it create E&O risk but also increases liability and puts the patient at risk (Ledlow et al. 2017).

9.10.2 Sustaining

Low volume is typically unattractive to the subcomponent manufacturer, especially if they have to sustain business over a long period of time. For example, an MRI machine may need an electronic component which is also used by a server. A medical device supply chain manager goes to a tier 1 contract manufacturer and forecasts a demand for 1 million memory modules over a 7 year period. In comparison a server manufacturer can ask 10 million memory modules over a 1 year span. This poses a significant challenge as the contract manufacturer will show a higher amount of willingness to accommodate the server customer than the medical devices customer. The contract manufacturer may be open to supporting the medical device customer as long as the server customer uses the same memory component. Now, as soon as the memory component goes through a version change and the server customer wants to change its design to accommodate that, the contract manufacturer would push the medical devices customer to discard the older memory-based design and adopt the new memory module. The medical device company has to bear that new cost even when they might not need the new module. The company either has to do a large one-time buffer stock build up where it buys sufficient amount of the older memory module to cover usage for a long period of time, or negotiates and pays a higher price for the contract manufacturer to sustain the product using the older generation memory module, or has to requalify the new part. The last option would mean taking the product through the regulatory compliance approval process again. This will not only cost the medical devices manufacturer money but will also take away resources from developing new products in order to sustain existing products. A supply chain manager has to effectively balance between cost and sustaining challenges to maintain business continuity and has to negotiate these options with the contract manufacturer and the component supplier.

9.10.3 Customer and Product Knowledge

At any level inside a medical device supply chain, successful supply chain managers often demonstrate strong customer focus and ability to look at products from the customer's perspective. They usually have a deeper understanding of the ramifications of any action that a supply chain would do to how a customer would operate and connect the dots and act accordingly. For example, when a manufacturer makes a million-dollar MRI machine, it may not always plan for the complications around the installation of a bulky machine. A hospital that has placed the order of one such machine could be in the middle of a metropolis such as New York City with narrow roads and busy traffic. The delivery and installation may require stopping traffic in the middle of Manhattan and making a big hole on the side of the building for the magnet to go through and be placed inside the building. Successful supply chain personnel often develop this customer perspective and bring it back to the business.

For an entry-level or a new employee, it is extremely useful to adjust and adapt by understanding and being cognizant to the other parts of the supply chain. It is important to understand the challenges and time sinks that other parts of the supply chain face. The person on the dock always feels that they are the last person to know when a huge or bulky shipment needs to be done or if something needs to be expedited, even though they are the people who are actually shipping the product out of the door. A product shipment can get delayed due to poor planning or due to lack of visibility at appropriate downstream levels. A successful supply chain manager takes an opportunity to walk the Gemba way (Six Sigma Daily 2018) and understand every piece of that supply chain and maybe even spend time working with all these pieces, whether that is materials management, floor management, warehouse management, logistics, etc.

This skill really helps to round out a very successful future leader within the supply chain organization. At the middle management level, one of the critical requirements is being able to manage various sorts of crises. There are so many different crises that occur in the supply chain, and those that are good at being able to communicate and keep a level head in a time of crisis tend to have long and successful careers. Again, in the supply chain, nobody really remembers you till something goes wrong, for the most part. And so, the fact is that things are constantly going wrong, and someone is constantly under fire for various things.

Often the daily operational decisions are taken at the middle level. Mid-level managers are the people who are typically responsible for getting stuff done and getting the troops to do what is necessary to move the organization forward on a day-to-day basis. They are the people who are solving problems. Keeping a level head and being able to see through the clutter, and navigating with long-term focus is key for having a successful career. Finally, at the highest level, supply chain leaders are expected to be collaborative. Supply chains in the medical devices industry can tend to be very insular organizations; however, strong leaders at the highest levels tend to be collaborative and create value for a cross-functional organization and improve continuously (Coimbra 2013). They often reach out across the organization to the sales, marketing, quality, engineering teams, etc. and really bring the company together in a way that can create positivity across the organization. These organizations tend to be the ones that more people like to work for and be inspired by and tend to stay in the company a lot longer.

9.11 CONCLUDING REMARKS

Overall, the medical devices industry's supply chain is unique in its own way. It is challenging and innovative as it brings contemporary technologies and techniques into the healthcare supply chain. It also provides a myriad of options for supply chain managers to grow their career and skills in various ways. The devices in healthcare are synthesized from all branches of science and technologies. Supply chain managers also enjoy a particular sense of pride and satisfaction at the end of the day as they are involved in making something which saves lives. Every product that ships out the door whether it is a mammogram, CT scanner, or an infusion pump is going to save hundreds of lives and actively help people around the world.

ABOUT THE CONTRIBUTOR

Upayan Sengupta is the Vice President and General Manager for Life and Health Sciences at UL (Underwriters Laboratories). He has been working in health care for more than a decade and has been in various leadership roles in managing global supply chain and operations teams for GE Healthcare. He is an electrical and computer engineering graduate from Purdue University and holds an MBA from Kellogg School of Management, Northwestern University.

REFERENCES

Abuhav, I. 2018. ISO 13485:2016: *A Complete Guide to Quality Management in the Medical Device Industry*. Boca Raton, FL: CRC/Taylor & Francis.

Amato, F., and Ezzell Jr, R. 2015. *Regulatory Affairs for Biomaterials and Medical Devices*. Cambridge, MA: Woodhead Publishing.

Boulton, G. Journal Sentinel. Sept 20, 2014. GE Healthcare's Waukesha business units at the forefront of 'big iron' https://archive.jsonline.com/business/ge-healthcares-waukesha-business-units-at-the-forefront-of-big-iron-b99349785z1-275905571.html/ (accessed December 21, 2020).

Coimbra, E, 2013. *Kaizen in Logistics and Supply Chains*. New York: McGraw-Hill.

Karuppan, C., Dunlap, N., and Waldrum, M. 2016. *Operations Management in Healthcare: Strategy and Practice*. New York: Springer.

Ledlow, G., Manrodt, K., and Schott D. 2017. *Health Care Supply Chain Management: Elements, Operations, and Strategies: Elements, Operations, and Strategies*. Burlington: Jones & Bartlett Learning.

Manz, S. 2019. *Medical Device Quality Management Systems: Strategy and Techniques for Improving Efficiency and Effectiveness*. London: Academic Press.

Medicare Payment Advisory Commission. June 2017. Report to the Congress: Medicare and the Health Care Delivery System. In Chapter 7: An overview of the medical device industry 207–242. http://www.medpac.gov/docs/default-source/reports/jun17_reporttocongress_sec.pdf?sfvrsn=0 (accessed December 21, 2020).

Min, H. 2014. *Healthcare Supply Chain Management: Basic Concepts and Principles*. New York: Business Expert Press.

O'Connor, C., and Periman, L. 2018. *The Healthcare Supply Chain: Best Practices for Operating at the Intersection of Cost, Quality, and Outcomes*. New York: GNYHA Ventures.

Six Sigma Daily. January 2018. What is a Gemba walk and why is it important? https://www.sixsigmadaily.com/what-is-a-gemba-walk/ (accessed Jan 5, 2021)

The Business Research Company. 2020. Intrado GlobeNewswire. https://www.globenewswire.com/news-release/2020/10/27/2114984/0/en/Global-Medical-Device-Market-2020-Size-To-Increase-Due-To-Rising-Infectious-And-Chronic-Disease-Cases-As-Per-The-Business-Research-Company-s-Medical-Devices-Global-Market-Opportuni.html (accessed December 21, 2020).

White, W. 2018. *Excellence Beyond Compliance: Establishing a Medical Device Quality System*. New York: Taylor & Francis.

10

Indirect Procurement in Information Technology

Adhiraj Kohli

CONTENTS

10.1 Objective 218
10.2 Overview of IT Procurement 218
10.3 Software Procurement 220
    10.3.1 Compliance 221
    10.3.2 Deliverables 221
    10.3.3 Preparation.. 222
    10.3.4 Removing the Bells and Whistles 224
    10.3.5 Data Privacy.. 225
    10.3.6 Open Source... 226
    10.3.7 Cloud Migration 227
10.4 IT Hardware Procurement.. 230
    10.4.1 Forecast...231
    10.4.2 Direct versus Distributors 232
10.5 Warranty Support...233
10.6 IT Professional Service Procurement 235
10.7 Case Studies..237
    10.7.1 Software License Distribution 237
    10.7.2 Hardware Asset Management 239
10.8 Organizational Leadership.. 240
10.9 Individual Performance... 242
10.10 Concluding Remarks.. 243
About the Contributor.. 244
References... 244

DOI: 10.4324/9780429273155-13

10.1 OBJECTIVE

In this chapter we will focus on the Information Technology (IT) procurement and its supply chain needs. Traditionally IT infrastructure consists of all technology products that a company needs for its internal employees' technology needs. For web-based companies, IT infrastructure also includes all the technology infrastructure that powers the actual product that the company offers through the internet. So, all social networking websites or e-commerce platforms are powered by their IT infrastructure and data running through their data centers. The procurement organization working to meet this need is often viewed as an indirect procurement team as they are procuring items which are not converted to products but rather used to meet the day-to-day business needs for the employees. This chapter will attempt to describe the nuances of an IT procurement organization.

10.2 OVERVIEW OF IT PROCUREMENT

For almost any company in the world, the IT infrastructure would be all the technology products that help their employees be productive, such as laptops, desktops, and wireless access points (which enable the office Wi-Fi connection and offices).

The IT needs can be categorized into hardware which would be the actual equipment for network, servers, data storage, etc. In addition there are software which are needed for productivity such as data management software, SAAS (software as a service) applications, or even software running within the data centers balancing the networking and traffic.

There are also various kinds of technology services which are part of IT procurement. These services could be pertinent to running the various networking and data-related technology such as domain name system lookup services (needed for routing emails and other data traffic inside the network) and content delivery network which delivers web-based contents to customers. Within a data center as well, there could be multiple technology services that a company may need specific to running the infrastructure (hardware and software).

IT procurement (Dovgalenko 2020) also includes consulting services which may be required for development of software applications or maintenance of certain processes. For some of these technologies, a company may require a specialized skill set for a short period of time. So the IT organization may have to procure some professional services or consulting services that are usually provided by consulting services vendors. In fact, some companies' business models revolve around selling their software products almost for free or at a really low margin and making money on services.

In fact, due to the recent explosion of availability of open source software, a lot of companies are moving away from enterprise software. Big companies like Red Hat, for example, are selling their software product for free and turning their main source of revenue into services or through providing support for the software.

If software is to be categorized into segments, one segment of procurement would be on the software which is primarily used for upgrading. This would include software used for upgrading the operating system and/or tools which are already acquired but need to be upgraded to the next version. These are primarily needed to run the business and to keep the lights on. This category consists of software that powers the actual infrastructure. For example, software for managing the wireless access points and software for managing all the endpoints such as each employee's laptops or desktops computers. These need to be constantly upgraded to keep up and are part of the basic technology infrastructure for employees.

The second category of software would be user requested or user focused. This would include all the SAAS applications that a company may have, such as messaging tools and conferencing tools which enable the employees to communicate. These are software where there are multiple options in terms of vendors and products. A company would typically buy some of these either based on what the employees have been requesting or what the leader or CIO feels will help improve productivity of employees and a service they believe that the company should provide. There are also large ERP suites like financial suites, human capital management (HCM) suites, customer relationship management, bookings and commission tools, etc. This category also includes cloud services needed for running the business or selling platform as a service (PAAS) products.

Finally, the third category of software is again a very specific category which may or may not be applicable to all companies, sometimes because of the way certain systems have been architected. This category includes middleware software or niche use-case software which are needed to ensure delivery of a particular functionality or service to employees because of the way the overall system has been architected. A lot of software in this category essentially provide a function of interoperability. For example, a company may have one type of cloud storage drive or messenger tool for its employees because it used one particular brand of products. Now let's say the company decides to switch to a different kind of cloud storage or messenger in order to reduce cost. However, it wants to transition slowly and continue to use the former suite of products for a long period of time for ease of migration for the employees. To ensure ease of migration the company may need a third piece of software which bridges the old and the new messaging platform or enables the two different cloud storage platforms to work with each other. In a multicloud environment, these software allow the company to balance between multiple cloud service providers without causing disruption and providing the employees with options.

10.3 SOFTWARE PROCUREMENT

For the first category, which is running the business, the procurement process is usually initiated by the internal IT organization as well as the procurement organization. The initial procurement process is initiated by engaging with the vendor after the completion of an NDA (nondisclosure agreement), POC (proof of concept), and technical qualifications of the software. The next step is to initiate a contract. This is where it's important to understand the details of the software licensing model and the company's business model (Guth 2007).

Different companies may have the same type of user-based licensing models, but their underlying business model might be different. For one company, the goal may be just to make money off the software, so as more and more users (employees) use it, they sell more units of licenses. For another company, the goal may not be making as much money on any particular software but each software sold to the customer increases the stickiness to their overall platform or their overall portfolio of products ecosystem. For example, the most popular office suite products or the most popular database companies are typically very large vendors who have many products. As a result, these vendors typically take a portfolio approach. So their goal could be to sell one popular software at a subsidized fee so that they can upsell, cross-sell, or just increase the overall proliferation of products from their entire ecosystems. Then these companies will be able to provide so much value addition that it's very difficult for the customer to cut them off and migrate away to competition at a later point. The stickiness or customer's inability to move away easily can also cause technical debt. Software needs to be constantly upgraded which requires budget and bandwidth. If an organization keeps running their business on older versions of software it starts accumulating technologi-cal debt. Often, the software companies enjoy revenue streams year after year due to this phenomenon. However, it usually gets disrupted by other compa-nies or by an in-house solution. There are some exceptions, like the financial or manufacturing enterprise resource planning (ERP) applications which are typically very deeply entrenched and tied to the finance and accounting pro-cesses so it's really difficult to get out, even if there are available disruptors.

10.3.1 Compliance

Understanding the business model is also important for negotiating the com-mercial side of the contract as it will enable a procurement manager to get the best possible value and pricing for their company. The final part of the contract involves negotiating clauses from the legal side, for example, clauses on data privacy. With all these new changes that have come from laws, such as the GDPR (Dibble 2020) and the California Proposition, compliance has also become increasingly very important. Software vendors who may have passed technical qualifications with flying colors and are ready to give amazing deals and prices, which would lower the total cost of ownership (TCO), are found to be getting rejected by big companies just because they can't meet certain data privacy guidelines like ability to delete employee personal data within 30 days of request. This is how stringent companies are becoming and how data pri-vacy and compliance with current and upcoming legalities have recently become very important for any software procurement process.

10.3.2 Deliverables

The key deliverables for a software procurement process are the license, the software support, and the maintenance that goes with it. The emerging Software as a service (SAAS) application business model is, however, chang-ing this approach. SAAS customers are now expected to pay a subscription for the software which actually bundles in the cost of the software support as well as the license. In these cases, the deliverable on the procurement side becomes figuring out the amount of subscriptions and cost per user for run-ning the application (Tollen 2015).

For the traditional license-based model, a procurement manager has to figure out how many licenses they have to buy and at what price. Again, the metric could be per employee license cost plus what the cost of the ongoing support and maintenance will be in order to keep the software running. This includes making sure the licenses are in compliance and getting support from the vendors in terms of getting new upgrades, patches, troubleshooting problems, user support, etc.

In terms of the licensing metrics, there are two sides, one of which is from the contractual perspective, which involves negotiating what the license actually provides. So for any software, the license could be the ability to use all features of a software on a per employee basis depending on the size of the organization. Depending on the software, it could also be pegged on the amount of data that all users ingest. Sometimes the count of license is dependent on the count of hardware needed to run the software. For example, if the underlying hardware is a computer and it has four processors instead of one then it will require four counts of licenses instead of one. Though it may involve loading the software in the computer only once, it will be counted as four licenses. This traditional model sometimes becomes very complex to manage across different vendors and a very large user base because each vendor has a different licensing model and in some cases even the same vendor can change their licensing model as time moves on.

So one aspect is what each license actually means and the other metric is the cost. A lot of the vendors have been moving from one licensing model to another, obviously sometimes depending on what makes sense to the customer and sometimes whatever maximizes their revenue. So, to handle this complexity and to simplify it, organizations tend to internally normalize across the software portfolio and look at a single dollar cost per employee metric for running the software and providing employees what they need.

In some cases, vendors are also flexible in changing their licensing model especially if the customer is a large organization with over ten thousand employees as they size up the opportunity and revenue potential. Vendors are also comfortable in changing their licensing model or creating a custom sku (stock keeping unit) or bundle of offerings as part of the license model and customizing their offerings into an unlimited enterprise wide agreement. This is often done in a bundled offering when the sum of each item's list price becomes cost prohibitive and their licensing model will not make sense at scale. In some sense, software price is often very flexible and a procurement manager has to really dig in and challenge the vendor to get favorable pricing.

10.3.3 Preparation

A procurement manager typically starts the process by first spending a lot of time with the internal requester to understand what is the basis of the requirement. Procurement is often engaged late in the game and brought in just to close the contract. Quite often, as a procurement manager engages and starts negotiation, they may find that engineering qualification has been completed and the vendor's sales team has had a certain level of conversations with the end customers. In these cases, the manager will already have a deeper understanding of the likelihood of the sale. In case of license renewal, the vendor may even have better data in terms of past license consumption and customer willingness to continue/any stickiness. In other words, a procurement manager may find that multiple parallel conversations have been happening between the vendor and various parts of the organization, and some conversations may even already have been held at an executive level/ some commitments may have been already made. The vendor manager's goal is to minimize cost and maximize the terms and conditions to the benefit of their organization.

So, in some sense, the procurement manager first starts the process by actually dismantling or undoing whatever has been done until that time or at least give that impression to the vendor to level the playing field. This is done to unsettle the vendor from negotiating from a position of power. Often the vendor uses earlier engagement opportunities with the end user organization to set an anchor on pricing, get more information about what can be up sold, get a sense of the amount budget earmarked for the purchase, and then tweak the pricing expectation accordingly. So, the first step for the procurement manager is to understand very clearly what is the requirement and what is the basis or the needs and wants of the requester? Does it include a certain buffer quantity of licenses? Does it include license or user base growth over the next couple of years? At the same time, it is also equally important to understand what is the time length that the customer organization is planning to use the software for.

Is it a software that the customer is just trying out and does the end user just want to see whether it's going to work? Is it going to be a short-term contract? Based on the term of the contract, once the procurement manager understands the requirements and also maps out the licensing model with the vendor, they can then come up with what is the ideal case for the company to commit to a commercial contract. This will provide them with a sense of what should be within the scope of the licensing terms, the count of licenses needed, and what are other contractual clauses (mostly around compliance and usage) that need to be negotiated.

A key aspect to be aware of is overage in use. The right count of licenses is important because buying more than necessary may imply over provisioning which will lead to waste. On the other hand, under committing will have the risk of overage and the vendor may penalize or have a different tiered pricing in those circumstances. From a license renewal negotiation perspective, if the customer has overage, i.e. installed higher count of licenses than permitted by the contract, then the procurement manager enters the negotiation with less leverage over the vendor, as the vendor knows that the product implementation has been a success, and it's been doing so well that there is overuse, and the vendor wouldn't have any incentive to give a better pricing. So, to that extent, the volume is what would dictate a procurement manager's leverage and pricing but at the same time, it's a double edged sword. If a procurement manager goes on the high side to get better pricing and in reality the utilization of the software comes lower than expected, then the procurement manager has probably overspent by a large amount (Hussey 2020).

10.3.4 Removing the Bells and Whistles

Prior to engaging the procurement manager, often it's noticed that a lot of preengagement has already happened between groups, such as an engineering team who was looking for a specific solution and a vendor who offers those solutions. Coming in fresh to the negotiation, a procurement manager often has to get their power and leverage back to negotiate with the vendor. There are situations where the sales team on the vendor side is aware that the actual customer inside the organization or the key organization leader from the technical side is completely sold on the product and there are no other competitors in the running. A supplier may feel that the deal is locked and can be less reluctant to change any price or add any other features without increasing cost. A procurement manager sometimes has to change or reset that expectation of the vendor. The vendor manager has to start from scratch and start from the base requirements with the vendor in order to get the best deal and do their job.

This is probably the most underrated part of the entire process. Let us consider an example: a customer needs feature A and feature A is part of many features in a software product. The supplier tries to sell him the software which has features A, B, and C but has one price for the whole package/bundle and is very eager to sell the entire product at a great discount.

Let's assume that the vendor has given the best possible price upfront and they have the resources and are fully committed to meet the customer's demands and make the procurement deal work. However, the question remains on who is going to check whether the item the customer is going to get is actually what they originally needed or something that the customer will utilize. The customer may not use features B and C ever. Suppliers will often over sell and customers can change the scope to get future proof of what they are procuring. Customers may also deviate from the original ask and wishfully get more features assuming that there is no major cost increase. Or a customer who is only looking for one particular feature is forced to take additional products, due to bundling. This happens when the customer is not knowledgeable and or hasn't really pushed the supplier. They have accepted and assumed that the price cannot be tailored to what they need versus what they are getting. So, even in a hypothetically perfect scenario when the price looks great and the vendor appears to be extremely cooperative, a procurement manager has to validate the very specific needs of the internal customers and compare the needs with what's being offered. For example, trying to renegotiate and tailor final costs based on actual feature utilization is an important task of the procurement manager. The manager also has to look ahead, such as a situation where feature C could become useful and it is important to have that option open. This is their fiduciary duty as well as making sure that requirements are not being overestimated because that is what drives the cost at the end of the day. If a procurement manager does not dig deeper and unbundle the offers and dismantles it to meet the core customer requirement, they are paying more for things that are unnecessary and not optimizing their costs.

10.3.5 Data Privacy

Data privacy has become a very important topic for negotiation from the legal side of software procurement. As GDPR and similar country-specific data privacy laws have come into effect, a company can get penalized by the European Union or other governments for a significant percentage of the total revenue if found in breach. For large companies such as fortune 500 companies, if the European Union decides to fine even a percentage of its total revenue, that is a huge amount. This has become a driving force behind a lot of negotiations as the customer organization would like to seek indemnity or protection from the vendor if the data violation is caused due to inadequacies in the vendor's product. Prior to laws like GDPR, vendors used to agree to signing up for damages capped to a certain amount typically equal to 12 months of the customer's revenue or attached to a certain amount of their revenue from the customer. That's why the challenge has been to negotiate a sufficient damage cap so that in case a data breach happens, the customer has enough protection in the contract to actually recuperate some of the costs.

Often vendors are not ready to sign up for such large liabilities, especially if the customer has access to certain personal data which most of these top companies do. The other aspect of importance here is because the IT world is moving to the cloud and data are no longer stored on premise, it is much easier for security breaches to happen. Potentially saving personal data of the employees and customers on a third-party data center which the customer or the software vendor does not own and controls is the other part of the complexity. The vendors are also unable to increase price in order to provide a higher damage cap, and so it has become a challenge to maintain balance, especially for smaller start-up companies or small to mid-sized companies who cannot potentially be responsible for the huge damages cap. Major video-conferencing companies who have captured so much market share are now facing privacy issues and security issues, and as privacy laws become more and more complex, these issues are sure to become a huge part of the industry.

10.3.6 Open Source

Another challenging area is dealing with Open Source software. As a new crop of companies are just providing services because their software is now free, they have come up with the solution framework of the software which they have contributed back to the community. And, now as that software gets adopted, a lot of other developers start contributing to the code base. As these codes mature to become enterprise-level tools, some of the original contributors start charging a premium for some of the enterprise-grade functionalities and features. Some of the companies have started to realize that they can potentially lock features or provide additional features in exchange for a premium. In general as an open source set of software matures, additional feature requirements start costing money. This also begs for a cost comparison with software which are not open source, as open source software products are increasingly becoming more and more expensive in the long run. In certain cases, a company may also evaluate a make versus buy which is comparing the cost of acquiring the software via open source or proprietary software versus the cost of developing the software in-house.

10.3.7 Cloud Migration

Most medium to large scale companies nowadays find it easier to migrate their infrastructure and applications to the cloud given the recent advancements and availability of many solutions and applications specifically designed and built to assist with the migration process. However, there is a critical decision point that companies face before undertaking this behemoth task.

It is to whether decommission their existing legacy applications and opt to either build from scratch or procure a third party SAAS offering to replicate the functionality of the legacy application. The other option is to simply lift and shift the legacy application on cloud infrastructure.

Both options have pros and cons and present different challenges for procurement and supply chain professionals. The first option allows for more flexibility in terms of addressing legacy technical challenges and also provides engineers an opportunity to leverage the latest features and realign applications towards a more service oriented architecture, which subsequently makes scaling and integrations with other applications easier. But this comes at a high cost both in terms of time and money.

For building from scratch, organizations will typically find themselves augmenting their engineering with professional services to speed up timelines. There would also be a time period of overlap when both legacy and new application would need to be hosted and supported during the migration and till the launch of the new application. This significantly adds to the costs of running the application.

Some organizations also take the faster route of opting for an existing SAAS solution. However, this usually carries a premium and the best way to minimize this premium is to compare the cost of licenses from a SAAS provider with the Total cost of operating and maintaining current on prem infrastructure. What companies often fail to take into account is the increase in warranty cost or the cost of support for hardware which most vendors increase for aging hardware to incentivize tech refresh by the customers.

SAAS providers also tend to push for factoring savings from not having to maintain backup and disaster recovery infrastructure but it is important to consider the licensing model and compliance rules of the software running on on-prem infrastructure as typically software vendors do not charge licensing costs for backup, test and dev environments. The savings from not having to maintain backup infrastructure are realized over a longer period of time and are further diminished due to tech refresh as most

Hardware suppliers either show lower marginal costs (e.g. lower cost per GB, lower cost per core) while also allowing for consolidation opportunities in order to incentivize customers to tech refresh or upgrade existing infrastructure.

Therefore, procurement and supply chain professionals must take into account a factor of savings realized from potential future tech refresh/upgrades and use it as leverage with SAAS providers when negotiating pricing as the goal should be to minimize the premium paid for faster go -live and also the higher costs of SAAS over a longer period of time.

If organizations do decide to opt for "moving" their legacy applications to the cloud either by refactoring or rehosting or a combination of both, then it is critical to consider some technical challenges and the impact to existing software licenses and compliance position.

Since software companies used to thrive on selling perpetual licenses with annual support fees or selling licenses based on different metrics, older or legacy license agreements and contracts typically did not account for a future where licensing could be consumption based or simply customers leveraging their software licenses in a cloud environment.

There are multiple challenges that may need to be addressed directly with the software vendor. Here are a few examples

1. Right to use: if licenses are tied to specific hardware or processors (e.g. SUN SPARC), what happens if they need to migrated to hardware running x86 or ARM.
2. Virtualization: if license is to be tied to one machine or one MAC address, what happens if the cloud environment only supports virtual machines (VMs) which can keep moving from one bare metal server (stand alone server) to another or it creates the possibility of assigning one license to multiple VMs on a single underlying bare metal machine, which could further exacerbate compliance violation
3. Change in licensing model: companies are actively switching to Subscription based licensing and use this opportunity to migrate customers to their newer licensing model. In these cases, they typically offer a one-time credit for existing perpetual licenses that needs to be heavily negotiated as it's a one-time deal and opportunity.

In all these cases mentioned above, it is crucial for supply chain and procurement professionals to take these issues into account and have a clear negotiation plan with desired outcomes that enable longer term benefits.

In addition, there are also technical challenges that may increase costs when doing such migrations. For instance, certain cloud providers either have limited options for hardware configurations or only allow for scaling up or down by pre-set factors.

In the case of migrating from one specific type of on-prem hardware to whatever is a closest match on the cloud provider catalog, there may be performance issues in the application and this may end up increasing unforecasted costs as either additional machines will need to be provisioned or additional licenses may need to be procured. It is very typical with software vendors with core based or processor based licensing that migrating to a different hardware with a potentially different processor and underlying instruction set, there may be a difference in performance which could be significant depending on the application and its use case. This would result in procuring additional licenses as the underlying hardware config would change and in some cases even increase cloud infrastructure costs if dedicated bare metals needs to be provisioned as these carry a premium over shared or virtualized resources.

It is essential for supply chain and procurement professionals to make a compelling case for suppliers to either allow for license usage post migration or define a licensing model that optimizes costs longer term post migration. The absence of any language that specifically does not allow for usage in a virtualized or cloud environment can provide leverage compared to having vague definitions and usage rights defined in legacy master contracts. If there are situations where there was either explicit language not allowing for migrating licenses or usage rights could be interpreted to also not allowed for migration, then the negotiations should be pivoted to what the longer-term future looks like for the supplier. Clear and succinct messaging to the vendor helps as it should be typically be on the lines of either partnering on this journey or subsequently losing out on a revenue stream.

The goal should then to be either establish a new contract or amend the existing one to alter usage rights or go the route of adding specific clauses ensuring that fees paid towards "migration compliance" would address all past and future issue pertaining to compliance and usage.

This would also ensure that these conversations do not resurface at the time of renewals or future adds.

Key challenges when interacting with external and internal stakeholders

- When usage rights or compliance position is not clear for software licenses, it is prudent to first establish internal requirements and

formalize the ask after having explored all avenues instead of sharing every bit of information as it becomes available. This would ensure
that the external stakeholders do not get confused by jumping between
options or anchored to a specific solution which may not be beneficial
to either party

- Internal stakeholders need to understand the true costs of migration
and the total cost as it compares to either a third party SAAS solution
or hosting an existing application on public cloud. This would also help
in negotiations as these data points can be shared with the supplier to
get align costs with budget.

- Internal stakeholders must also be made aware of potential technical
challenges in terms of performance and to address this it is best to bundle some professional services hours at the time of purchase as most
suppliers are incentivized to make the migration work and understand
the leverage that comes with the initial purchase. These "free" hours
and consultations typically go away in subsequent renewals.

10.4 IT HARDWARE PROCUREMENT

Procuring various kinds of hardware is the second important aspect of IT
procurement, even though the types of hardware are simpler in comparison to software licenses. The broad categories are computers which include
laptops for employees and back-end servers or computers for corporate
data, and networking which includes routers, switches, wireless infrastructure, load balancers, etc. to connect the intranet (inside the company) and
internet (outside the company). There are also storage servers where all the
data are stored for all the applications to run the day-to-day business. Most
of the hardware would fall into these three categories; however, there are
also network security-related hardware and software which are important
for IT. This includes firewall machines and other software and services
which ensure that whatever traffic that's being transmitted across the network inside the company's network or outside into the cloud is secure and
encrypted. If the data are going to the public or private cloud, organizations
have to deploy the security components and buy certain licenses from their
firewall providers to make sure that those entry points into the public cloud
are also secured.

10.4.1 Forecast

With software there is no cost of scaling out, but with hardware there is a cost of scaling out because of its lead time. Multiple software copies can be produced and spun up to the customers in an instant. If a software contract is made for an unlimited license, then the new licenses can be granted to new users in an instant. However, a new server cannot be purchased, built, delivered, and installed in an instant. Hardware has a lead time which needs to be accounted for. In addition, there is also the challenge of actual physical storage or space needed for hardware. It is not economical or feasible to stock an unlimited amount of equipment for data centers or offices because not only is there an upfront purchase cost but the additional cost of storing the hardware as well. So, in terms of the supply chain, the forecasting and placing of the orders becomes really important, especially for those categories of hardware which are driven by the applications that are running on them.

The hardware associated with the core infrastructure for running applications may be slightly less critical if it is internal to the organization providing some functionalities to the employees. Employees may or may not see some degraded services. However, if the hardware is for actual products or used in situations such as running an e-commerce platform or a social network, then proper forecasting is really critical. It carries a much bigger risk because if it is not done properly and the organization does not have sufficient hardware and cannot scale out as needed, it would have a direct impact on the revenue. So, it is critical to understand the drivers behind the forecast. Typically for any internet-based platforms, the underlying infrastructure forecast of servers, networks, etc. are usually driven by the volume of internet traffic. As a supply chain manager, it is important to understand how much traffic is growing annually. For example, the annual growth of metrics such as the number of people concurrently present in a web site or the number of search queries done by customers per second are key drivers behind the infrastructure forecast. The second important item is to have some level of understanding of the infrastructure architecture. For example, an increase in traffic will imply a need for more computer servers; however, the servers need to be connected by networking gear. Therefore the forecast for network gear is related to the forecast for computer servers. Certain architectures also call for redundancies through additional hardware which would also drive up the forecast. Overall, there are quite a few drivers which makes forecasting a complex process especially for companies that have a strong online presence. It becomes sometimes difficult to predict what exactly the demand will be and a supplier manager may have to manage those peaks and valleys in demand using business acumen and historic data.

10.4.2 Direct versus Distributors

Supplier managers typically procure directly from suppliers or indirectly through supplier's distributors and resellers. In order to get the most discounted prices, going direct is often more advantageous. However, going direct is often a function of the volume of business. An organization may have a higher volume of business in one geography and have a much smaller volume of business in another geography where the number of users are much less. In such cases, a supplier manager may procure the same product through different channels: one direct and one through a local distributor. For example, a US-based company bringing in infrastructure gear in Brazil will require help from a local distributor due to complexities in import, taxations, and compliance laws. However, for the domestic business, the US-based company may deal directly with the manufacturer because of the economy of scale of the US, European Union, and Asia-Pacific based businesses.

Supplier managers for data centers typically deal directly with suppliers because data centers require large amounts of infrastructure. Typically distributors or resellers add their markup which increases the cost. Supplier managers can negotiate and establish a cost plus model, where the distributor or reseller can only add a fixed percentage or a known amount of markup, but at large scale it will still add up to a large amount of money. Therefore a cost plus model with a distributor or reseller may be still higher than the target cost. As data centers expand, keeping the marginal cost of scaling has to be kept under control, which becomes a supplier manager's key responsibility. The cost per user has to go down. Therefore, it's very important to squeeze each and every cent out of the hardware that is being procured. So, the best way to go about this is to have a direct relationship with the supplier and in some cases, even their suppliers or subsuppliers and component manufacturers as well. This gives better control on the pricing and the markups that are being paid. It also gives predictability into what is going to be the lead time or delivery time so that the supply and size of the fleet can be managed accordingly. In this ecosystem, resellers now have pivoted to providing additional services such as having warehouses to store additional inventory to minimize the lead time. They can provide other services such as rack and stack or other integration services for additional revenue opportunities.

10.5 WARRANTY SUPPORT

Along with purchasing hardware or software the next important spend that a supplier manager has to manage is warranty support or maintenance cost. After purchasing thousands or millions of servers, networking and other infrastructure gear, some of this gear will inevitably break down in the course of time and thus need to be repaired. The manufacturers typically charge a certain amount of money for covering the repair cost, which they may offer for 3 or more years after purchase. This charge covers cost and labor, which means the cost of the replacement part and the cost of visiting the customer site and replacing it.

In case of servers particularly, a lot of the subcomponents are commoditized and easily available for procurement. In such cases it is quite possible to figure out what the cost of repair and maintenance is. One can look at what are the parts which typically breakdown and look at the list of parts that the suppliers are providing to fix the broken hardware. If the cost of the components that typically fail are known and can be stocked up based on the failure rate, a supplier manager can create a self-support model by purchasing and stocking spare parts and by building an internal team of technicians to do the physical repair. This is known as the self-support model. However, this approach only makes sense at a certain scale of a data center as only when there are thousands of very similar pieces of hardware whose failures are similar and replacement parts also similar.

In a situation where the self-support model is not viable, the supplier managers would have to work with the original equipment manufacturer to negotiate the warranty cost. However, even if the self-support model is not chosen in a specific case, a supplier manager can still model the cost of part and labor in the previous paragraph and use it as a benchmark for negotiation. Let's say the part and labor cost comes around 2%–5% of the hardware acquisition cost, then this number could be used by the supplier manager when they are putting a warranty support contract in place. This warranty support cost, however, includes an additional item which is the service level agreement (SLA) that is needed. SLA is an additional cost driver beyond the parts and labor cost. For example, a platinum-level support SLA may require the manufacturer to fix the broken machine within 24 hours or the part may have to be shipped within 24 hours. For such SLAs, the cost of warranty will be significantly higher. In comparison if instead of 24 hours, the duration is increased to 5 business days or a silver-level SLA, the cost of the SLA will drop significantly. For a supplier manager, it is very important to optimize the SLAs. This is where knowledge of the architecture or software stack is important. Depending on where each hardware sits in the stack of the product architecture, the SLAs need to be designed. For example, there could be a lot of redundancy designed in, so a server can easily fail and it will not impact the application. There could be certain components which have so much redundancy built-in that one doesn't need the platinum-level SLAs of having the vendor show up within 24 hours of reporting an incident. However, there could be other areas which may actually need that.

Optimizing SLAs can drive significant savings when a supplier manager is negotiating the cost of maintenance and support for hardware. They need to understand the level of SLA and also check internally to understand whether it's actually needed by having historic data of actual utilization. This, however, gets a bit challenging for categories such as network or storage, where the SLA includes cost of maintaining or supporting the software that's running on the hardware. In most of these cases, the software that typically runs on the router or a switch or a network attached storage device is not as rudimentary as a firmware. These have become sophisticated to a point where it's almost like an entire operating system. In other words, without the proprietary software, the hardware is pretty much unusable. However, as soon as the warranty includes the support and maintenance of software, the cost of support shoots up. The vendors obviously lean toward the pricing for maintenance and support per software and may charge up to 15%–20% of the acquisition price of the software. As the hardware becomes more complicated, this is a key factor which drives the cost of maintenance of the hardware.

Supplier managers may also try to optimize warranty cost by breaking down the support into hardware support and software support. They could ask the manufacturer to break the invoice for the procured product as a piece of hardware and a piece of software needed on top of that. This may produce an advantage that the total warranty cost is now 5% of the hardware cost and 15% of the software cost instead of a uniform 15% of the total cost. This approach sometimes works with the vendors, as quite often the manufacturer publishes updates and features during the course of the lifetime of the product which can be applied without any modification to the hardware. So essentially the hardware never changes or needs any repair while the software upgrades continue to happen. However, one may also find that vendors may come back and reduce the cost of the purchase price of the hardware so much and increase the cost of software so high that the savings are negligible.

Though this approach is a bit challenging, vendors are increasingly more open to decoupling the software from the hardware, mainly due to competitive pressure from open source hardware and software. Open source is an effort where companies are essentially sharing their design to the world for anybody to copy and reuse. For example, instead of purchasing a branded server or a network gear with a proprietary design and proprietary operating system running on it, it is possible to use any of the open source designs. Just by joining the open source consortiums which are a collective of companies who are freely exchanging information around hardware and software among each other, the companies can avoid a certain amount of cost. Today those open source designs can be modified and be built by any generic manufacturer. This has also proven to be a point of leverage with the suppliers.

This has changed the market dynamics as manufacturers who used to sell hardware and software as a bundled solution are now amenable to sell their software on generic open source hardware. This allows the manufacturer to at least maintain the software revenue stream instead of losing both. In parallel, the open source software are also reaching a point of sophistication where even the small to mid-sized companies with limited engineering resources have been able to take advantage without compromising data security and functionality. As the margin has declined and hardware business has become a zero-some game, the hardware manufacturers are focusing more on software and services companies.

10.6 IT PROFESSIONAL SERVICE PROCUREMENT

Professional services are needed to run various kinds of operations (Clifford 2016). These are services which a company often needs either on a temporary basis or it calls for skill sets which are not considered as core for the organization. This kind of service can be also utilized in locations where the company does not have a presence and needs help to scale. So instead of hiring full time employees, a company uses a professional service and outsources the overall task to another company. The professional services company provides either the service in a turnkey manner where they are accountable for fulfilling an overall set of deliverables or they could be responsible for just providing a specific number of headcounts.

Some of these use cases are highlighted in this section. One such use case could be doing various kinds of technical and physical services for running a data center such as running all the networking cables in the data centers to provide connectivity between the servers, networking gears, etc. Another use case could be procuring commodity items in mass quantities such as power cables, patch panels, and office supplies for the entire organization. There could be more complex use cases where technical experts are needed to do complex maintenance work in the data centers. For example, if a self-sparing model is implemented by a company, then there is a need for technical professionals to do the actual repair work. A supplier manager can go directly to some of the third-party maintenance providers who can be made responsible for maintaining the spares and doing the repair work. Some of these types of third-party services also often use procurement for the spares as a service. Professional services are also used extensively for IT software development where a large set of people may be needed to do the initial development and release of the software and subsequently a small team needs to be kept for maintaining the software.

There are multiple ways a supplier manager manages professional services. The easiest way to negotiate would be to establish a rate card for various types of expertise and skill levels. The supplier manager would typically try to establish a low hourly rate or a monthly rate. At a certain volume the rates are more dictated by the market condition. The scope of the work plays a key role in determining the right type of contract. For example if a perpetual maintenance work is the primary scope, then perhaps a rate card-based contract is better suited. However, if there is a project with a clear set of objectives, a statement of work (SOW)-based contract is better suited. In the latter the professional services company is responsible to deliver or meet a certain set of goals within a stipulated set of time with a stipulated amount of resources. A supplier manager needs to understand whether the proposal makes sense or not and also negotiate how to manage if the scope gets changed by the customer during the course of the project or if the vendor fails to meet the requirements within the limits of the resources defined.

For open source and nonessential IT software projects, often the companies feel that their internal engineers are burdened with other projects and hence do not have the bandwidth, and what they need is a one-time lift to build out a certain software feature. There are now a lot of developers out there who have entered this market and a supplier manager has to establish a SOW. One key requirement for the supplier manager is to ensure that the ownership of the end product or the IP should belong to the company, so that later the developer cannot claim the product as their own or the company does not have to buy licenses from them to use in the future.

10.7 CASE STUDIES

These are two examples where an IT procurement manager can make an impact. The first example is around understanding software licenses and utilizing a divestiture as an opportunity to deep dive and taking stock of a portfolio of licenses. Often software procurements are done based on assumptions and these types of situations produce new cost-optimizing opportunities. The second example is how a supply chain manager can help in creating an asset management program to create a predictive demand with a supplier and obtain a better cost and lead time (McLachlan 2018).

10.7.1 Software License Distribution

A company was going through a divestiture event and split into two separate entities. The challenge was to divide the licenses correctly between the two entities. However, in order to do that, first an accurate count of all software that has been deployed needs to be established and then, the next step would be figuring out a way of allocating those licenses.

Step one required developing a knowledge of the entire software count inside the company across all the vendors involving all the various license models that existed. The next step was building out an exhaustive list of software, publishers (software manufacturers), counts of licenses in use versus actually granted, versions of the software, ownership status, limited or unlimited count of licenses allowed, expiration date or contract end date, warranty support and maintenance terms and conditions, assignment of license, etc., which essentially created a license manifest. The list of entitlements would also articulate what software was owned, which ones were fully paid for, and how each of them was being utilized and by which parts of the organization. This would enable the supplier manager to accurately set out the contracts and the future entities could get their shares allocated without inviting any of the vendors to perform an audit. Vendors often took these kinds of events as a golden opportunity to audit and charge huge sums of money for potential overages or even assignment rights.

In some cases, a contract was set up in a way where one cannot legally assign the licenses to another entity whether it was the parent entity or the child entity during a divestiture. This problem was solved by first developing an understanding of the scope of the software, whether it was tied to specific types of hardware and what percentage or count of the licenses were actually utilized. In addition, it was important to identify whether the software was fully owned by the company. Certain software once purchased become perpetually owned by the purchaser and certain software were only owned for a limited time period. In addition the license count would be different. For example, one software counted each employee as the basis of license, another may be using the number of computers, and a third one could be using processors as the basis. The overall license count for all software was then converted to a common per employee basis. The supplier manager built this final data set to calculate the final license entitlement positions accurately.

The above process was done based on contractual terms and conditions. For example, the contract had listed 50,000 licenses. The next step was to understand the utilization. In order to do that, each and every computer needed to be checked to determine how many of those software were actually running. Then only it could be compared whether all 50,000 licenses are being utilized. If less than 50,000 instances are actually existing then it would be underutilized and if more were in existence that would be a case of overage.

Detection and getting the correct count of software in every computer was a technologically difficult job. Though the computers were all connected through the network but checking for instances of a targeted list of software actually deployed in a networked environment was very complicated. In theory sniffing out all the computers by connecting to each and every one of them via the network and detecting the existence of various instances of all types of software was possible by using a sniffing tool. However, to run such a tool also posed potential security risk of being able to access all computers through a single sniffing tool. Also searching computers isolated in various network domains or environments and identifying the targeted list of software required a lot of tweaking of the sniffing tool. A hybrid approach was taken with a mix of detection by sniffing tool but restricted to some of the domains, by in-house home brew scripts and by manual calculation to come up with an accurate utilization list. The entitlement count calculated from the contracts and the instances counted are put in the manifest. This is effectively a rudimentary software asset management process. This comparison provided an accurate assessment. Getting this state of accuracy was hard and once achieved needed to be maintained and became the basis for accurate divestiture.

10.7.2 Hardware Asset Management

The second use case was created to establish a process for optimizing purchase of endpoints or laptops. The challenge was that the demand for laptops was unpredictable which made it difficult to put an accurate monthly budget for laptops. Employees were eligible for refreshing their old laptops after 3 years. However, certain employees would come in because they were eligible per the refresh policy to exchange their laptops after 3 years, but some would not. In addition, there were laptops which needed to be refreshed on an emergency basis due to incidents like broken screens or other problems. These factors made forecasting new laptops needed per month quite unpredictable. This also made it difficult to keep a sufficient amount of inventory of new laptops available to meet unpredictable demands and also made price negotiation difficult as the volume would significantly vary month over month.

So there were two challenges, first how to lower the cost and second how to create a process for managing to refresh the assets more efficiently and establish a supply chain process. A number of solutions were put in place to meet the overall challenge of cost and ensure that end users did not get impacted by insufficient inventory. To reduce cost, first the difference between leasing versus buying was explored.

A significant portion of the laptops were needed for the US-based employees. Instead of purchasing laptops, leasing laptops for the US employees made more sense. The overall lease model was established with almost 0% financing. A 4 year lease was put in place with an option of either returning the laptop or purchasing the laptop at fair market value at the end of lease, i.e. after 4 years. If an employee decided to return the equipment after 3 years, per their internal refresh policy, the company would effectively return the laptop early to their supplier and would end up saving anywhere from 25% to 30% of the original cost.

Second as a process, a single source of hardware assets was built which kept track of all the laptops coming in for refresh. From a process standpoint, a single source of truth for all laptops was established which kept track of new laptops coming in and old laptops being returned. In addition to managing the broken laptop replacements, the laptops were purchased with blanket warranty that included accidental protection. So this use case no longer impacted the budget or took from the inventory. The supplier was now responsible for taking care of all broken laptops under the lease. The asset tracking database kept track of all such incidents and also tracked laptops for which the warranty coverage had been exercised.

This took care of the bulk of the budgeting issue. For geographies outside the US, where the necessary volume was much lower, the process was kept simple where a reasonably small amount of inventory was sufficient to meet all replacement requirements.

So overall a hybrid model was established to optimize savings and supply. Due to the inventory management system, it became easy to predict how many laptops were up for refreshment due to the end of lease or close to the 3 years mark. So from a supply chain standpoint, demand became more consistent without the spikes which were previously caused by warranty issues. From a budget point of view the extended warranty was bought from a separate budgeting account. Overall instead of setting a uniform solution, an optimized hybrid solution was implemented.

10.8 ORGANIZATIONAL LEADERSHIP

For a lot of organizations, the IT supply chain leadership can be distributed. The responsibilities of managing cost and contracts could be distributed between multiple teams. For example, sometimes a leadership team of a particular organization may do initial negotiation and then request the legal team to close a contract. So for example, an engineering database software could be negotiated by the director or VP of the engineering organization. However, as the organization grows, a formal procurement is typically put in place, even though the key decisions were often made by the same leadership team. As an organizational leader of such a procurement team, gaining trust is important to be the actual business owner instead of leaving the decision-making in the hands of the stakeholders. The stakeholders start relinquishing authority to the organizational leader when they start seeing value. Quite often the procurement team is viewed as a blocker instead of as an advantage. The perspective changes when the leaders demonstrate ability to manage the business with an efficiency level which the stakeholders could not achieve working individually. The key factors for establishing efficiency are staying ahead of demand, i.e. not letting procurement get into a stage where business is about to get impacted, ability to complete price and other legal negotiation quickly, i.e. not dragging on and taking resources and attention away from the stakeholder's organization, and getting more out of the budget but optimizing spend, i.e. not exceeding the allocated budget and pushing for savings to stretch out of the budget.

Large organizations tend to become more reliant on enforcement tactics to force the stakeholders to allow the procurement management to manage all aspects of procurement. While this ensures compliance to the documented acquisition process, a leader should understand that this is often viewed as bureaucratic red tape by the stakeholders. Stakeholders hate losing agility and getting stuck with the processes while waiting for the products that are necessary tools to get their work done. A good leader understands this side and ensures that procurement processes are not becoming too overweight. Quite often if that doesn't happen, the stakeholders start their own procurement process and start to bypass the procurement organization. In a way, a leader has to maintain a balance between maintaining broader financial goals, compliance requirements, minimizing risk of litigation for the organization, and ensuring efficient closure of contracts to enable the business to function.

Depending on the size of the organization, an IT procurement leader often recruits experts instead of generalists. However, this approach has a challenge as the generalist does not get the opportunity to understand the nuances of a specific product, its market trends, alternates, etc., and what exactly are the stakeholders' wants and needs. For example, a generalist may have to close a database contract one week followed by a travel service contract in the next. Instead, the primary goal is to quickly negotiate based on historic price data point and focus on the payment terms. Hiring two specialists such as one who is an expert on technology practice and understands database software, license models, and another who is an expert in travel services who understands loyalty points management, airlines industry etc. and can produce better optimization for the entire company. This, however, depends on the scale of the organization and the leader understands when to choose what level of expertise based on available resources and need.

10.9 INDIVIDUAL PERFORMANCE

An individual contributor in an IT procurement team needs to understand contracts, and payment terms and conditions first. Depending on the organization size, a procurement manager needs to understand their portfolio and stakeholders. The key leverage for negotiation is the volume, understanding the forecast, and actual demand, which is achieved by tracking demand and volume through the life cycle of products. Depending on the product, a procurement manager should be able to develop knowledge to understand the engineering requirements of their stakeholders. Similarly if the procurement manager is managing some other area, they need to develop a deeper level of understanding of the products, why the product is needed, and what are the drivers for that necessity.

IT procurement is typically budget driven, even where the budget often has to be used up and never typically returned back to finance as unused. While cost savings is important, since the budget is never returned (typically), cost avoidance is more important. For example, a company allocates $1 million for acquisition renewal of software licenses for 10,000 licenses. Now say that the supplier manager manages to create cost savings by reducing the cost to say $900K for the 10,000 licenses. Then $100K is the cost savings. However, another approach is that the supplier manager negotiates 15,000 licenses with $1M. This could be a better option where effectively the company avoided spending the cost for 5000 licenses, provided that there is an increase in internal demand in the near term.

A successful procurement manager also provides a companywide perspective to its stakeholders. For example, a software could be used by three different departments or subentities even though none of these departments have that same global view. A good procurement manager unites the stakeholders and drives the optimal decision for the company.

Business leaders are often not aware of commercial or legal nuances and here is where a procurement manager is expected to add in their value. They have vast experience in understanding how vendors try to lock in a customer and how to avoid those traps and future proof the contracts so the stakeholder can engage or disengage with relative ease. Relationships change and business conditions change and a vendor manager should understand this very well. Often a procurement manager will get into a commercially emotional situation where the business is either attempting to go in partnership with a vendor very fast and has already tied their success metrics with that event or wants to exit from a relationship immediately based on an executive's wish. Both of these situations will create pressure on the supply chain, and a procurement manager has to handle this situation and do what is right for the overall organization.

Quite often stakeholders look for additional intangibles like additional training, additional features but not necessarily have the right amount of budget. By using their negotiation power, a procurement manager can make their customer happy and successful. At minimum, a vendor manager should read legal contracts prior to meeting with their stake holder as the stakeholders rarely read the contracts. A minimum expectation from a procurement manager is to have curiosity to learn the nuances of the business side or technology side and also have the curiosity and initiative to understand the nuances of the commercial and legal terms. As the stakeholders change and the procurement manager moves on in their careers and changes jobs, the contracts that are left behind should be able to continue protecting the business and allow it to run efficiently. A procurement manager should always be aware and accountable of the legacy they leave behind.

10.10 CONCLUDING REMARKS

IT procurement is one of the most common procurement organizations as almost every modern enterprise needs it irrespective of its end product. Unlike a manufacturing supply chain organization, due to the diverse nature of goods and services required by an enterprise, its procurement team has to deal with a diverse set of products, vendors, and internal customers. The categories are broad and underlying technologies are often too diverse. Quite often developing an understanding of the asset footprint is key to get a picture of what does an enterprise actually own, what exactly it needs, and how it does monitor.

The internal stakeholders are often end users of technology and view the procurement process as bureaucratic and blocking them from doing their daily function and any time spent in it has a risk of business interruption. While choosing a technology and negotiating contracts with the vendor, it becomes the IT procurement leader's responsibility to protect the organization from business, legal and financial risk, without interrupting the business function that depends on that technology.

ABOUT THE CONTRIBUTOR

Adhiraj Kohli is a technology business leader at LinkedIn where he manages procurement of infrastructure-related technologies. He has in-depth experience in leading business operations and indirect procurement. As a business leader, Adhiraj negotiates contracts with cloud-based companies, acquires web-scale infrastructure and IT hardware, and sets up post-acquisition asset management practices. He has an MBA from Boston University.

REFERENCES

Clifford, D. 2016. *An Introduction to Service Integration and Management/ Multi-Sourcing Integration for IT Service Management*. Cambridgeshire: IT Governance Publishing.

Dibble, S. 2020. *GDPR for Dummies*. Hoboken, NJ: John Wiley & Sons.

Dovgalenko, S. 2020. *The Technology Procurement Handbook: A Practical Guide to Digital Buying*. London: Kogan Page.

Guth, S. 2007. The Vendor Management Office: Unleashing the Power of Strategic Sourcing. Morrisville, NC: Lulu Press.

Hussey, J. 2020. *The SAM Leader Survival Guide: A Practical Success Guide for Software Asset Management Professionals*. Lakewood, TX: Technology Vendor Management Educational Services.

McLachlan, P. 2018. *Pocket CIO – The Guide to Successful IT Asset Management: Get to grips with the fundamentals of IT Asset Management, Software Asset Management, and Software License Compliance Audits with this guide*. Birmingham: Pact Publishing.

Tollen, D. 2015. *The Tech Contracts Handbook: Cloud Computing Agreements, Software Licenses, and Other IT Contracts for Lawyers and Businesspeople*. Chicago, IL: ABA Publishing.

11

Powering the Future of the Energy Industry

Bradley Andrews

CONTENTS

11.1   Objective 245
11.2   Overview of Energy Sector Supply Base 246
11.3   Macro Trends Impacting the Industry Landscape 247
11.4   Digital Supply Chain 248
11.5   Shift in Supply Chain Management 249
11.6   Shift in Relationship Management – the Seller's Perspective 250
        11.6.1 Battle between Close Alliances and Hard Competition 252
11.7   Case Studies 253
        11.7.1 Valves that Did Not Need to Be Replaced 253
        11.7.2 Parts that Did Not Need to Be Produced 254
11.8   Tribal Knowledge and Data Transparency 256
11.9   Organizational Leadership 257
11.10 Individual Performance 258
11.11 Operational Excellence 259
11.12 Concluding Remarks 261
About the Contributor 261
References 262

11.1 OBJECTIVE

This chapter will focus on the supply chain in the energy industry. Energy industry includes all the companies that are involved with all aspects of energy production, delivery and sale. This includes fossil fuel exploration, refining, global delivery and sale at different stages of refinement. Besides petroleum, gas and coal, the energy sector also includes generation of

DOI: 10.4324/9780429273155-14

electricity from energy sources such as solar, wind, nuclear, geothermal and hydel power.

There is also an ecosystem of EPC (Engineering, Procurement and Construction) companies (Dutta 2020) who have the necessary knowledge and capability to undertake complex resource and capital-intensive construction projects to build the necessary energy-producing infrastructure. For example, an EPC will work with an oil production company to build a refinery or an EPC may work with a government to build a thermal power plant for electricity generation. EPCs are typically responsible for providing engineering design service for the installations, procuring necessary infrastructure items needed for the installation and then managing the installation and construction project to build the final plant. A large EPC typically manages a lot of projects globally and hence has an arm dedicated to managing the supply chain for procurement, logistics to ensure delivery of the goods and services needed for the undertaken projects.

11.2 OVERVIEW OF ENERGY SECTOR SUPPLY BASE

The energy space is a multitrillion-dollar supply chain. It's essentially broken into three groupings. One is spare parts, equipment and technology, whose supply base includes companies such as General Electric, Schneiders, Siemens and similar big multibillion dollar global monoliths around the world who offer operating technology in the energy sector. Then there are the service group of companies, and that ecosystem comprises information technology companies such as IBM and Amazon, who provide IT services. There is also a whole suite of technology services companies, and those are broken into the small niche management consulting servers right up to the biggest engineering firms in the world, such as Worley or Fluor Workgroup to the biggest technology services companies like, Halliburton and Schlumberger. And then there is also another group of companies which includes the commodities and is referred to either as the inputs or the outputs. So at one side of the spectrum would be the input side, for example, bitumen which turns into crude oil, which turns into polyethylene, which turns into a derivative of plastics. Similarly on the output side there are companies who trade energy and resources like power or other resources such as oil derivatives, gas or minerals. Effectively these are the three buckets: the technology companies, the services that use the technology and then the products. So it is a multifaceted multitrillion-dollar supply chain that essentially allows human beings to get the energy and resources that are needed to run the world.

11.3 MACRO TRENDS IMPACTING THE INDUSTRY LANDSCAPE

There are three main macro trends happening in the energy industry. The first one is globalization and that is going to shift around COVID and some of the risks that are being noticed in the supply chain around that. The energy companies, the products they use and where they sell to are all on a globalized matter (Fukuizumi 2020).

The second is an unstoppable trend which is the energy transition. Due to global warming, societal pressures, new banking pressures and shareholder advocacy, the energy, chemicals and resources world need to provide the same amount of energy but with more restrictions. The world's electricity use is only going to go up. The world's energy production is only going to go up as we continue to pull human beings out of economic and energy poverty (Sarkodie and Adams 2020). But we have to do it with much less carbon footprint and with much more of a sustainable approach so that in 50 years if we are looking at the Paris Accord or IEA, we need to have half the emissions, but we need twice the energy. So there's this massive energy transition that's happening in our industry and that's touching everybody. From the miners who are mining resources to companies who are refining and processing resources and finally the companies who are generating and distributing energy, the entire ecosystem is impacted by the energy transition.

The third is digitization and automation of processes: how do we move things to the cloud and how do we do bots to do robotic process automation? But as we think about automation and remotely automating plants, we have to think of the cyber security aspects which is not just protecting data but protecting facilities from preventing someone from hacking in and taking over a power plant.

So we're dealing with those three macro trends which are unstoppable trends. Besides them, there are short-term trends which affect the supply and demand. For example, this is the only time in modern history where both the supply side and the demand side of energy has been pressurized at the same time. Typically, when you have excess supply, future oil contract price drops. As demand increases, contract prices also increase as dictated by the economics of demand and supply. However, this is perhaps the only time in history when demand has come down and there's an oversupply which is leading to current crazy outcomes such as negative oil prices for futures contracts. Apart from these, the industry always had ongoing macro trends arising due to geopolitical issues, trade wars and weather-related impacts. There are always macro or micro forces in play, but in reality we are primarily driven by those three unstoppable global trends (Deloitte 2020).

11.4 DIGITAL SUPPLY CHAIN

The supply chain for the energy sector is globalized. EPC companies serve a handful of customers whose operations are spread across the world. The corresponding supply base is also equally globalized as the suppliers typically have a worldwide network of factory, fulfillment centers, distributors and logistics providers. Due to constant improvement in industrial electronics, materials and engineering designs, more efficient machineries are being produced by the market leaders which are procured at the lowest possible cost to maintain the pace of energy transition.

The underlying supply chain management process which has stayed unchanged for years is now ready for a significant change (Martinotti et al. 2014). There's a difference between digitizing a manual process and a digital process. In this chapter we will mainly focus on the digital aspect of the supply chain and the big shift that's occurring. Existing enterprise resource planning (ERP) systems and the procurement modules that go along with it are sequential workflows that have been digitized. In other words supply chain digitization often means migrating paper or personal computer-based supply chain documents to a giant database which keeps track of all the supply chain-related data and approvals in one place and makes them available across the organization. It does not necessarily change how procurement is done instead it has taken the existing process and moved it from paper and pen to the computer.

For example, think about any popular procurement platform and what it does during a buying event. It has been built to basically follow the traditional procurement process where a supply chain manager's workflow involves prequalification, evaluation and purchase. First, it helps the supply chain manager to reach out to suppliers to get three competitive bids in a buying process of prequalification: shortlisting, going out and getting expressions of interest, doing a technical bid evaluation, doing a commercial bid evaluation, making a recommendation, negotiating the outcome and making the purchase. There is nothing digital about it other than the flow of information is done in a digital way through a computer and all data get deposited in a giant database for easy access and analysis.

The e-commerce world has developed into open marketplaces on cloud environments where the buyer and seller not only interact directly but they actually change positions and apply elsewhere. The new trend is to explore whether we can take that learning into the business-to-consumer or consumer-to-consumer world where we are seeing popular global marketplace platforms which are enabling both domestic and cross border trade via online shopping. Can this approach be adopted in the industrial world? And that is the real shift and the excitement we're seeing and it is turning into a true marketplace, not necessarily a linear kind workflow-based traditional supply chain.

The second is given the pressures of this industry partnerships, preferential contracting, smart contracting the overall profit margin is shrinking. Earlier, the profits used to be so large that everybody could get a slice of the value adds and still make their profits. Now, with profits being squeezed all the way through the supply chain, it has forced a reduction in the number of intermediaries and pushed for new ways of creating value. Some of the values are achieved via more shared risk-reward models, more long-term supplier agreements or a multitude of different tactics.

11.5 SHIFT IN SUPPLY CHAIN MANAGEMENT

Twenty years ago, the EPC industry started to learn from defense and aerospace and the automotive industry of how to procure in a different way. It followed by creating category managers whose job was to lower prices and they did a really good job. Some of the big customers hired or developed deep expertise on understanding the cost structure of their suppliers. Armed with that information the supply chain category managers would go to the suppliers and dictate the price for the amount they would buy. And what that has done is driven a systematic approach to the supply chain whereas it squeezes the suppliers down to a certain point where they can't get cost reduce anymore and the EPC supply chain has started to observe to see failure in the supply chain because of it. A really good example are some of the engineering services companies (sellers), some of which have fallen into financial challenges because the supply chain (buyer) on the buy-side has pushed them to taking high-risk lump-sum EPC contracts than their balance sheets can hold. This increased financial risk for the seller and impacted their liquidity and cash reserve (Jacoby 2012).

So what the supply chain (buyer) did was they did their job almost too well, too clinical to use their leverage too much and actually hurt the overall business in the long run. What they also did was they moved away from sustaining the relationships and became very clinical. They tried to commoditize everything down. Now, there are certain things that can not be commoditzed. Human efforts are tough to commoditize. Now, you're seeing a reversal of that as the executive is starting to drive the decision-making rather than just the supply chain. The supply chain will go back and try to get their target price and drive the lowest cost. That may be expected in a capitalist world. However, now we're seeing a shift to value-based pricing, more of a long-term risk-sharing business model. Therefore, for the supply chain expert, it's actually more important to develop their skill sets toward business thinking, how to translate risk-reward and how to calculate the time value of money into a broader sense rather than following a very deliberate and cost-centric process. That process produces a very desired outcome and even though that individual outcome might be the most optimized, the overall outcome may not be in the long-term viability of the supply chain. Supply chain experts are going to get out of just following a rigid process to actually having to do some strategic thinking and creative decision-making which will also make the supply chain role more exciting (Steinberg 2017).

11.6 SHIFT IN RELATIONSHIP MANAGEMENT – THE SELLER'S PERSPECTIVE

Managing the relationship between a customer and its supplier is a two-way responsibility. Customers need to recognize the following three factors when managing relationships with a supplier. First, there is a shift in supplier's attitude. Suppliers are telling their customers now that they are not interested in the working-for approach. Instead, they are interested in the working-with approach. This is because when the profit margin for a business gets ground down to the extreme, the supplier starts failing to give any return to its own investors and thereby losing focus on the customer (One Pull 2020). So suppliers are pushing back on the idea of being commoditized and on giving out the best price, instead they are walking away from the business and expressing disinterest in business. As businesses like to be more efficient and continuously improve, there is a shift from working-for to working-with the supply chain.

Often supply chain folks fail to look or negotiate based on value. In any negotiation, the quickest resolution to negotiation is where both sides have access to equal or symmetric amounts of information. When you have asymmetric information, then it creates doubt, it creates distrust and it actually gets worse outcomes. And what supply chain has traditionally done is they have filtered and prevented suppliers from getting context to their own business. For instance a supply chain organization may often attempt to manage its suppliers' interaction with its internal customer such as engineering or design executives. Sometimes they will tie the access in lieu of any kind of cost-related benefit that has to be provided by the supplier. This prevents the suppliers from understanding the context of the engineering organization. What are their challenges that need to be solved? Where are their projects going? What are their strategies? Subsequently it prevents the supplier from adjusting its own business (Donnely 2020).

This kind of behavior by the supply chain organization arises from a belief that relationship between the supplier will lead to worse price and therefore needs to be always managed by the supply chain. When a supplier has good relationships with a customer, the customer ends up getting the best service and best value for the price it pays. When push comes to shove, that particular customer can lean on the supplier to take more risk or allocate more resources or even lower the prices. For those customers, it is easy for the suppliers organization to extend more favorable terms or remove internal barriers within the suppliers organization. The supply chain has to change and open up the asymmetric information, allowing the corporate strategy to flow because blocking information flow just creates a supply chain in silos.

The third issue is creating more synergy in the broader supply base. The supply chain organization has to bring its entire supply chain together and not through only unilateral discussions but through multilateral discussions. In order to build a best in class solution in a cost-effective and time-efficient manner, it is important to understand the strengths and dependencies of the suppliers between one another. A solution may require multiple EPCs to work together. Though the EPCs compete with each other but the customer company can make them work together. Let's say the solution requires a component or software from one EPC which the second EPC will use to develop the end product. Creating a collaborative workshop between the competitors and making them review and answer to a common product goal will break down barriers and benefit the end customer. This transforms the supply base into a multilateral supply chain. In summary we discussed three concepts: working-with the customer rather than working-for, enabling access to information to provide more context to suppliers and having a multilateral approach to the supply chain. This is quite contrary to the traditional hub and spoke kind of information disbursement model where the supply chain is at the center and the vendors are held at arm's length away from each other.

Each concept along with the traditional way of managing a supplier relationship has its value and supply chain managers have to match their business tactics to the choice they pick. However, when a supply chain organization is being overly systematic in grounding down the supply base it will be difficult for it to transition or transform its supply base when needed. When they want to transition their core business or during a supply chain disruption it will be more challenging to get cooperation from the suppliers. The supply chain has to live in the bed they have. For example, a cost-conscious supply chain organization has a supplier which also matches its corporate culture and way of working with the same expectation all the way up and down the line. That particular supplier will be very steady with their strategy and is going to be the low-cost provider. However, they will not be changing tactics. So as a supply chain organization you cannot be agile and forward-thinking and have the partnerships antiquated and one-sided. The supplier may not put its investment without a strategic imperative, otherwise it will put themselves at risk.

11.6.1 Battle between Close Alliances and Hard Competition

In the EPC world, there's always been this battle between close alliances and hard competition. It has ebbed and flowed, where there have been alliances between the supply chain organization and suppliers resulting in sole-source programs which are run based on metrics or key performance indicators (KPI) or gain or profit sharing business model. In this approach there is a high level of trust and relationship between the organization and the customer and is not always trying to get competitive bids for every program and grinding the supplier. Instead the customer is sourcing supply solely from one specific partner but monitoring the overall supplier performance based on metrics such as cost performance, speed of engagement and risk taken by supplier to keep the relationship in check. This typically brings the two companies together. Sometimes these alliances go on for a long period of time. However, whether it was going on for the last 20 years or last few years, there is always someone who comes in and questions the sole-source arrangement and pushes for the introduction of additional suppliers to drive more competition. Typically this takes the alliance out of status quo and often leads to a change in the supply chain.

Supply chain organizations sometimes have a bias that it does not get competition and lowest price if it wants to go down on an alliance path or have a closer relationship based on a shared value model. Therefore it leans toward the approach of grinding everybody down via three competitive bids in a buy and systematically negotiating down costs. However, there has been case studies which demonstrate that you can do both. This is where a supply chain organization needs to transform because it can get so much from a buyer and a seller through an alliance model. In this approach both parties understand each other's values and visions, and they get great corporate support easily as there is shared risk and reward but the supply chain can still build competition within that. The supply chain organization can create two or three alliances with a set of suppliers and this way supply chain not only increases competition but also incentivizes the competing companies to work together based on their strengths. This approach needs a little bit more sophisticated management but the benefits is everybody gets well aligned.

11.7 CASE STUDIES

These are two important examples where a supply chain person irrespective of the level can move the needle by delivering tens of millions of dollars of savings. The first example is about developing supply chain understanding: do they really need to make that buy? The cheapest buy you can ever make is the one you don't have to make. This requires understanding the data around the demand side for the supply chain.

11.7.1 Valves that Did Not Need to Be Replaced

In an organization, the maintenance and repair side of operations comes to the procurement and requests for 10,000 units of valves. Typically supply immediately initiates the necessary sourcing activity to procure the valves.

It requests for necessary written approval from its customer and from finance to obtain the necessary budget. As soon as the procurement request and underlying requirements (part number, quantity and deliver location) come in, the buying process starts. But there's a disconnect here between what they're buying and what maintenance really needs. Lot of times, especially with automated procurement systems being in place to communicate between departments, the work flow is also coded into the procurement process. The order for 10,000 units could be the result of a very regimented maintenance log which dictates the requirement of replacing valves every 6 months. This often happens because sometimes we design-in logics and algorithms in our systems which starts dictating the procurement process between organizations in silos. There have been cases where upon investigation it was found that the supply chain is buying valves for a warehouse that's already got too many of them and for replacement of valves that no longer exist.

This type of situation when the supply chain could be buying replacement parts for something that no longer exists happens because the data cleansing has not occurred at the maintenance level. Often to accelerate and automate processes, these work orders are cascaded down between various organizations and in parallel approvals are also generated. However, the former may be executed by one team who is responsible for the job while the approval may be done by another team who is accountable for the process. The separation and lack of communication between the two often results in such a situation. Now here's an opportunity to create value. Supply chain needs to challenge and ask the business to explain where the inventories are going and link the inventory to actual operations needed. Supply chains need to find out what's actually happening in operation. Also, as the supply chain investigates and discovers the correct data of what is actually needed or not needed, it also needs to figure out where the discovered data needs to be stored going forward. The physical data on assets and the tracking mechanism need to come back to complete the loop which got broken. Supply chain too often just looks at their own wedge and not looking at the total life cycle of that part, which by the way, they're accountable for as they go in. They bought the assets for the company which they used shareholder money to buy, they have an obligation to understand the values.

11.7.2 Parts that Did Not Need to Be Produced

We have taken the supply chain and compartmentalized it and specialized it so much that information is segmented and you just know a limited amount only. We have done that because we want to turn human beings into robots but we have neglected to do what human beings do best, which is think, analyze and question.

A supply chain manager manages procurement of valves as a supply chain category. Every year he has to procure valves of different types, say small, medium and large, based on the amount of fluid flow, and for each type, there are valves of different specifications like one with diameter dimensions in the metric system (centimeters and millimeters) and the other in imperial system (inches and foot). The supply chain manager for valves, who is also referred to as category manager, manages a large amount of specification data as part of his role. During the course of work, the category manager is asked by operations to procure 16 different variants of a particular type of valve. These variations were created to optimize engineering solutions needed for specific projects taken over the last 7 years. The category manager managed the suppliers, obtained quotes, measured delivery performance, benchmarked price, tracked quality and managed multiple suppliers to continually source all 16 different types of valves. However, what is often left unaddressed is why do we have 16 different versions to begin with and what's preventing it from getting reduced to three? It is the category manager's responsibility to challenge operations and explain the increase in level of complexity and difficulty in managing and procuring between 3 and 16 variants.

A category manager may continue to procure 16 variants instead of stepping back and questioning and comparing the engineering specifications. It is their job to explain the advantages of standardization and reduction of variants from business and production point of view instead of just perfecting the engineering design. Often the suppliers themselves will propose ideas for consolidation as it is also advantageous for them in terms of manufacturing and providing warranty support (less amount of spares to hold). A category manager should be able to articulate to engineering that having small volumes of many variants, large volumes of a few are always better for getting better price and also better for continuity of supply.

This kind of situation happens because often the supply chain is engaged at the end of the design and procurement process. By the time a category manager is engaged it could be already too late to influence engineering as the solution design time has already been invested. Ideally the supply chain team should be initiated by presenting how the price will trend down based on increasing volume on a few variants instead of many. Instead we can buy what we can design, the direction should be we should design what we can buy and here is the data supporting the approach.

Challenging is not always easy as engineering will often team up and push back very strongly and even belligerently about the importance of uncompromising engineering integrity. Engineering naturally looks for perfection. The classical argument against this is always cost savings and continuity of supply. Money talks and so does an argument laid down by the category manager articulating the amount of savings that could be obtained by changing the design.

It is not at all uncommon for supply chain managers to find cost savings opportunities by discovering millions of dollars' worth of purchased items being unusable and sitting idle on the warehouse shelves depreciating. The part may have become inactive and due to communication gaps are being procured and then having no other option but to be scrapped at pennies to the dollar. Similarly hundreds of millions of dollars could be saved around designing and what is available to buy rather than buying what was designed.

11.8 TRIBAL KNOWLEDGE AND DATA TRANSPARENCY

In the previous section we talked about the broader responsibilities of a supply chain manager as they need to dive deep and explore and validate the real need as opposed to immediately responding to the need communicated to them. However, in order to ask those probing questions and to follow-up and solve cases similar to those mentioned above, how does a supply chain manager do it? Typically this is driven by their curiosity or based on their past experience and tribal knowledge. Often because of their curiosity, a supply chain manager would dive deep and unearth the gap and solve the problem and become a hero. That's one way of doing it but the other way should be a systematic approach. Instead of being solved by a supply chain manager who is a go getter and decides to push the envelope, the supply chain organization should be closer to the parts of the organization where this transparency exists so that the details behind the demand jumps at you and you are more inclined to solve it as part of your natural DNA instead of solving such issues in a situational way.

In order to ensure data transparency, companies cannot hold their data in silos as we need to have cross organization data transparency. It could be access to a data repository or it could be access to groups that are doing data analytic or data science or data amalgamation but it has to be done centrally or has to be done horizontally. In absence of that, the natural trend is people hold on to their own data and they are not set up to share across the organization. To knock it down, a successful supply chain leader often creates centralized data strategies. For example, when the information cuts across the organizational boundaries, the disconnects start to reveal themselves. You may find a large amount of valve already sitting in the warehouse and another set of valve with a different specification may be sitting at a forward buffer location in the maintenance spare pool, and finally at the actual site yet another variant is being used. Then you compare with the quantity and specification you have been given to procure; does it align with different sets of valves across the organization?

Nowadays we can apply data cleansing techniques, anybody can do this now. Any company that is worth their weight would have that but it's got to be integrated as it's all about slicing the data and you do it in a central way, otherwise you will get things that are disconnected. To get actionable information, the data need to trend toward a certain insight. It happens when the supply chain data have been linked with operational and financial data otherwise it does not give you the value you need. Over the last 3 years new techniques have emerged and overall data transparency has taken a massive step forward and cloud has made everything more accessible (Veridian 2020).

11.9 ORGANIZATIONAL LEADERSHIP

This section is dedicated to leadership and demonstration of leadership attributes. This is written from the perspective of a leader in the energy sector and what they look for in someone in the energy sector or in someone looking to break into the energy sector and what people in the sector also view as optimal leadership.

Leadership is not dependent on an individual's style of communication. There are leaders who could be an extrovert or introvert, or an analytical person, or a visionary. The key is authenticity and communicating with a clarity of intent in a convincing manner. A leader needs to ask themselves, can I be very clear with my intention whether that is a vision or a direction or a response. This is consistently required at every level. Communicating effectively the intent is really very important for supply chain professionals because they need to understand why they are challenging and where this fits in the grand scheme. Depending on the audience, they need to be able to see the big picture through the lenses of business, operations or engineering and try to drive accord between them. Leaders need to understand the system or the end-to-end process that they are working in. Within the context of that process, they have to be able to present their ideas with simple and clear communication. They have to explain why it is important for the overall process or intended business outcome. They have to think in an agile manner.

Apart from the clarity of communication there are some basic tenets expected from a supply chain leader. Some of these have no room for error, for example, ethics and treating people inside and outside the organization in the right way. Companies have ethical guidelines but leaders often go above and beyond. Leaders also demonstrate a higher level of resiliency. Mental resilience is very important because by definition as you become a leader, things are going to shift on you and your performance.

11.10 INDIVIDUAL PERFORMANCE

For an individual contributor, their performance is often measured based on two things: their own personal actions and the environment in which they are put in. Some people may perform incredibly well with bad results because the environment wasn't there and vice versa. Managers may want to understand as the environment changes, what does the individual contributor do? How do they act in distress, how do they act in good times or bad, what are their thinking style switches based on environment changes and how does that lead to resiliency? Are they just a person who gets really steadfast in their thinking and unable to adapt? They have to be able to sense, understand what's going on and then be resilient in their mind set.

In summary, the key leadership attributes are clarity of intent and being able to communicate, understanding the intention and communicating it and demonstrating resilience. All the other factors to be able to personally lead depend on the behavioral side of a person and also depend on the team. Second, a young supply chain professional often in their aspiration for professional growth tries to emulate their organizational leader. In this regard it may be important to remember, it is hard to become someone through emulation instead and it's more important to become the best of themselves. The third one is developing cultural sophistication. We live in a global world and any time we make assumptions about anything, it's usually to the detriment of our thinking. Cultural sophistication means that one is aware of having conscious and unconscious biases while doing business in a different country and culture. Cultural sophistication is not knowing but more of developing the pragmatic humility that we do not know and therefore being aware that our biases are really going to hurt us if we are not conscious.

11.11 OPERATIONAL EXCELLENCE

Organizations often go through changes like becoming centralized or becoming decentralized or regional based. They also get organized based on functions and typically people move to matrix organization because of behavioral and cultural problems hoping that the organization chart will change those. Typically, it does not, as you just get into silos in different ways. Businesses often restructure, say from being organized by regions to being organized by functions, but still fail to materialize the expected efficiency. We have to match our organization to our strategy and culture of the organization. In order to do that we need to understand our market, our strategy, our operating model, the business culture and then address how the organization chart is going to take into account these factors. Often when two businesses are combined, a leader may want to quickly create a new organization chart by smashing and combining multiple existing organizational charts. This would fail unless it is done carefully after considering the vision of the new organization with strategy, market, business model and culture.

The very last thing we do is going to put names in boxes. A good organization knows how to operate but a great organization knows how to operate themselves. A great organization knows what they are good at and what they are not good at. It is not possible to change a culture directly but you can change the environment and culture will change but you can't determine what that will be. It's a group of people doing habitual things. Great organizations and great leaders understand their strategy, understand the culture they need and the operating model they want and then everything flows from that and the organization chart comes out last.

For example, to address a market that is highly competitive one needs to be aggressive on price. Having a customer-centric organization is not going to work. A lean operating team with six sigma training who can meticulously control processes to get every penny out is needed for operational excellence. Therefore, the operating model will look much less aligned with a customer outcome but more aligned with the production side. The key person in this organization will be the head of production.

In comparison, a market that's emerging will call for a different strategy. More flexibility will be required which will call for more flexibility in the organization. So the organization chart is going to look different, probably with four or five different leaders who can deal with more ambiguity. A more culturally agile organization is needed which is also customer centric to define and develop the business together. You want to be very close to the customers and what they are doing and how they are changing and you want to be part of their change so therefore an organization will be aligned with the customer segments. They key person here will be the head of all the projects who will act like a general manager (GM) owning the profit and loss (P&L).

However, it is not always possible to be both a customer-centric and a production-efficiency organization. A customer-centric organization is going to actually change your own business to help the customer. On the other hand a cost-sensitive business cannot change the price and expect the customer to adopt. So the point of views are different. Great companies realize that different parts of the market cycle and different parts of their strategy need different operating models and need different types of organization.

For example, new emerging energy businesses such as offshore wind and hydrogen need small start-up types of business. In comparison a mature and low-margin business such as refining has to be lean, cost optimized and automated as it's very competitive. Then there are big oil and gas businesses where a customer-centric model is needed to support them. This demonstrates that even within a single business, three operating models could be required to support three different types of projects with three different strategies.

A leader needs to be familiar with multiple business models and environments. Without that experience it will be hard to be successful, as a leader has to recognize their game plan or their whole approach has to change. It has to fit into the business they are serving for.

This is where resiliency in recognizing the trend in their environment becomes important. Resiliency is not waiting for something to happen to you, resiliency is something that is about to come to me, what am I going to do, good, bad or otherwise and so that's the real key and then be able to see that. For supply chain executives having multiple perspectives is very important. Past experience in running projects, owning operations are some of the key ways to bring in the perspectives from multiple angles. Then it becomes easier for them to wear multiple hats as the situation demands. The clarity of intent and ability to communicate in an authentic manner comes when you have lived it or at least immersed yourself in it.

11.12 CONCLUDING REMARKS

The EPC supply chain has worked too efficiently over decades and has optimized operational cost in the most efficient way to its own detriment. Its profitability has gone down to a very low level. As the world tries to get out of energy poverty while lowering carbon footprint from burning fossil fuel to alternate sources of energy, the responsibility, however, lies on the shoulders of the EPC companies to carry out this transition. In order to respond, the EPCs are innovating in more automation and moving into the digital supply chain.

There has been a shift in perspective in the EPC supply chain. The supply chain is focusing on reducing financial risk and increasing value. The value-based approach which requires strategic thinking from the supply chain leadership ensures better alignment between customers and the EPCs via more symmetric communication, goal alignment and cultural synergy.

ABOUT THE CONTRIBUTOR

Bradley Andrews is a visionary and transformational leader in the digital supply chain arena. He is currently a president at Worley and leads large-scale energy procurement and construction projects globally. He has set up and led numerous commercial strategy and contract negotiation for resources and energy industry projects and operations. Bradley is passionate about reinventing the future of projects and industry through the energy transition, how these industries work with communities to mitigate climate change risk, lead best practice in project delivery and technology advancement and unlock tangible value in the digital transformation. Bradley is a Professional Geophysicist with a Bachelor of Science in Geophysics from the University of Calgary, a Master's Degree in International Business from Haskayne School of Business, and has completed post-graduate study in Executive Leadership in major projects at John Grill Center in the University of Sydney.

REFERENCES

Deloitte. 2020. Future of energy – Deloitte. https://www2.deloitte.com/global/en/pages/energy-and-resources/topics/future-of-energy.html (Accessed December 24, 2021).

Donnely, P. 2020. Business trends: EPC 2030: Five vital characteristics that will define the EPC firm of tomorrow. Hydrocarbon processing. https://www.hydrocarbonprocessing.com/magazine/2020/january-2020/trends-and-resources/business-trends-epc-2030-five-vital-characteristics-that-will-define-the-epc-firm-of-tomorrow (Accessed December 24, 2020).

Dutta, S. 2020. 15 largest EPC companies in the world - Yahoo finance. https://www.yahoo.com/entertainment/15-largest-epc-companies-world-115523006.html (Accessed December 24, 2020).

Fukuizumi, Y. 2020. 3 trends that will transform the energy industry. World economic forum. https://www.weforum.org/agenda/2020/09/3-trends-transform-energy-industry/ (Accessed December 24, 2020).

Jacoby, D. 2012. *Optimal Supply Chain Management in Oil, Gas and Power Generation.* Tulsa: Pennwell Corporation

Martinotti, S., Nolten, J., and Steinsbø, J. 2014. Digitizing oil and gas production - McKinsey. https://www.mckinsey.com/industries/oil-and-gas/our-insights/digitizing-oil-and-gas-production (Accessed December 24, 2020).

One Pull. 2020. 5 problems facing EPC companies - one pull wire. https://onepullwire.com/news/problems-epc-companies/ (Accessed December 24, 2020).

Sarkodie, S., Adams, S. 2020. Electricity access, human development index, governance and income inequality in Sub-Saharan Africa. *Energy Reports.* 6: 455–466. https://www.sciencedirect.com/science/article/pii/S2352484719310443 (Accessed December 26, 2020).

Steinberg, H. 2017. *Understanding and Negotiating EPC Contracts.* Oxon: Routledge.

Veridian. 2020. Data and the supply chain: What is a data warehouse and why is it important in supply chain management? https://veridian.info/data-and-the-supply-chain/ (Accessed December 24, 2020).

12

Driving Automotive Long-Haul Growth

Michael Cupo

CONTENTS

12.1 Objective 263
12.2 Overview of the Automotive Industry 264
12.3 Macro Trends Impacting the Industry 264
    12.3.1 Global Footprint 265
    12.3.2 Service Orientation 265
    12.3.3 Remanufacturing 266
12.4 Supply Chain Management 267
    12.4.1 Products 267
    12.4.2 Forecasting 268
    12.4.3 Product Mix 269
    12.4.4 Internal and External Manufacturing 271
    12.4.5 Supply Base 271
    12.4.6 Price 273
    12.4.7 Spares 273
    12.4.8 Aging Inventory 274
12.5 Organizational Leadership 275
12.6 Individual Performance 276
12.7 Concluding Remarks 277
About the Contributor 278
References 278

12.1 OBJECTIVE

In this chapter we will focus on the supply chain side of industrial automotives. We will focus on heavy-, mid-, and light-duty (HMLD) automotives which also include the high horse power machinery manufacturing industry. This is a mature vertical where the products and prices are quite stable

DOI: 10.4324/9780429273155-15

and typically where the supply chain processes have been well optimized both in terms of cost and execution.

12.2 OVERVIEW OF THE AUTOMOTIVE INDUSTRY

Overall, this segment consists of engine manufacturing for large trucks and also industrial equipment such as tractors and similar large vehicles. There are also other segments in the industry, such as turbochargers, filtration products, and emission systems. Some of the largest companies competing in that space are Cummins, PACCAR which owns popular brands like Kenworth, Peterbilt engine or trucks, as well as Caterpillar, Volvo, Navistar, John Deere, Kubota, etc. Some of these companies make the final vehicle and some of them make essential components such as engines that go into the vehicles and hence are more focused on the power train than the actual build of the vehicle. The power train goes into anything from a bus to on-highway vehicles to tractors, excavators, and even to large generators that can power hospitals.

12.3 MACRO TRENDS IMPACTING THE INDUSTRY

One of the key trends that makes HMLD a challenging environment is its dependence on the economic cycle (Albrahim et al. 2019). The customers of this industry typically look to invest in engines or trucks or tractors based on speculation of economic growth. Unlike an auto dealership where one can walk in to buy a car, a customer cannot just pick up a phone and call a company to get a large excavator and instead has to plan this well ahead of time. So, if the stock market seems to indicate that there's going to be an increase in gas prices or mining of certain types of materials, several months ahead of that, companies need to look at their fleets. They need to evaluate if they have the ability to flex their capacity once this demand begins to increase and get out there and do the mining work that will be needed. That takes months of planning ahead of time and understanding what is happening in the larger environment. However, planning for these machines when the demand arises tends to be very cyclical. Because of that, the second trend among the customers is to do with less when the demand cycle is on the downward trend. When demand is down, a customer who has say a thousand trucks in their fleet, rather than buying the products, may scavenge through those products. They may have downtime on some of their trucks and they will pull parts off of those trucks rather than buy a new engine and make do. But when the trend changes and the market is expected to grow, then the same customer will start making investments to procure infrastructure and machines that can help them grow.

12.3.1 Global Footprint

The HMLD is a global marketplace with a lot of international competitors (Wong 2018). There are some manufacturers who have very strong regional presences like one of the large truck company primary manufactures for the US market. However in general, the manufacturers need to have a supply chain that is global to compete in this market. They need to have not only global manufacturing footprints to keep costs where they need to be but they also need to be able to produce certain types of products for different types of markets. Like in Australia, for example, they have these land trains that have one truck that pulls five or six different containers behind it on the highway which requires a certain type of truck engine that can produce that much power. But then in Europe, for example, due to narrower roads, there is strong demand for much smaller vehicles. So, the global manufacturers need to be able to offer a range of different-sized products that can handle the different types of markets that they are in. In addition, these are big and heavy products, and therefore suppliers try not to transport them any further than they have to. That requires the ability to manufacture the products in that region with as little movement as possible.

12.3.2 Service Orientation

The HMLD industry has a distributed revenue model. There may be a limited amount of up-front profit in the manufacturing of large engines, but most of the revenue that is made comes from the service of products that are sold afterward. So, in terms of supply chain, the supply of spares to support the after sales service is key. Cyclical demand and a long life cycle creates a unique supply constraint which is typical for this industry. It is always going to be very constrained. Just like any supply chain, the inventory needs to be as low as possible to meet a certain level of service level agreement (SLA) requirements. With different product offerings, product offerings based on geography, and different models being created every year, a manufacturer can end up having a massive issue with parts proliferation. They have to stock a lot of different parts to get the after-sales service revenue. Also depending on the type of machines, the product can last for a really long time, such as diesel engines which can run for three decades and have millions of hours of operations put on them. So, the difficulty that has to be dealt with is these parts are very big, they're very expensive, and they have to be planned ahead of the market demand. So, if the customers are planning for a certain amount of time ahead of the economy, the manufacturer has to be prepared to meet that demand otherwise they will lose the competition. So, a manufacturer has to have a lot of flexibility in manufacturing to support this environment and tackle constraints based on how much inventory they can carry.

So, the supply chain organizations for these manufacturers have to accept a certain level of risk. For example, say a company has a 100,000 orderable parts or stock keeping units (SKUs). However, it may not be possible to carry supply for all of those active SKUs, so the supply chain may have to figure out within the business what level of risk to be taken to prevent loss of sales. So, the company has to make a strategy that's going to say the service level to the customer may be 95%. That means the customers' orders would be delivered as soon as they order 95% of the time. This is not an easy task by itself as one has to be able to look at historic order data, understand the life cycles of the products, select what parts are going to help meet that need, and then also make a reasonable strategy to understand what parts are going to be sacrificed that are going to have long returns to service and are not going to impact the customers in such a way that their engines are not going to function and will ultimately hurt their business.

12.3.3 Remanufacturing

As a significant portion of the revenue comes during the long life cycle of the HMLD machineries, and manufacturers are very focused on staying close to their customers and having a long lasting relationship. One way of achieving this is through a process called remanufacturing (Nasr 2020). In this process a manufacturer offers to buy back an engine block or essential part of the machine like the injectors and replace them. For example, if a customer buys a battery at a car part store for $108 and after using it the battery is returned, the customer gets back $8 for the core part. This core part can be reused and that's why the manufacturer buys back the core part. In a similar fashion, an HMLD manufacturer can provide an up-front discount for the core part which is the engine block. This ensures the customer will go back to the manufacturer when the engine needs to be overhauled. This is a win-win situation for both parties as the customer's up-front procurement cost is discounted, plus the repair cost and overall operating cost goes down, and for the manufacturer, they get more revenue by buying back old engine blocks and selling remanufactured engine blocks. Ultimately, the customer effectively ends up paying for the depreciation on the part.

The manufacturers often use this strategy as a bank. For example, say an engine is $50,000 with $10,000 offered as a discount for the engine block which is the core part. After using it for several years, the customer returns the engine block. The manufacturer may allow the customer to take some other part like remanufactured pistons or another injector instead of a remanufactured engineer block if the customer so desires. So, it's like a bank and the manufacturer ends up keeping your customers locked by providing value and instead of having to give the $10,000 as cash they can instead give away a part that costs them less than market value to produce. This approach also allows the manufacturer to compete against brands that are making generic parts or fake parts.

In addition to these there are other trends which are shaping this industry. Electrification of vehicles is driving innovation. There is also a massive push toward lowering emission (Evans et al. 2020) and sustainability (Khadikar et al. 2020).

12.4 SUPPLY CHAIN MANAGEMENT

12.4.1 Products

In the HMLD vertical there are many kinds of engines supporting different kinds of vehicles, different types of power generation tools such as generators and turbines, and filtration and emission systems. While it may be relevant to understand how the products work, for a supply chain manager it is key to be able to understand the forecast trends. In order to do that, it is essential to understand the product life cycles. One also needs to have an understanding of basic supply chain concepts like full length and safety stock levels, basic materials requirement, planning models, and an understanding of lead times (Iyer et al. 2009). These are some of the supply chain aspects that one needs to understand beyond any knowledge of the products themselves. Lot of HMLD manufacturers have very strong vertical integration which means a lot of manufacturing is owned by the manufacturers themselves and not by third-party suppliers. Foundries and manufacturing sites, as well as all of the key major components that go into the engine, are often produced by the companies themselves. They're not farmed out to somebody else, because manufacturers want to protect their proprietary technology as much as possible and keep an advantage over competitors. But at the same time, there are still going to be a whole wide array of different types of suppliers that a supplier manager has to manage. So, one needs to be able to manage internal suppliers, where one has to be able to negotiate internally with their own company and drive their messages to be successful, as well as manage a large number of external suppliers. There are various categories of external suppliers or commodities ranging from metal blocks, electrical boards for power control, filtration, chemicals, fluids, paints, and anything else that's going to go into a vehicle. The supply chain manager has to have the ability to work with each of those different types of components. Typically, the commodities are divided into castings that are dealing with metals; chemicals which deal with lubricants, coolants, etc.; and electrical which deals with power boards, controls, etc.

12.4.2 Forecasting

Typically, there's a lot of forecasting challenges with the products which impacts the forecast of the subcomponents. There are a lot of sources of forecast volatility. It could be due to seasonality, change in Environmental Protection Agency (EPA) standards, etc., but volatility is a big part of the industry.

For example, filtration parts have massive seasonality because a lot of recreation vehicles (RV) customers use their vehicles only in summer. The demand of such items will slow down as the customers use the vehicles less in winter. So, there are some parts where the demand may increase to three or four times in the summer as compared to what the rest of the year's demand is.

There are filtration products which also may be tied to the United States EPA emission standards and other similar worldwide organizations. When a new model comes up which meets a new EPA standard, the demand growth follows the product life cycle and product proliferation. As the model is tied to certain EPA standards, they are going to follow a certain path of growth. Usually, the demand starts low, when the customers are just building up their fleet, but then over time, the population of engines in the market starts to grow, and the replacements start to happen. So, demand may change from say, 50 pieces a month, to all of a sudden a 1000 pieces a month for that specific model. This may happen very quickly, say in 3 months, when thousands of the vehicle model have been sold and are now in need of replacement filters. It's very difficult to predict when that actually happens based on the popularity of the vehicle and shelf life of the filter. Additionally, because this growth goes with a model year, once the demand on a model starts to die out, the demand for the filter dies quickly as well.

There are also use cases where the demand is sporadic due to the proliferation footprint. So, there could be pieces that only get ordered two or three times a year, and the amount of pieces in this category might be three to four, twenty, or even a hundred. So, a supply chain manager needs to understand the different types of forecasting models. For parts which are active and selling steadily, a moving average-based forecast should be sufficient. However, if there is a strong seasonality then specific seasons over multiple years need to be compared. For example, questions arise such as when the seasonal behavior for winter starts and ends, when it peaks, and how does a particular winter peak compare with multiple winters. A supplier manager needs to forecast in such a way that it's covering that demand and needs to be able to determine if the demand is increasing. They may need to use exponential smoothing or similar techniques, otherwise forecasting models are just going to look at upward demand and drive excess inventory instead of smoothing the demand into a decline after the peak. So there are multiple ways to forecast which will help in optimizing the amount of on-hand inventory.

12.4.3 Product Mix

For a supply chain manager, understanding the product mix is very important. Let us look at the product mix from two dimensions: cost and demand volume velocity. Assume there are five different categories of parts, A, B, C, D, and E, with A being the most expensive unit price range and E being the cheapest unit price range. For example, say A is in the $50,000–$100,000 category whereas E is in the $1–$100 price range. For each category to indicate demand volume velocity, there are A1–A5, B1–B5, etc. All the 1 category (A1, B1, C1, D1, E1) parts are high volume and fast selling products and

A5–E5 are slow selling but expensive parts. The supply chain manager typically has to optimize the inventory to minimize the risk of a large inventory cost and balance it against lost revenue opportunity.

This is achieved by optimizing service level for each subcategory depending on price and revenue impact. So, maybe A1 parts are very expensive, but by selling only a few of them, the manufacturer is getting a large amount of revenue. However, an E1 part may also be sold in millions of pieces and fetch a larger amount of revenue overall. So the best way to optimize revenue and risk is to attach different SLAs for each category.

A1 parts are going to be the most expensive inventory but they are moving the fastest and an A2 or A3 part would be something which still has a lot of money associated with it but does not move as fast as an A1. A C1 part on the other hand would not be that expensive for the manufacturer to carry and also moves very fast. For fast moving parts, the inventory turns faster and hence a supply chain manager can set a high SLA to fulfill the demand with the subcategory 1 parts. As the parts become cheaper, like a D1 or E1, the servicer could also put at an even higher level SLA as they are even more inexpensive to carry. A supplier manager optimizes the service level associated with each category to maximize revenue and customer satisfaction. For example, a D1 and E1 revenue opportunity should never be missed and the SLA could be set to a high value like 98% which means 98% of the time the customer demand for such parts will be fulfilled immediately. Whereas for an A1, an SLA of 95% is appropriate as it is expensive to carry and moves quickly, but for an A5 which is very expensive and hard to move, carrying a lot of inventory is not advisable. A risk averse supply chain manager may put a 75% SLA for an A5 which means 25% of the customer will have to wait for their demands to be fulfilled. However, for an E5 due to its inexpensive nature higher amount of stocking could be acceptable. For each category starting from A1 to A2 to B1s and E5, they all have their own service level targets and the aggregate of all those targets should determine the overall service level across all the categories.

A supply chain manager cannot just rely on the system to do this kind of optimization. They need to develop some level of understanding of the demand side such as why parts move the way they do. As someone who is responsible for this, one needs to understand what are the most important parts, who are the customers that buy them, why do they buy them, and what do they use them for. One needs to be able to explain why a part is moving faster than others. Similarly on the supply side, they need to understand their suppliers production capacities as well as ability to carry inventory of finished and raw materials. A supplier manager needs to understand every single thing about these parts and if they just rely on the system's analytics then they end up drawing the wrong conclusion. For example, the system may signal to reduce inventory for an A1 type of part since they are very expensive and that is where all the inventory dollars are currently stuck. However, that may not be the right thing to do because as a supply chain manager, you may know the reason why there is so much supply right now because demand is seasonal and is about to start picking up rapidly and so there should be plenty of supply to prepare. So, one cannot blindly trust the system-based signals and needs to know what is happening to the products and why.

12.4.4 Internal and External Manufacturing

As this industry is quite vertically integrated, it manufactures a big portion of the subcomponents or critical items on its own rather than relying on a third-party supplier. As a result of this, a supplier manager may have to also manage supply from their own factories. This is critical, especially at a manufacturing level. For example, the engine forecast should be tied to a bill of materials where some of the materials are going to be coming from external sources and some of them are going to be coming in from internal sources. And obviously, as the supply chain manager, you could be blocked from optimizing sales by not having engine blocks available to sell. This should theoretically never happen; however, it does due to real world situations and error. A supplier manager needs to understand that they should never be blocked by a tiny component that could have just been overstocked.

This is why it's so critical to understand what parts one has to have per forecast, what parts one can get away with having extra spares for, and which parts one can procure only at lead time. All of these demand streams should eventually be tied to the ability to burn down the supply over time or being able to write off a certain extent in the event demand disappears. However, the forecast that drives the demand should be based on the actual build capacity. So, it doesn't matter if one can procure 100,000 engine blocks if a factory can only build 10 trucks a week. Hence the forecast is bounded by not only customer demand but also based on build capacity.

12.4.5 Supply Base

Like any other industry, the supply base for HMLD vehicles is a mix of suppliers having different levels of engagement. There are a certain number of strategic suppliers at one end of the spectrum and more transactional suppliers on the other end. For categories such as screws, gaskets, and rubber hoses, a supplier manager has to manage supply by providing forecasts. The engagements are low touch and a commodity manager can manage a larger number of suppliers. Usually there are multiple suppliers in such categories so if any particular supplier's supply starts going down, business is usually redirected to other suppliers as these are typically transactional relationships.

On strategic commodities such as electrical components, the categories are critical and expensive. These strategic commodities are more complex and highly engineered and could be more customized. Managing these suppliers requires high touch as a supplier manager has to communicate on a more frequent basis and review forecasts. These suppliers need to understand upcoming demand shifts so that suppliers can react and plan in advance. So a supplier manager has to be in touch with marketing groups, sales groups, engineering groups, etc. and understand the driving factors such as when big sales are coming in. These suppliers need both short and long-term forecast guidance and a supplier manager should be able to provide guidance from one to multiple quarters out. In exchange, the supplier manager should demand a supply plan from suppliers in response to the guidance given and assurance of supply.

An experienced supplier manager typically manages a portfolio of suppliers or a portfolio of categories of products. A large HMLD manufacturer can have hundreds or even thousands of suppliers managed by only a handful of supplier managers. The natural tendency for a supplier manager is to focus on relationships proportional to the spend. However, even suppliers with low spend but with high product proliferation end up demanding a significant portion of a supplier manager's attention. Certain suppliers are also extremely important even if their product is cheap or not as complex but crucial to the entire build process or in high demand. Proliferation happens when the parts provided by a supplier are used in many HMLD products and over multiple generations. This can happen even if the supplier has only a few part numbers or variants but the part is used in many places. Each of the variants need to be stocked in various amounts in various locations, which consumes inventory space and needs to be monitored and managed. One may end up spending the same amount of time trying to fix a problem for a supplier who is supplying a few gaskets or with a supplier who has a wide portfolio of products. It is especially challenging when there is a decade's worth of material for different products that a supplier manager has to manage.

12.4.6 Price

In general pricing in this industry is pretty stable as the end products (HMLD automotive products) do not fluctuate a lot in price. For example, the diesel engine is very mature in terms of its design and development and unlike industries such as the electronics industry, there is no rapid change in terms of capacity, speed, etc. Consequently the price of the parts needed to produce the engine have also stayed quite stable and hence any sudden shift in technology or price is not really expected. However, there are times when the price for a part can increase but that may be primarily driven by scarcity when the part is no longer in production. A supplier manager from an automotive manufacturer may have to work with a supplier to make such kinds of components which are no longer profitable but still need to be produced as part of warranty service to a customer. In such a situation, a supplier manager may have to face an uphill battle to negotiate with a sole-source or a single-source component supplier. The supplier is aware that it is the only supplier and it may increase the price. So as a supplier manager one may have to take a decision to procure that specific part even though it may not be profitable but necessary. Overall when these parts are combined with other components, the business should still be profitable.

Often a supplier supplies multiple parts and one particular part may have a cost increase, but other parts may still have room to negotiate price to offset this increase. When a supplier finds its customer being out of options and tends to increase the prices of unavailable parts, as a supplier manager one has to check whether the price hike is reasonable. The supplier may also incur a cost to restart production of the part solely for a single customer and hence may incur higher set up cost which to some extent justifies the higher margin. There is a fine balance as this could be a situation where the supplier is actually doing a favor to the manufacturer and is diverting its resources from more profitable parts of its business to manufacture the one off part based on a customer's request or it could be an opportunity for a supplier which it is then leveraging to increase its own profits. However, apart from such situations, the prices of automotive parts are not expected to wildly fluctuate and while they do change every quarter, that is more market-driven.

12.4.7 Spares

As the HMLD vehicles are expected to have a long service life, some of the parts inside are expected to be changed in a periodic manner. This makes supply of spare parts a key part of the business which a supply chain organization has to sustain. A significant portion of the revenue from a vehicle comes after it is sold, as parts such as filters, belts, and brake pads need to be changed every so many thousands of hours or miles of operation (Brandt and Springer 2015). There is a cycle for all of such parts that are out there inside the vehicles. There could be a customer who may have say ten thousand of certain types of engines, and every so often, the customer would need a service package of parts. It is a supplier manager's responsibility to ensure supply for these service packages of parts in a cyclical manner. There are generators out in the field which could be even more than 30 years old which need to be sustained. Sometimes procuring such parts could be very challenging if not impossible. HMLD products are designed in a very rugged manner so that they last for a long time, and so finding a part for a 20- or 30-year-old machine that is not even in production anymore is the kind of problem that a supplier manager is going to face. And it's a real challenge as there is typically a huge number of parts spread across a large network and some parts that may not even still be manufactured. A supply chain team for a typical HMLD manufacturer may have around 50,000 part numbers in the system and probably a couple of million pieces in the warehouses to support spares inventory.

12.4.8 Aging Inventory

As a large amount of spares needs to be managed, a challenge which rises out of it is age of the inventory. Though this is common across most industries, due to the long service life of the products and complexity of parts in the automotive industry, a supplier manager has to procure and manage a large inventory to support the parts. Also as the portfolio of HMLD models increases, a supplier manager has to have more and more parts that they have to support and they end up with a growing bulk of inactive inventory that also sits there. Often there could be a situation where there is a large inventory worth millions of dollars but without any demand for it. A supplier manager still has to maintain it because they need to be able to support the demand if and when it comes in. However, a large inventory also means a large amount of cash which is stuck and cannot be used to buy additional inventory. So the amount of on hand but inactive inventory limits the amount of active inventory that can be procured. A supplier manager has to deal with these two opposing forces and make it work. On one hand they have to carry the inventory to support any kind of demand, but on the other hand there could be a strong business goal to reduce the inventory and increase delivery and free up cash.

The challenge for a supply chain manager is to determine whether to write off old and inactive inventory or not to backfill slow moving inventory. For example let us assume, 10 years ago, 100% of the inventory was active and was worth $50M. This means there was clear cyclical demand and as the inventory depleted a supplier manager would backfill the inventory with new inventory. As the inventory got sold, cash became available which was used to procure new inventory. Now, go forward 10 years, 50% of that inventory is active which means $25M worth inventory is sitting there unsold and $25M as well is no longer available to procure new inventory, but in the past 10 years an additional number of parts worth $50M have also become active. So effectively as a supplier manager, one has only $25M to procure the total active portfolio of parts. An additional amount of cash has to be added to take the inventory level to $100M to cover the newly active parts. However, the key point is as the portion of the inventory becomes inactive the amount of cash gets tied to the inactive inventory which results in a decline in cash available to cover an increasing amount of active parts. One option is to write off a portion of the cash by liquidating inactive inventory. In order to minimize that, a supply chain manager needs to work toward maintaining or improving the delivery to deplete active inventory as that is the only portion of inventory which is moving. So a supplier manager has to use their business acumen and balance between which portion of the inventory that has already sold will not be backfilled due to a decline in interest and what portion of the inventory should be refilled and to what extent. Managing cash to buy the right amount of inventory and to shape the inventory amount between aging parts and new parts is a challenge which a HMLD supply chain manager needs to manage.

12.5 ORGANIZATIONAL LEADERSHIP

Typically the HMLD supply chain team is a global network of people. So the leaders of such an organization are responsible for putting together and managing a global network of manufacturing and distribution. They should have the ability to understand contracts and negotiations from the purchasing side. They need to understand the supply chains and what are the relevant issues. A great understanding of the supply chain network is a must as they need to be able to determine where the manufacturing and distribution centers should be located to maximize profit and optimize movement of goods and services. Another key requirement is to be able to quickly root

276 • Becoming a Supply Chain Leader

out the cause of a situation from a diverse set of cross-functional teams and have the ability to draw out real answers from people. A leader is expected to dig into the specific situation and then to effectively lead through the problems and to guide and help people solve problems so that their supply chain organization, in the end, will be effective.

12.6 INDIVIDUAL PERFORMANCE

Companies often provide fringe benefits to their employees. For example, a hotel company may offer a discounted rate for rooms or a tech company may offer free food which sometimes makes these industries more attractive to prospective employers. The HMLD industry on the other hand is typically very frugally responsible and does not typically offer such amenities or fringe benefits. These companies are typically going to be very tight with the money that they spend. However, this industry is also very mature and process oriented where a new college graduate can get in without having a lot of background on the products themselves. It really lends itself if one has a knack for numbers and a knack for the supply chain. The ability to grow quickly and to be able to take on a lot of responsibility that one wouldn't normally gain as quickly is a possibility here. In this segment, a lot of supply chain managers come from diverse educational backgrounds and as long as you have a willingness to learn, like being creative, and are curious, you will be able to adapt quickly. In addition, a knack for data gives a huge ability to move very quickly because you will grow into managing hundreds of thousands of part numbers and similar volumes of data to track and manage in order to run the business. A supply chain manager has to have the ability to learn how to work with large data sets generated from resource planning systems and pivot and extract relevant information to make decisions. Hiring managers may look for someone who can demonstrate that they are interested, curious and creative and have some basic ability to work through problems.

In an interview, a hiring manager can describe a situation and ask the candidate how they would solve it and then observe the candidate's problem-solving skills. A good GPA always helps to get in but developing a business acumen and ability to find data-driven solutions in a complex situation will help a supplier manager to grow. It is also common in the industry to grow upward if one starts in the supply chain as it first gives a good understanding of the business and its revenue challenges. Then as a supplier manager's knowledge of the products increases, it really lends itself to being able to move up and to take on furthermore responsibilities. As a career path it offers the experience that is needed for people who are managing and leading the organization. For any specific manufacturing company, growing organically has an advantage as it is often very hard to find that level of talent and knowledge from other industries. A company often does much better to hire and train people than if it tries to try to get people who have been trained in other companies.

In terms of growth a supplier manager needs to be able to understand forecasts and the supply chain concepts, as well as understanding the suppliers. They need to develop the necessary technical and analytical skills to be able to pull data and put it into a meaningful manner to drive decisions. Next comes the ability to put the decision and supporting data into a presentation and articulate to leadership. So, as a senior supply chain person, one needs to be able to understand the supply chain, work with data and be able to talk to your leadership about what the problems are. At the next level as a supply chain person becomes a manager, it is very important to be able to train the people who are working under you. For that, one has to know their supply chain really well and needs to have tools and teams who would ensure clear flow of validated data which needs to be turned into reports to non-supply chain leadership.

12.7 CONCLUDING REMARKS

The industrial automotive supply chain has to sustain long life cycles of the vehicles and machineries. The sales volume of a particular generation of industrial vehicle model depends on speculation of potential economic growth and that introduces volatility in forecast. Similarly post-sales spare forecast depends on demand materializing from the footprint of the fleet still active. As 50% of the profit margin comes from post-sales support, the manufacturers have become more and more service oriented by providing different SLA-based support plans to its customer base.

The challenge in the automotive supply chain arises from maintaining a high inventory of spare parts due to the large product mix of various generations of vehicles and machineries. This puts the supply chain at excess and obsolescence risk and significant effort is spent to analyze how much to continue holding and how much to liquidate. Liquidation of inventory also requires more investment to build new inventory as new generations of vehicles and machineries get to production. The automotive supply chain is a mature vertical which is very cost focused and frugal; however, it provides a lot of opportunities for professional growth.

ABOUT THE CONTRIBUTOR

Michael Cupo is a senior supply chain leader and has worked in multiple supply chain verticals. His expertise includes material planning, supply chain forecasting and demand planning for automotive and electronics manufacturing. Michael is an expert in developing analytical models for simulations in inventory management and forecasting using statistical methods and Lean principles. Michael is a business graduate from the University of Memphis.

REFERENCES

Albrahim, M., Zahrani, A.A., Arora, A. et al. An overview of key evolutions in the light-duty vehicle sector and their impact on oil demand. *Energy Transit* 3, 81–103 (2019). doi:10.1007/s41825-019-00017-7 (Accessed December 26 2020).

Brandt, F., Springer, M. 2015. The Next Horizon for Automotive after-Sales https://www.oliverwyman.com/content/dam/oliver-wyman/global/en/2015/jul/Oliver-Wyman-41-43-Automotive-Manager-2015-After-sales.pdf (Accessed December 26, 2020).

Evans, D., Evans, D., and Williamson, A. 2020. *The Road to Zero Emissions: The Future of Trucks, Transport and Automotive Industry Supply Chains*. London: Kogan Page.

Iyer, A., Seshadri, S., and Vasher R. 2009. *Toyota Supply Chain Management: A Strategic Approach to Toyota's Renowned System*. New York: McGraw-Hill.

Khadikar, A., Buvat, J., Nath, S. et al. 2020. The Automotive Industry in the Era of Sustainability. https://www.capgemini.com/wp-content/uploads/2020/03/The-Automotive-Industry-in-the-Era-of-Sustainability.pdf (Accessed December 26 2020).

Nasr, N. 2020. *Remanufacturing in the Circular Economy: Operations, Engineering and Logistics*. Hoboken, NJ: John Wiley & Sons.

Wong, W. 2018. *Automotive Global Value Chain: The Rise of Mega Suppliers*. Oxon: Routledge.

13

Managing Hyper Growth in the Cloud

Prasad Sabada

CONTENTS

13.1 Objective 279
13.2 Public and Private Cloud... 280
13.3 Overview of the Cloud Service Providers and the Market..............281
13.4 Challenges of the Cloud Service Providers 282
13.5 Supply Chain Management.. 284
    13.5.1 Forecast and Demand Volatility.. 285
    13.5.2 Suppliers... 287
    13.5.3 Relationship... 288
    13.5.4 Execution Challenges... 290
13.6 Individual Performance..293
13.7 Organizational Leadership.. 294
13.8 Concluding Remarks.. 296
About the Contributor.. 296
References... 296

13.1 OBJECTIVE

In this chapter we will focus on the Cloud computing business and its supply chain. Due to rapid growth of the Internet and social media, there has been tremendous growth in the computer infrastructure and the Cloud business. As the entire world is spending more and more time on-line with their personal digital devices such as phones and tablets, and corporate IT equipment such as laptops and computers, the underlying connectivity and digital data are increasingly being provided by Cloud-based services built on millions of computers, storage devices, and networking equipment. This chapter will attempt to describe the nuances of a Cloud supply chain organization and its underlying opportunities and challenges.

DOI: 10.4324/9780429273155-16

13.2 PUBLIC AND PRIVATE CLOUD

The Cloud computing business is also referred to as hyperscale computing business (Miller 2019). Hyperscale indicates an architecture where more computers can be added to scale up as demand grows. We often hear two common terms: Public Cloud and Private Cloud. Every company stores a lot of data and that data are often stored in a storage software called a database or a database engine. The function of the database server (running that software) is storing data, for that data to be accessible, and provided upon request. Like any other software, the database engine also needs to run on a computer which has adequate storage space to store the data, has the processing power to write the data to the storage device like a hard drive or a solid-state-based storage device, and the ability to fetch data when requested. This is a simplistic view; in reality this can range from a simple server hardware and software running on it to being as complex as clusters of computers distributed globally and connected in a way that they never fail to do two essential functions: writing the data and fetching or reading the data.

In other words, the database software needs to be hosted on a hardware server and the hardware and software combined is often referred to as the database server. There are popular brands of database software companies who sell their software both as a product or as a service. A Fortune 500 company may choose to run its database as a stand-alone product on computers within the four walls of its own enterprise. This is referred to as a Private Cloud as the company has to own both hardware and software, keep them in their premises or own data centers and manage the Private Cloud. Or the same company may decide not to own the computer for hosting the database or any other hardware and software. Instead they can use databases as a service from the database software manufacturer. The manufacturer of the database software may have put up the necessary computer infrastructure to host the database and granted access to the customer to access the databases. In a similar fashion the database software company may host databases for different companies and the databases virtually coexist in a secured manner in the computer infrastructure which consists of myriad of computers. This is an example where the Fortune 500 company is using a Public Cloud to get databases as a service. The Fortune 500 company (that owns the data) and provides to the end-user is choosing between a Public and Private Cloud implementation, or a hybrid using both types of Cloud to make the data accessible to the end user. Well implemented, the end user is oblivious to the implementation choice made by the customer company.

13.3 OVERVIEW OF THE CLOUD SERVICE PROVIDERS AND THE MARKET

According to ZDnet the top five Cloud service providers are Amazon Web Services (AWS), Microsoft Azure, Google Cloud Platform, Alibaba Cloud, and IBM. All these companies are large caps, i.e. with market capitalizations over tens of billions, Cloud services is one part of their overall business (Dignan 2020).

These large-scale Cloud services companies are also referred to as hyperscalers. These are providers of Public Cloud services where they host public company websites, governmental agency websites, e-commerce websites, social media sites, gaming sites, etc. They operate as Cloud service providers for the branded websites. They are said to operate at hyperscale as they have built millions of square feet of data center space and deployed millions of computers that run concurrently consuming hundreds of megawatts of energy. Data centers are the building blocks for the infrastructure and where all these computers reside. These companies primarily provide their clients with compute power, storage capacity, and network to provide web based access. So these companies are also referred to as infrastructure as a service or IAAS companies. Their customers use these computers as hosts and build their own software utilizing the computer and storage (Jamsa 2013).

Major social media companies, e-commerce marketplaces, ride sharing companies, etc. often have the economy of scale to build their own data centers and enjoy more autonomy and profit margins. Most of these companies do not sell computer and storage services, instead they develop a platform and provide customers means to build web applications on these platforms using all the tools that have been provided as part of the service. These companies are known as platform as a service (PAAS) providers. There are companies who go one step even further; they pre-build an application and provide access to the application via the web. So instead of accessing a website, the customers get the experience of using the software through their Internet browser, for example, companies who are offering services to create documents online, storing photos online, or doing spreadsheet based calculations online. These are providers of software as a service or SAAS.

The major IAAS players do not necessarily restrict themselves and have also been offering PAAS and SAAS. Independent of the business model, the fundamental underlying infrastructure of the hyperscalers are the same, which is a massive global footprint of data centers. For example, Google maps its 20+ data centers in its website showing most of them in US and

Europe, 2 in Asia and 1 in South America (Google 2020). There are over 500 data centers globally and the number is growing rapidly.

The number of data centers and their sizes in the US and Europe are expanding rapidly to support global growth of the Internet. With the data centers being the eventual recipients of all the infrastructure items, infrastructure assembly and manufacturing facilities, warehouses, etc. are also growing at the same rate. The manufacturing facilities for final assembly of the server racks and warehousing facilities to hold the finished goods and the subcomponents are built strategically in close proximity to the data center clusters to improve delivery time and minimize logistics cost. Imagine the number of servers and other infrastructures needed when a billion people concurrently search online and everybody gets their answers within milliseconds. Imagine the amount of storage that will be needed if every minute 400 hours of video gets uploaded and then viewed by people all over the world. The number of searches and the number of video uploads are only increasing. In order to support this kind of infrastructure growth, the hyperscalers require a massive supply chain capacity. Another key indicator of this scale is the power consumed by data centers. In 2017, US-based data centers consumed 90 billion KWh and the worldwide data centers (Danilak 2017) consumed more electricity than the entire nation of the United Kingdom. Now imagine how many computers, storage devices, networking gears, and cooling systems are needed to consume that kind of energy.

The current stay-at-home economy that is being perpetuated by the COVID-19 pandemic is only accelerating this trend. To support this overall demand, the hyperscalers spend more than $100B/year on capital expenses (Capex) to sustain and grow the infrastructure.

13.4 CHALLENGES OF THE CLOUD SERVICE PROVIDERS

The Cloud services industry is a relatively young one. Each of the hyperscalers had a different journey to get to the Cloud services space. One of the hyperscalers, for instance has a massive customer base for their productivity suite for e-mail, documentation, presentation, etc. To offer the productivity suites in a SAAS model it had to migrate the suite of software to the Cloud and subsequently the underlying infrastructure needed to be built. Another hyperscaler, needed to build its own global infrastructure to support its global online e-commerce platform. A third hyperscaler, had developed a portfolio of products ranging from productivity tools like e-mail, video sharing, and search engines which required a relatively similar underlying infrastructure globally.

The hyperscalers typically have a strong track record of providing search or e-commerce or other capabilities as a service and typically have a lot of technical innovation at all layers of the software stack. They would essentially offer search or e-mail, chat, storage, or other services for free and monetize through advertising. In order to draw in more customers they typically develop a multitude of products which allows them to understand the end users more intimately. This, however, requires them to house a lot of data which further drives their need for more infrastructure. Innovation also drives unpredictable demand. Some innovations become instantly popular and immediately require massive infrastructure deployment to support meteoric global adoption. Despite the challenges with such growth at scale during huge adoptions, hyperscalersalways strive to be the first to market. These launches are completely different from product launch in a traditional products company based on a relatively predictable ramp plan based on historic data on customer take rate. However, the massive infrastructure also produces the economy of scale where a small cost saving change in design in a server has a multiplicative effect. For example, when a hyperscaler with a plan to purchase a very large footprint of computers makes a small improvement in the hardware design by removing an unnecessary component, it accumulates to a huge dollar value benefit in the total cost of ownership (TCO) of the incoming fleet (Linthicum 2016).

Most of the hyperscalers entered the infrastructure business driven by their own challenges and needs and lack of any externally available solution. During the course of building their own data centers, they faced a completely unsolved set of challenges such as how to design a scalable and global infrastructure where the whole world can demand services such as searching or shopping concurrently. They learnt how to deal with unpredictable demands as customer adoption which does not follow brick and mortar type business models. A relatively unknown application can become a phenomenon overnight, and figuring out how to monetize an application which can draw a huge customer base if it is made available free of cost globally is a large problem. Figuring out how to deal with Moore's law in the electronic industry, where computers double their performance every 2 years and so old computers need to be constantly replaced, is yet another challenge. The list of opportunities goes on, including how do you optimize supply chain cost of all hardware manufacturing and maintenance cost of the hardware so that capital expenditure is optimal, and how do you help deliver software applications to customers through the web. In addition, due to homogeneity of the large-scale footprint, any small efficiency improvement has a large dollar impact and advantage against competition. As the hyperscalers solved these problems for their own business by investing in their own infrastructure in a massive way, the hyperscalers opened up their infrastructure and offered their solutions to others. This gave an opportunity to the hyperscalers to monetize their investment in two ways and design know-how through the Cloud business.

Due to their huge economies of scale, the hyperscalers have become one of the largest customers of their own suppliers. In any industry, when a particular customer becomes the number one customer of its supplier, the customer not only presses to get best possible price, but also presses to influence supplier operations to get more efficiency. For example, it would not be considered a surprise if a hyperscaler reaches deeper into the supply chain, negotiating the price of raw materials and associated overhead cost with their suppliers' supply base. So much so, the supplier is paid only for its value additions, i.e., only its labor and innovation (R&D). In addition, economies of scale and supply chain de-risking mandates could also result in vertical integration at the hyperscaler, in full or in part. It is well known that the hyperscalers do design and manufacture servers, networking equipment and subcomponents (Merritt 2019).

13.5 SUPPLY CHAIN MANAGEMENT

All of the hyperscalers need server products such as computers, networking hardware such as switches and routers, storage products such as media storage, solid-state, and hard disk drive based storage, and infrastructure products such as power and cooling systems, generators, and battery backup units. In each of these categories, they buy various types of products in enormously large quantities. In each of those areas, innovation occurs at different pace which in turn drives demand as they strive to become more efficient by replacing old, relatively inefficient, infrastructure with new infrastructure. Typically each of these categories has its own supply chain team managing end-to-end life cycle for these families of products. However, at the end of the day all these categories are connected, as all these categories of products have to converge inside the data center, the end location

For example, a common issue in data center supply/demand management is availability of infrastructure items such as space and network versus availability of racks of computers. For instance, computer manufacturers face significant challenges if they get stranded with inventory on hand when a the data center runs out of space or network or power capacity and the hyperscaler refuses to accept delivery of server racks. Similarly a constraint in the supply of networking gear may delay a data center expansion even when computers, space, and power are available. So constraints in one supply chain for any of these categories could have a ripple effect resulting in stranding of products and a halt in production in the other categories.

13.5.1 Forecast and Demand Volatility

One of the key challenges is the forecast, as demand tends to be quite volatile. Meeting demand becomes challenging as often the demand spikes suddenly. There are three main drivers behind demand: new features, organic growth, and refresh and expansion. When a web customer releases a new feature, there is an unpredictability in customer adoption. Say an e-commerce company which sells shoes also decides to sell socks. Or say, a hyperscaler offers a new type of storage technology which allows streaming of movies in high definition for its customers. In the first use case of socks as offering, the e-commerce company has to now store pictures of many kinds of socks and suggest matching sets of socks to the customers while they browse for shoes. This would of course require more computing and storage capacities. It has to remember more data about the customer's preferences of colors, patterns, size, inventory, etc. when a customer now buys a pair of socks. Similarly, for the second instance, the hyperscaler has to now build a different kind of network as the difference between ordinary and high definition means more data needing to be sent to the customer and provide a better viewing experience.

As hyperscalers operate globally to support customers who can come from any part of the world, releasing any of these features would mean having a lot of infrastructure ready and in place to launch the feature. Now depending on success, as the feature is being rolled out, more servers and infrastructure will be needed. This kind of global feature roll out creates demand spikes and introduces volatility. A supplier manager needs to understand the demand drivers behind the commodity.

The second source of demand is organic growth. We have heard of days like Cyber Monday which has become a phenomenon much like Black Friday. Black Friday is the day after Thanksgiving holiday in the United States when traditionally people begin shopping for Christmas. It is a huge day for retail business. Cyber Monday which is the Monday after Thanksgiving has become a similar phenomenon for online retailing in the US. On Cyber Monday, the overall Internet traffic at e-commerce sites increases as a people visit the online retailer to purchase items of their choice. As more people go online concurrently it increases the need for organic growth of the infrastructure. Organic growth of demand can have seasonality and is more predictable than the previous demand driver.

The third driver is refresh and expansion. Due to Moore's law computers double their performance every two years and therefore old computers need to be constantly replaced. In addition, the servers and networking gears have subcomponents which start failing in about 3–5 years. Due to both of these reasons, the servers, storages, and networking gears need to be replaced in about 3–5 years. Many web scale companies also financially lease their hardware to reduce operating cost which requires returning the financially leased hardware after 3 years. The new hardware is also typically more power efficient. For example, while 10 computers can support 10 customers simultaneously today, 3 years from now, 7 computers will be able to do the same. So the new computers are 30% more efficient. Now assume the power consumption of old and new computers are both 1 KW. Therefore the 10 old computers consumed 10 KW and the 7 new computers would only consume 7 KW. So one way of looking at it is if the computers are replaced there would be monetary savings due to reduction of 3 KW power consumption. Another way of looking at it would be if we get 10 new computers which would keep the power consumption the same but would now support an organic growth up to 14 customers. Similarly, 7 computers will occupy less space than 10. So refresh drives multiple types of cost savings.

The data centers gain efficiency by refreshing their hardware and a supplier manager needs to understand the return on investment (ROI). However, often a tech refresh is not as simple as rolling out an old server and rolling in a new server. As the old servers have active customers, the refresh requires an expansion. A new infrastructure cluster needs to be built, then customers from the old server needs to be migrated to the new, and only then the old infrastructure can be decommissioned. Though modern infrastructure management software is making this "lift and shift" process more manageable, usually infrastructure refresh takes several months if not years and this causes uncertainty in demand. Also, as mentioned earlier, a constraint in any of the infrastructure can cause delay for the entire expansion process. For example, a constraint in networking gear availability will delay the build out even though the servers and storage gears are available. A supplier manager therefore needs to understand the constraints of other commodities as it impacts their own commodity demand. A certain amount of volatility is also introduced because of constant innovation.

A data center has three independently evolving technical arms: the server, network, and storage. Therefore, it is always a challenge to ensure a particular combination of the three which can interoperate with each other perfectly and that requires rigorous design and testing. This often introduces delays and creates uncertainty in demand fulfillment lead time.

In addition, there is also seasonality, with industry trends like high amount of enterprise migration to the Cloud, geopolitical or other disruptions related dynamics change such as the present pandemic which has led to mass adoption of online education, which creates and influences the demand for the cloud infrastructure.

13.5.2 Suppliers

From the supply standpoint, a Cloud supply chain organization has to deal with different types of suppliers which calls for various types of relationships. There are different product categories across different supply chains each having its own ecosystem and characteristics. However, it is fair to say overall, the semiconductor is the biggest industry that supplies into the computing space. Semiconductor chips such as CPU and memory are not only used to build the computer but also used to build the essential subcomponents such as solid-state or spinning media-based storage drives, and power supplies (Jhonsa 2020).

The semiconductor industry is in a consolidation trend. The semiconductor manufacturers are dealing with complex, expensive, and critical technology investment to maintain a competitive edge in their respective areas. Each of the products require complex hardware and software integration to get the best possible performance. If the performance is not up to par, the manufacturer's particular generation or batch of production can get rejected. However, if the performance and quality are acceptable, the design win at the hyperscaler creates a certain level of stickiness for the semiconductor manufacturer. The hyperscaler will continue to use a particular type of microprocessor, graphical processing unit, storage drive, or memory across multiple versions as it requires them to make heavy engineering investments. Migration from one architecture to another is expensive, so any engineering design decision tends to lock a product to its particular generation of components (Kannan and Thomas 2018, Trendforce 2019).

There is also a lot of customization via application-specific integrated circuits (ASIC) or customized semiconductor chips. The ASIC market is heavily consolidated as there are only few players, especially the high-end ASICs used by hpyerscalers. As a result,a hyperscaler may choose to start designing and building ASICs in-house. This vertical integration typically moves the supply chain continuity to the next level, which means ensuring capacity at the chip fabricator level. Since the hyperscalers operate at a massive scale, they may now have to negotiate with foundries to ensure capacity at semiconductor wafer level to get their demand met. All these examples of consolidation occurring in the semiconductor space implies that hyperscalers are heavily dependent on a handful of suppliers for the critical technology areas and do not have too many alternates to choose from. Other than semiconductors, the compute infrastructure also requires fiber optics supply which is also consolidating but to a lesser extent and still somewhat fragmented. Other than the electronics parts there are mechanical parts like the enclosures, wires, cables and connectors, etc., also characterized by few big players in each category who serve the entire IT industry including hyperscalers. The same story is true with the printed circuit board (PCB) where while there are more players, the supply sources becomes fewwhen it comes to building multilayer PCBs or with specific substrate material for faster and more powerful computers.

13.5.3 Relationship

Due to architectural stickiness as described earlier, relationships with suppliers span over multiple generations of products. In other words a hyperscaler may buy products spanning over multiple generations of subcomponents to support its own multiple generations of platforms. The platforms as well as the subcomponents could be in various stages of their lives. If both the hyperscaler and the manufacturer does it right, then both the platform and underlying subcomponent should be coming to market and going to end of life synchronously. However, that is rarely the case and impossible to do with so many critical semiconductor-based subcomponents. There is always a lag due to significant design, testing and qualification time required by a hyperscaler to design in or adopt a particular generation of subcomponents. This leads to an offset where a hyperscaler is ramping up but the manufacturer wants to wane off one of the underlying components.

Relationships and volume of business play a big role in decision to extend production life of the component so as to sustain the platform at a reasonable cost. A Cloud supply chain manager needs to manage across this multigenerational relationship. It is also important for them to assess if the supplier has the necessary deep pockets to continue with Capex investments that build manufacturing capacity, reflecting their commitment to the space (O'Brien 2014).

Another variable in relationship management is to assess the suppliers' innovation capabilities and whether that matches with the DNA of the hyperscaler. Since the suppliers are limited and strategic, most of the supplier relationships in this space are long-term partnership-based engagement. Although there could be some areas like PCB or low-cost components which are mainstream where pricing and availability are more important, whenever the design becomes more complex, for example, the situation demands a more complex PCB, a Cloud supplier manager has to build a relationship with suppliers who are willing and capable of handling the design complexity and have the capacity to do so. Due to massive scale, the hyperscalers often become customers who control significant revenue of their suppliers which necessitates the Cloud manager to be fair to the supplier and keep an the eye on the supplier's financial health. However, at the same time, a supplier manager needs to drive efficiency, especially with pricing.

Hyperscalers are also very innovation driven and often come up with the next coolest idea which would give them a quantum leap, and deliver improvement in TCO. However, they want their ideas protected. During this process, the supplier manager has to work with the suppliers to make them develop certain capabilities to produce this next coolest product. Developing capabilities is a critical part of managing supplier relationships. Often suppliers may not have the key technological capability or background that the hyperscaler is looking for in the product they need. At the same note, it may not make sense for the hyperscaler to develop it in-house as it may not align with their core competence or it has to be embedded and be an integral part of the supplier's product. In such a situation a hyperscaler may consider it to be critical enough to invest and get it developed at a supplier. In other words the hyperscaler asks the supplier to learn something new and add to the product. Even though it creates a future dependency on the supplier for supply, this may put an upfront cost burden on the supplier. The supplier has to be able to codevelop the product, come up with creative ways to finance and amortize the development cost, and recover the development investment later during the lifetime of the product. Quite often, the supply base members for hyperscalers are expected to have more of these engagements which also deepen the relationship.

The microprocessor space is heavily consolidated with two main ×86 architecture-based suppliers. ×86 is a microprocessor architecture which is dominant and covers almost the entire server market. Companies have attempted to enter but often have not continued due to the high barrier of entry as it relates to the microprocessor space (Morgan 2020). There is a similar high barrier of entry in the artificial intelligence (AI) and machine learning (ML) spaces. As AI is transforming the industry slowly, it is becoming an integral part of the offerings in enterprise Cloud services. The AI/MS has a high dependency on the software framework which runs on the hardware. Different hyperscalers are trying to lock their customers with the specific architecture of hardware and software framework. As a result of this a Cloud supply manager in all likelihood has to maintain a long-term relationship with a fixed set of hardware and software providers which are strategically chosen due to the hyperscalers architectural choice. In this space, graphics processing unit (GPU) manufacturers are having a leg up as current x86-based devices do not have the architectural advantage to be able to handle AI/ML workloads well in comparison and found to be not customer's preferred choice yet. This is perhaps one area where in spite of a high barrier of entry, a lot of venture capital investment has been pumped in. This has launched a lot of new start-ups, and some of them will emerge as potential suppliers who will join the hyperscalers supply base in future and may compete or replace existing suppliers.

Maintaining technological advantage is a key concern of a hyperscaler. A supplier manager has to consider a make versus buy option to protect this advantage especially when building a new relationship with a start-up instead of developing a technology in-house. Supplier managers have to be able to assess the risks involved in both in-house versus buy from a start-up. For make, the opportunity cost for the engineering resources need to be calculated as the engineering resources should put in efforts which will produce the maximum benefit either monetarily or technologically. A supplier manager has to consider all the variables that need to be evaluated when looking at start-ups.

13.5.4 Execution Challenges

Overall there are a spectrum of suppliers, and the size of the business plays a key role in managing the relationship and getting attention. From a market standpoint, the hyperscale vertical is growing rapidly and the current state of economy is providing additional tailwinds.

For the hyperscale business there is also a high barrier of entry as scale really matters in this space. To become a Cloud service provider from scratch, one will be hard pressed to build their own Cloud service infrastructure. More and more companies are likely to partner with one of the major hyperscalers, especially in the early stages of their companies. For a new entrant, they have to have a preexisting scale and the critical mass in terms of their supply capabilities and the willingness to take the challenge.

The business dynamics in the Cloud infrastructure vertical often lead to a few key execution challenges. Challenges may come from complexities arising from dealing with multiple types of supply chains or due to the introduction of untested partners to satisfy innovation pressure. Problems may also arise from the launching of new services without any historical pattern to rely on for demand forecasting, etc.

Due to the complexity of dealing with multiple types of supply chains, concurrent flawless execution is critical without which inventory can build up in one part of the supply chain while waiting on other parts to catch up (Sanders 2019). To support the unpredicted surges in demand, a massive amount of coordination is needed across Cloud infrastructure supply chains which includes space (building), power, cooling, and the IT hardware supply chains. Managing with such high volumes of inventory and dealing with how real-time consumption at the data center will be forecasted is critical. This is probably the biggest challenge that exists in this space. Hyper growth or sudden lack thereof can swing the problem in either direction. It can exacerbate the problem either on the physical infrastructure side which could be ready and taking in megawatts of stranded power while waiting for design of the network and computer to finish or it can swing toward the IT side where the servers cannot be installed and cannot support the demand as space or power is not available.

The suppliers also become victim to unpredictability or downstream execution of the Cloud providers. For example, the semiconductor segment of the Cloud supply chain consists of more mature suppliers. Typically they look at the macroeconomic trends, macro industry trends, and the average need across multiple industries to project their capacity requirements are likely to be across multiple years. Based on that they have a multiyear investment model for their foundries which they use for longer term planning. These investment plans to build capacity or to ramp-up a technology transition from one generation to the next are often done based on averaging the demands over a time period. They often fail to meet the demand surges from the hyperscaler. Due to a single hyperscaler's massive demand, a surge can move an entire supply chain into shortage and the whole industry can get constrained (Clarke 2019).

A parallel can be drawn to a movie released on the Internet. A relatively unknown movie released on the Internet may have a few thousand concurrent viewers in the first few days. As the movie becomes more popular, the movie starts trending toward the top of the movie watchers search list or recommendation list and more people start viewing it. The number of concurrent users start increasing to tens of thousands as the more viewers start to come from different parts of the globe. If we look under the hood, the backend infrastructure also starts scaling in parallel to support the demand. To support the increase in the number of concurrent viewers, more computer servers need to be able to stream the movie. In addition, in order to provide a uniform viewing experience, more servers in different geographies need to be spun up to support local viewers. So in the first few days the movie could have been streamed from say a few servers from one US-based data center but within a week the movie on its way to become an Internet block buster may demand hundreds of servers in data centers spread across different parts of the globe. This can happen for movies, games, and various kinds of apps released through the Internet, news, and social media. Due to this extreme velocity between launching to erupting into a global phenomenon, the hyper growth in the underlying infrastructure has the ability to overwhelm the Cloud.

In addition to these issues, challenges also come from internal processes. Cloud service providers are relatively young and often the scaling happens too fast. This essentially means that supply chain capability upgrades may have not kept up with what the business requires. Being always in hyperscale mode, all the resource and energy have been spent in firefighting and chasing point problems, and so supply chain IT systems and tools, manufacturing resource planning (MRP) systems, etc. are always playing catch up. Typically there is a lot of human glue, manual processes, and "dive and catches" to keep the supply chain running. Of course when people are holding the overall system together, the system works at human speed instead of at the speed of light. Essentially this can create a drag on the Cloud supply chain execution velocity. Cloud providers typically have their roots in software development and often underestimate the power of industry standard MRP tools and systems of records. When it comes to making in-house versus buying tools, they tend to be overly self-reliant on home-brewn tools. These tools may not be at industrial strength, and so they could struggle by depending on people resolving issues to keep the business running.

However, there is also a silver lining with this strategy. As the infrastructure demand is all internal, i.e. inside your data centers, it is not directly catering to the broad set of end customers across a variety of industries. Cloud supply chains have a large appetite to take risk and if one part of the supply chain gets stranded, they can redeploy unused computing capacity from other data centers in a second as it is all virtual and transparent to customers. One of the additional advantages is products are also released silently. Usually customers are not even aware when a new feature in an e-commerce platform gets released. The user experience continues to get better. So, in other words, a hyperscaler can launch a product and iterate as many times as they want instead of spending a lot of hours to make the product squeaky clean from day one, because at the end of the day, they are optimizing the TCO. For example, they are not necessarily looking to hit 99%+ quality with their first release and the products can improve while in production.

13.6 INDIVIDUAL PERFORMANCE

The primary measure for a supply chain professional's performance in Cloud supply is their ability to continuously learn and deal with ambiguity as there is no cookie cutter role in a Cloud supply chain organization. Every season a new challenge will come up from either external or internal sources. It could be an external market constraint or an external product transition but it could also be an internal software or hardware delay or an internal supply constraint from another category. The challenges are endless and of course due to volatility of demand, the forecast data is not reliable. How does one grow in their career in this industry? If one wants to go up the ladder, one needs to show that they are operating at the next level, and that is how most of the companies typically tend to look at someone who is ready for promotion. If this person is operating at the next level then the person deserves to work at the next level.

From a knowledge standpoint, if a person is handling projects with the right kind of complexity or degree of difficulty, and demonstrating the right kind of leadership to make the products successful, they will be noticed. Especially when the products are impactful and they play a pivotal role in the making of the overall product. As one grows professionally, leadership becomes more important because it's not just about what one does, but it's about how one makes the team better including sharing of knowledge. A leader is expected to mentor people, call out gaps, and articulate what needs to be done to fill said gaps.

Collaborators often get noticed, especially those who keep their leadership informed. People with a can do attitude are often given responsibilities that their superiors think that they can deliver. People who are willing to do grunt work and ready to do heavy lifting and not just focus on shiny new projects for visibility are respected especially when issues are ambiguous and resources are stretched thin due to hyper growth. The ability to support one's opinions or point of views with data and the ability to do data mining and analytics will always enable a supply chain manager and their peers to be successful. The ability to build tools whether via writing queries or putting things in spreadsheets is always recognized by leaders.

13.7 ORGANIZATIONAL LEADERSHIP

From an organizational standpoint, perhaps the most important thing which makes an organization better is having the right culture. That's the most important thing. The success of a well-run supply chain is measured by its success in demand fulfillment, carrying reasonable amount of buffering, sourcing at best in class cost, maintaining product quality at acceptable levels, an appropriate level of risk taking, assessing risk and balance against the opportunity cost, enabling innovation, having the right level of tooling, etc. Given the short history of the Cloud supply chain companies, a lot of these attributes are often in various states of maturity and need to be driven.

A supply chain organization needs to be in continuous improvement mode to attain some of these capabilities. To achieve supply chain excellence in all these different areas, an organization needs to develop competence in collaboration across all the different functional areas. Often sourcing, planning, manufacturing, etc. get into their individual silos as they are organized and expected to execute like a stack of chevrons. However, in order to drive operational excellence there needs to be a tremendous amount of collaboration, especially in areas which have not stepped up to the scale of the hyper growth business yet. Given the complexity of coordinating multiple supply chains operating at disparate speed and lack of any single tool which can effectively help with coordination across the supply chains, there has to be a culture of transparency and collaboration. There needs to be a team that measures success based on success of the project and the success of the team rather than individual success. Operational excellence is incumbent on everyone even though it places a bigger burden on execs.

Because of dealing with the global footprint, Cloud service providers have to deal with communication across geographical boundaries, time zones, and perhaps even across the familiarity and comfort of English or other languages. Communication in all shapes and forms is therefore of paramount importance and it has a major bearing on how an organization is built and the culture it develops. The organization is also shaped by the talent it hires. If it is recruiting an individual contributor, a manager, or an executive it needs to bring in talent with the right domain expertise based on the need. However, it also needs to ensure onboarding people who have the proven ability to learn. They may be lacking in one area but one cannot compromise in their ability to keep learning.

The leaders such as managers and executives need to be recruited based on who can demonstrate leadership skills which are complementing the existing leadership teams. In other words, bringing in people from different environments or maybe even from different industries often helps with infusing new ideas. While supply chain skills are transferable across different industry verticals, having an open mind to brainstorm without being biased based on past experience is often a recipe for success in an environment which is continually evolving.

A manager is closer to overseeing execution and delivery based on the key metrics for their function-specific teams. An executive in comparison is more agnostic to the function or functional groups that they are leading and delivers to broader supply chain metrics to make the overall supply chain successful. A leader whether a manager or an executive is responsible for putting the right incentive in the right place to have the right culture. They create a culture by what type of behavior they reward and what kind of people they are promoting. They gain trust by being fair in terms of how impacts are being assessed. An executive is expected to be able to look around the corner, anticipate the future, and drive supply chain excellence projects.

A leader also builds for the future and hires based on future needs and is able to develop talent and keep the talent pool motivated and excited. A leader provides psychological safety for their team for its success. Failure comes hand in hand with taking risks and it is a leader's responsibility to create a safe environment where it is ok to fail as long as the team takes a reasonable amount of risk and learns from their failure. A leader is expected to set clear goals and make responsibilities clear and align the team to broader organizational metrics. They also very importantly need to promote equitability and create an inclusive improvement.

13.8 CONCLUDING REMARKS

The Cloud supply chain is in hyper growth mode which empowers its supply chain to manage massive amounts of spend with a diverse set of suppliers ranging from semiconductor to generators and building construction. This also makes the overall lead time management quite complex as they are interdependent and a shortage in any particular category can create a cascading effect.

Volume of business plays a big role in managing relationships. The world of Cloud is extremely innovative which is also expected from its supply base. It's constantly driving innovation to leverage the economy of scale from its growing global footprint. For the web scale Cloud providers, lost opportunity cost is unacceptable as their inability to provide infrastructures prevent their customer from scaling for growth. To minimize lost opportunity cost, the volatility risk in forecast is often mitigated through appropriate buffering strategy by the Cloud infrastructure providers so that the overall supply lead time for their customers stays manageable.

ABOUT THE CONTRIBUTOR

Prasad Sabada is a Global Operations Executive with expertise in scaling cloud infrastructure business into hyper-growth phases. He has set up and led numerous operations and supply chain teams for delivering data center and IT hardware businesses. He is an expert in leading sourcing and supply chain management teams globally. Prasad has an MBA from UC Berkeley and is a graduate from the Indian Institute of Technology.

REFERENCES

Clarke, P. 2019. Industry is putting the brakes on memory Capex. EE News. https://www.eenewsanalog.com/news/industry-putting-brakes-memory-capex (Accessed December 26 2020).

Compuforum 2019: The Big Future of Data Economics—TrendForce Takeaways. Trendforce. https://www.trendforce.com/presscenter/news/20190531-10142.html (Accessed December 23, 2020).

Dignan, L., 2020. Top Cloud providers in 2020: AWS, Microsoft Azure, and Google Cloud, hybrid, SaaS players. Part of a ZDNet Special Feature: Managing the MultiCloud. https://www.zdnet.com/article/the-top-Cloud-providers-of-2020-aws-microsoft-azure-google-Cloud-hybrid-saas/ (Accessed December 26, 2020).

Jamsa, K. 2013: *Cloud Computing: SaaS, PaaS, IaaS, Virtualization, Business Models, Mobile, Security and More*. Burlington, MA: Jones & Bartlett Learning.

Jhonsa, E. 2020. Cloud Capex looks poised to grow strongly in 2021. The Street. https://www. thestreet.com/investing/Cloud-capex-poised-to-grow-strongly-in-2021-intel-micron-western-digital (Accessed December 24, 2020).

Danilak, R. 2017. Why Energy Is A Big And Rapidly Growing Problem For Data Centers. Forbes Technology Counis. https://www.forbes.com/sites/forbestechcouncil/2017/12/15/why-energy-is-a-big-and-rapidly-growing-problem-for-data-centers/?sh=3e3873325a30 (Accessed April 20, 2021)

Google. 2020. Discover our data centers. https://www.google.com/about/datacenters/locations/ (Accessed May 31, 2021).

Kannan, H., and Thomas, C. 2018. How high-tech suppliers are responding to the hyperscaler opportunity. Mckinsey. https://www.mckinsey.com/industries/technology-media-and-telecommunications/our-insights/how-high-tech-suppliers-are-responding-to-the-hyperscaler-opportunity (Accessed December 23, 2020).

Linthicum, D. 2016. Cloud computing's elusive total cost of ownership. https://www. Cloudtp.com/doppler/Cloud-computings-elusive-total-cost-ownership/ (Accessed December 24, 2020).

Merritt, R. 2019. The whales changing the industry. EE Times Asia https://www.eetasia.com/ the-whales-changing-the-industry/ (Accessed December 24, 2020).

Miller, R. 2019. Who are the data center's industry's hyperscale players? Data center Frontier. https://datacenterfrontier.com/data-centers-industry-hyperscale-players/ (Accessed December 24, 2020).

Morgan, T. 2020. Throwing down the gauntlet to CPU incumbents. The Next Platform. https://www.nextplatform.com/2020/02/11/throwing-down-the-gauntlet-to-cpu-incumbents/ (Accessed December 24, 2020).

O'Brien, J. 2014. *Supplier Relationship Management: Unlocking the Hidden Value in Your Supply Base*. London: Kogan Page.

Sanders, J. 2019. DRAM chip prices plummet due to Intel CPU shortage, prompting retail sales. Techrepublic. https://www.techrepublic.com/article/dram-chip-prices-plummet-due-to-intel-cpu-shortage-prompting-retail-sales/ (Accessed December 24, 2020).

14

The Logistics of Cooperative Supply Chain in Dairy

Andrew Bawden

CONTENTS

14.1 Objective 299
14.2 Overview of Global Dairy Market 300
    14.2.1 New Zealand 300
14.3 Overview of Dairy Product Producers 301
14.4 Macro Trends Impacting the Industry 301
    14.4.1 Economic Trends 302
14.5 Supply Chain 303
    14.5.1 Global Producers 303
    14.5.2 Seasonality 303
    14.5.3 Logistics 304
    14.5.4 Product 304
    14.5.5 Cooperative 305
14.6 Case Studies: Ultra-Heat Treated Product 306
14.7 Individual Performance 307
14.8 Organizational Leadership 308
14.9 Concluding Remarks 309
About the Contributor 309
References 309

14.1 OBJECTIVE

This chapter will focus on the nutrition industry, specifically the dairy industry. Dairy is a global industry and is unique as the top producers are not identified as corporations but as nations. New Zealand is one nation which is one of the world's top exporters of milk and other dairy products.

DOI: 10.4324/9780429273155-17

This chapter will focus on global consumption trends and how this industry is evolving. Unlike other chapters, we will focus on a business model where the suppliers are also owners.

14.2 OVERVIEW OF GLOBAL DAIRY MARKET

Most dairy products in the world are consumed within the region or country in which they are produced. Due to the highly perishable nature of milk, it is consumed or processed very close to where it is produced. A large quantity of trade, in general, is informal and through short supply chains, i.e. procured through local producers. Even in case of formal trade, most of the quantity of milk does not cross international borders. So, the global trading of dairy is done by countries based on the surplus as compared to the country's required needs. Countries typically serve their home market first and then look to export the surplus. Milk is viewed as a relatively affordable source of important nutrients and proteins in the diet by most communities. There is also protectionist legislation and food security which influences this market.

The share of globally traded dairy products is about 8% of the overall production. As the total volume of international trade in dairy is small, it is considered to be a thin market. As a result, it is kind of susceptible to imbalances in supply and demand which can easily disrupt the market.

14.2.1 New Zealand

When it comes to international trade, the top four "regions" cover 70%–80% of the total dairy products that are globally traded. Out of which, New Zealand is the equal largest (along with the entire EU), accounting for 30% of the global export market as of 2020. So, for New Zealand, in quite a stark contrast to other milk-producing countries, less than 4% of its milk produced is consumed within the country. Around 95% of New Zealand's dairy production is actually exported. New Zealand, with a relatively small population of under 5 million, produces more than 21 billion liters of milk per year. Its customer client base is therefore predominantly offshore and mostly, but not exclusively, includes countries that are not self-sufficient in their dairy needs. The global dairy production has increased around 60% from 2000 to 2020. A number of factors are behind this including rise of middle class in third-world countries such as India, China and Brazil, which raises the affordability and therefore the demand of nutrition products such as dairy.

So, total production grew from 559 million tons in 2000 to 831 million tons in 2017. As of 2020, the top-producing regions were Asia at 31%, European Union (EU) at 21%, and North & Central America at 15%. In terms of specific countries, the largest milk producers continue to be India at 141 million tons and the US at 93 million tons. So, New Zealand has a milk production of 21.3 million tons and accounts for just 3% of global milk production but is a larger player in terms of total export volume (Shadbolt and Apparao 2016).

14.3 OVERVIEW OF DAIRY PRODUCT PRODUCERS

In countries that have large populations and large production volumes as well, such as India, dairy production is often very localized and very regional (GalketiAratchilage and Marocco 2020). It is almost direct from farm to consumers. This is what causes the informal dairy trade. On the other hand, New Zealand is very much on the formal side as it exports 95% of its production. The formal dairy market is extremely fragmented as there are many regional producers. From a revenue point of view, the leading producers of dairy products globally are Nestle, Lactalis, Danone, Fonterra, Friesland Campina, etc. These companies compete with each other for retail shelf space in certain markets as well as forge strategic alliances in certain markets and even supply each other with raw materials. Often the milk producers are cooperatives. For instance Fonterra is a cooperative and is owned by the farmers. Cooperatives are committed to procure all of the milk produced by its farmer-owners. Some of the dairy product companies have their own branded products and have positioned themselves as branded companies. Some of them are more likely to be the provider of raw materials to other companies and have positioned themselves as an ingredient company. There have been significant mergers, acquisitions, and strategic alliances in this market as the dairy product manufacturers are jockeying for positions in the growth of global dairy trades.

14.4 MACRO TRENDS IMPACTING THE INDUSTRY

Manufacturers get to their customers through primarily three channels in the formal market. The primary channel is via selling through a number of food retailers by putting branded products on shelf space. The second channel is by supplying to the foodservice industry and the third is by serving the food processing industry.

Looking at dairy consumption by region, Asia and Europe are the largest dairy consumption regions, accounting for 43% and 26% of the world dairy consumption, respectively. In terms of global consumption trends, the Asian market, in particular, is quite interesting and plays a significant role as the largest customer base.

In Asia, rising per-capita consumption levels and growth in wealth have been driving a strong growth in demand for dairy products. This is driving both domestic supply and import as well. The growing middle class is driving the need for more high-quality protein. As the middle class emerges, it is seeking different food options, and potentially more westernized food options, such as pizzas, coffee, etc., which have plenty of dairy products in them. With growing wealth, the market is driving the need for more high-quality protein, which is fueling the overall growth of the dairy industry (Benham 2020).

14.4.1 Economic Trends

Of the overall amount of dairy products that are traded through the formal trade process, the products can be classified into several categories. Liquid milk accounts for almost a third of the formal trade and is the largest segment, closely followed by butter and cheese. Milk powder and skimmed milk powder have a much smaller share but still are equally important for longer shelf life. With the growth in the global economy and particularly with the growing Asian economy, it's expected that the share of milk that passes through the formal trade will increase in the future.

Europe and US markets are both quite mature but also a bit different in terms of being predominantly more self-sufficient. The dairy that is produced is also consumed within that region, and so they're not as reliant on importing dairy. Asia certainly is a focus in terms of a region where there is a gap. In 2014, Asia had 38.4% share of world production and a 42.9% share of world consumption. In other words it had a self-sufficiency rating of 90%. They were the only region actually to report a decrease in self-sufficiency. So that suggests that there's a growing population which the local or regional market is unable to keep up with and that requires the need for more imports (Shadbolt and Apparao 2016).

14.5 SUPPLY CHAIN

14.5.1 Global Producers

From a food supply chain perspective the dairy product manufacturers are reliant on exporting and traders. Dairy is one of the most regulated agricultural products, and suppliers have to predict where the demand is going to come from among all the regions (Wallace et al 2018). There is competition from the regional manufacturers and hence there are a lot of region-specific dynamics including different regions requiring different products. For example, Russia is one of the largest importers of cheese and butter oil. China, on the other, hand is one of the largest importers of whole milk powder, skim milk powder, etc. There are also other regions which tend toward certain product combinations over others. So, manufacturers have to really keep their eyes on the global and regional trends as compared to supply and inventory levels, and then really tailor their production to meet these requirements.

14.5.2 Seasonality

From a supply perspective there are also certain unique challenges. For example, New Zealand's production has a strong seasonal influence which is known as the milk curve. The uniqueness of the New Zealand dairy supply is that it's predominantly pasture based and cows are free range and not housed. So the production amount changes with the climate. It peaks during spring which is between the months of September to November when pastures have maximum vegetation. The production continues through the summer months of December and January but starts tailing off as New Zealand goes into winter. In the winter the production gets quite close to zero. Only a few farms actually continue to produce milk through the winter months. This poses a unique capacity challenge.

The manufacturing facilities are built to deal with the maximum capacity which happens only during spring. Therefore, while built to handle the peak capacity, for most of the year the manufacturing sites operate below their full capacity. The milk production curve creates an uneven production volume which the manufacturers have to plan ahead for and manage in order to meet the regional demands at different parts of the world. In addition to optimizing the output volume, there is also fluctuation in the dairy prices which needs to be accounted for. The supplier managers need to understand the demand forecast and price forecasts and balance that with the milk curve as the factory output changes with the season. Since dairy is a predominantly perishable product, a lot of the production during the peak period has to be turned into powders such as whole milk powder and skim milk powder which are long shelf-life products (Dani 2015).

14.5.3 Logistics

On the shorter shelf-life products such as different kinds of butter and cheeses which require refrigeration, the supply chain becomes more specialized in terms of logistics (Kroll 2015). They require a specialized supply chain of refrigerated containers and cold storage warehouses both where they are produced in New Zealand and also in the countries where they are exported and sold.

Milk as a raw material does not really travel far, whereas the finished products are actually shipped out globally. The factories in New Zealand are essentially aligned to supply as best as possible for the farms where the milk is picked up but even that has its challenges. New Zealand has rugged terrain where the farms are located. Since the dairy industry is full of cooperatives, however isolated the farms are, the cooperative is committed to picking up every farmer's milk and processing it.

Managing end-to-end refrigeration from procurement to processing to production and then transportation and warehousing is challenging and all has to meet stringent regulatory standards. The supplier manager has to understand the complexity of the logistics as well as the regulations and requirements of exporting dairy products internationally.

14.5.4 Product

The main ingredient for all the dairy products is of course the raw milk. There are a lot of complex manufacturing processes to process and upgrade that milk into a range of different products like cheese, powdered milk, butter, etc. The next important item of procurement is the packaging. There are not many other ingredients anywhere close to the volume of milk that are need processing. Some dairy product manufacturers require other items like cocoa or other secondary products to create the end products such as chocolate or other milk-based products;

however, these needs are second at this stage to packaging and creating the finished product.

14.5.5 Cooperative

As some of the manufacturers are cooperatives, the suppliers of milk do have that stake in the business to balance between supply and maintaining the price. The local dairy farmers have to keep the price of the milk viable for the end product to be competitive on the global market. So in a way the dairy farmers expose themselves to be impacted by the global price of dairy. The price paid to the farmers are not really negotiated like that between a customer and supplier as they are the same. So the manufacturer looks at the base commodity prices that are being traded around the world and then forms a view of what the milk price or the supplier price will be given back to the farmers (Stevenson and Lev 2013).

The other complexity is the prices attained on these markets, which acts as the reference, comes from different specialized ingredients or finished products. For example, the price obtained for cheeses, skimmed milk powder, and whole milk powder are all different; however, on the input side the farmers are producing the same milk for all these products. So it can be a bit of a complex mathematical problem to figure out what the milk's solid or raw price will be. That is why the prices obtained at the Global Dairy Trade Auctions act as the best reference price for the cooperative farmers, especially in New Zealand. Being the largest exporter, the auction doubles as a sales channel also. The auctioning platforms are used to trade different commodity groups such as whole milk powder and skimmed milk powder. So this allows the manufacturer to be equitable to the farmer suppliers.

There is also a unique and natural tension between the farmers and the owners of the company as they want to be able to make decisions which drive equity value. So the farmers have to balance between the profit margin at the raw milk level and also the profit margin at the final product level as they are doubling as owners. However, the prices of milk powder and other dairy-based commodities are determined by the global trade which limits the price flexibility. This is one of the reasons why cooperatives are moving away from just selling the base commodity ingredients and moving toward more valued-added, highly specialized ingredients which are more profitable, such as butter fats, a type of protein which is also referred to as milk solids (Pieters et al 2017).

14.6 CASE STUDIES: ULTRA-HEAT TREATED PRODUCT

This is an example of adapting the supply chain to accommodate for a change in consumer preference, and how this impacted on the logistics and other processes. This requires an understanding of the shelf life of an ultra-high-temperature (UHT)-treated milk product and the type of packaging it would require.

UHT-treated milk is a longer shelf life liquid milk product, which is achieved through a sterilization process by heating the milk to a level where it's sterilized and can be stored without any refrigeration for about three months in a sterilized carton. This process increases the shelf life and reduces the perishable nature of dairy. However, after opening the carton, the milk behaves like regular freshly processed milk and will perish within the normal course of time.

The challenge came about due to a change in consumer preference to a protruding "Helicap style" cap vs the existing flat "Recap style" cap that previously existed on the carton. it. This created a supply chain challenge as the rectangular cartons now have protruding knobs instead of flat rectangular tops. Flat rectangular tops allowed stacking two pallets one on top of another, but with a knob sticking out as a result of the new packaging style, putting a pallet over another would crush the protruding cap and damage the cartons.

This required a review of the whole delivery logistics from packaging to loading on pallets to warehousing as now double stacking was no longer viable. The entire process was reoptimized so the cost of transportation and storage could remain the same. The previous state allowed double-stacked pallet loads into a 20-foot refrigerated container. This needed to change to a new pallet format which was single-level, but in a 40-foot container. The solution needed to be reconfigured at the manufacturing level, so the robotic system could load a newly designed pallet configuration. On top of all of this, the warehouse had to be reconfigured. For transportation, the route had to be rechecked as 40-foot containers are larger and heavier and each country has different road weights limits, and even within a country, different routes have different road weight limits. So there was an end-to-end impact all due to this relatively simple and small change in packaging. However, this end-to-end impact resulted in additional optimization as it was now feasible to fit more products in a single container. The total container volume was also reduced with a new pallet format which in turn reduced freight costs, including land freight savings, ocean freight savings, and other freight related charges.

The overall process needed a lot of cross-functional collaboration. Different pallet configurations and container configurations were tested. The marketing teams and sales operations team needed to be aligned as one had the pulse on the consumers and the other ensured successful delivery in the new format. The overall package development process needed to be program managed with supply chain, manufacturing, transport, and warehousing both domestically and internationally.

14.7 INDIVIDUAL PERFORMANCE

For an entry-level supply chain manager in the dairy industry or for someone coming in from a different industry, the ability to get stuck in and solve problems while being agile and adaptive is very important. It is also equally important to be aligned with the company values. As a farmer cooperative, dairy manufacturers are often very value-driven, and they do have this need to not be seen as a corporation that just spends money loosely. They do have that responsibility to their farmers and to be responsible and make the best decisions for them. So one needs to align with the company values first and foremost, but as someone new in the industry, the willingness to solve problems and get stuck in is very important.

At the next level, senior management's ability to break down the barriers between different business units within the larger organization is critical. With large organizations, there is this tendency that sometimes different arms of the organization can stay in their silos. Employees often just stick within their specific swim lanes and get specifically tied to the manufacturing side or the logistics side. It is important to understand that the challenges are often cross-functional, across different business units, and so it's important that someone has that ability to break down those barriers. The dairy industry also has a unique difficulty, as on the supply side it may depend on a domestic market but on the sales side the manufacturers are global. New Zealand may procure raw milk only from the farmers of New Zealand even though it's a global exporter. So it's important for a supplier manager to have that global view and the ability to work and collaborate across different markets with different people and the need to be able to communicate and work with people from all ranges and positions in the organization. So, one should be able to talk to people who are working on the shop floor, a warehouse or manufacturing environment, and all the way up to the executive suite.

14.8 ORGANIZATIONAL LEADERSHIP

From a general manager or a senior leaders perspective, to be successful, one needs to be able to manage multiple teams efficiently, especially if the teams are geographically diverse. For running a large organization, it is important to be able to demonstrate the following four skills.

Number one is leadership fundamentals or the ability to manage teams across different geographies and the ability to develop this in others. Ability to do coaching, performance management, and to get results through others becomes a lot more important than just being able to manage a team within a narrow radius. It is critical for a senior leader to be able to talk about purpose and strategy creation and lead projects that impact a lot of different business units. Ability to get people on board across those business units is a communication skill that is fundamentally important.

"Mana" is a Polynesian world which in the Maori language means having that presence and that respect that comes with both age and experience. It is a combination of knowledge, respect, and communicating with influence, stakeholder engagement, and management that a leader needs to be able to demonstrate (Moorfield 2020).

Number two would be a deep understanding of the specific operation, such as looking after the warehousing division or the distribution team. It requires having a deep understanding and being able to handle the day-to-day operations and the challenges that the individual warehousing managers would face and then knowing the levers that one can pull to manage and to transform performance around those specific operations.

Number three is the commercial acumen and typically when one gets to a very senior role, one is managing a large core center and teams of hundreds of people. Often some of these teams are logistics or teams which do not generate any profit. A senior leader needs to be always scrutinizing for performance optimization and sustainably increasing productivity over time.

Lastly, number four is delivery. This is about sustained delivery through others over time and to being able to gain trust of others and to be viewed as someone who is going to be able to deliver results time and time again. This comes by demonstrating one's ability to lead, influence, and engage and by delivering exceptional results which is essential.

14.9 CONCLUDING REMARKS

The dairy industry is enjoying a growth in demand driven from countries whose per-capita consumption is increasing while their demand is continuing to outpace their domestic supply. New Zealand with its huge surplus exports more than 90% of its dairy production to meet this demand. Since the primary raw material for the dairy industry is a single component, i.e. milk, the supply chain for this vertical revolves around processing and delivery logistics of milk. While the acquisition cost of milk depends on the milk curve and international trading prices, the dairy cooperatives are looking at additional profitable alternatives. New process innovations are extending the life of milk and creating more derivative products such as butter fat, and these efforts are increasing the profit margin for the cooperatives.

ABOUT THE CONTRIBUTOR

Andrew Bawden is a supply chain leader at Fonterra where he manages sourcing, partnerships and global supply chain development in the dairy industry. Andrew has extensive knowledge and experience of the global dairy market and its supply chain including sourcing from cooperatives situated in New Zealand. Andrew is a graduate from the University of Auckland.

REFERENCES

Benham, L. 2020. Five trends shaping the global dairy industry in 2020. Foodbev Media. https://www.foodbev.com/news/5-trends-shaping-the-global-dairy-industry-in-2020/ (Accessed December 24, 2020).

Dani, S. 2015. *Food Supply Chain Management and Logistics: From Farm to Fork*. London: Kogan Page.

GalketiAratchilage, U., and Marocco, E. 2020. FAO (Food and Agricultural Organization of the United Nations) Dairy Market Review, March, 2020. Overview of global dairy market developments in 2019. http://www.fao.org/3/ca8341en/CA8341EN.pdf (Accessed December 24, 2020).

Kroll, K. 2015. The dairy supply chain: From farm to fridge. Inbound Logistics. https://www.inboundlogistics.com/cms/article/the-dairy-supply-chain-from-farm-to-fridge/ (Accessed December 24, 2020).

Moorfield, J., Te Aka Online Māori Dictionary. 2020. Mana. https://maoridictionary.co.nz/word/3424 (Accessed December 24 2020).

Pieters, L., McMahon, M., Schunck, P. et al. Deloitte. January 2017. Global dairy sector – Trends and opportunities. https://www2.deloitte.com/content/dam/Deloitte/ie/Documents/ConsumerBusiness/ie_Dairy_Industry_Trends_and_Opportunities.pdf (Accessed December 24 2020).

Shadbolt, N., and Apparao, D. 2016. The International Food and Agribusiness Management Review, January 2016. *Factors influencing the dairy trade from New Zealand.*

Stevenson, S., and Lev, L. 2013. Center for Integrated Agricultural Systems, UW-Madison College of Agricultural and Life Sciences, March 2013. Values-based food supply chain case study: Organic valley. https://www.cias.wisc.edu/wp-content/uploads/2013/04/rb80organicvalleyfinal041813.pdf (Accessed December 24, 2020).

Wallace, C., Sperber, W., and Mortimore, S. 2018. *Food Safety for the 21st Century: Managing HACCP and Food Safety Throughout the Global Supply Chain.* Hoboken, NJ: McGraw-Hill.

15

Data Analytics in Supply Chain Management

Tony Chen

CONTENTS

15.1 Objective 311
15.2 Sources of Data in Supply Chain 312
15.3 Challenges with Managing Uncertainties 313
15.4 Opportunities 314
15.5 Vendor Selection Using TCO Model 315
15.6 Vendor Selection Using Hypothesis Testing................................320
15.7 Logistics Challenge Solved through Network Optimization and
     Linear Programing 321
15.8 Risk Management through Probability Simulation 323
15.9 Inventory Management Models 325
15.10 Machine Learning-Based Forecasting..326
15.11 Concluding Remarks...328
About the Contributor 329
References...329

15.1 OBJECTIVE

In this chapter we will focus on usage of data in the supply chain. By analyzing data we get critical trends which help us in making decisions. Data analysis provides us with directional guidance for various kinds of problems. In this chapter we will go through a few examples of how data analytics can be used to select a vendor, or to solve logistics challenges, to assess and minimize supply chain risk, to manage inventory, etc. We will discuss what data can be used to model and evaluate the problem and what kind of conclusions can be drawn by using various kinds of analytical techniques (Robertson 2021).

DOI: 10.4324/9780429273155-18

15.2 SOURCES OF DATA IN SUPPLY CHAIN

In this section we will discuss availability of data in the supply chain. We will discuss how we get data from various sources, and how it helps us drive decisions. Data does not tell us what to do but it helps us in deciding it. In this chapter we will consider various examples of data and what decisions we try to drive out of it. Let us start from a state where we assume we are looking into a business which is already in production. Therefore we can infer that a few essential data sets such as quotations from a vendor, purchase orders (PO), invoices, and forecasts are available. These data sets will be required for doing the analytics (Chen 2019).

POs are documents that every company has when it tries to procure any goods and services. We are starting with an assumption that we have access to the PO data which consist of description of the item to be purchased, associated part number which is a numerical code for the item, quantities, unit price, total quantity, timeline for expected delivery, and applicable terms and conditions. Similar to POs, on the vendor side, a complete sales record should also exist for the goods or services sold to customers. Typically the price is based on a quotation provided by the vendor and the PO is issued by the customer to the supplier. After the transaction gets completed, the supplier sends an invoice to the customer based on the sales record. The invoice contains the final cost and payment terms. Typically an end-to-end process consists of a customer requesting and obtaining a price quotation followed by the customer issuing a PO which is then followed by delivery of the goods and invoice from the vendor and payment. The entire communication creates a set of data as mentioned above.

However, in order to fulfill the PO, one important question to ask is how does the supplier prepare to fulfill its obligation? How does the supplier itself procure material from its upstream vendors to build the product that has been ordered via the PO? In a coordinated supply chain, normally a communication gets sent out prior to placing the PO by the customer. This is known as the forecast, which informs the supplier about the potential amount that the supplier needs to prepare for to fulfill the future POs. This information provides directional guidance to the supplier in order to plan ahead regarding manufacturing capacity, raw materials, logistics, etc.

The durations of supply chain activities are also important. For example, a medical device manufacturer that produces flu detection instruments first purchases subcomponents from its vendors, and after receiving these parts, the manufacturing begins and is then followed by shipping. The total process takes 3 months and is often referred to as the lead time of the product. Depending on the uncertainties in availability of the subcomponents and dependencies on these activities, lead time can vary. Understanding lead time is critical; from a manufacturer's perspective, a subcomponent supplier with shorter lead time means more flexibility and less costly to adjust to changes for the manufacturer. On the other hand, from the supplier's perspective, giving that shorter lead time flexibility to the customer (manufacturer) is accompanied with higher levels of inventory and/or production capacity investments (more costs).

Along with the forecast, POs, and lead time there are certain contracts which have the guiding terms and conditions around the three elements. For example, the forecast is a potential guidance on what happens if the manufacturer builds the device and the hospital buys less or more quantities than the quantity communicated through the forecast. These contracts are between the hospital and the medical device manufacturer. It may involve commitment of a certain price in exchange for a certain order quantity or commitments on a shorter lead time if the quantity purchase is within the forecast. In addition there could be additional terms and conditions around safeguarding both the customer and suppliers against various eventualities. The forecast data typically consist of part numbers, part description, and quantities needed over a timeline. It could be quantities needed per month, could be per week, and could be per quarter, depending on an agreed upon purchase frequency. It may have additional information such as location where the goods will be needed to give guidance on delivery logistics.

15.3 CHALLENGES WITH MANAGING UNCERTAINTIES

In the context of supply chain, uncertainties are often reflected in the volatility of demand, price, and/or duration; the higher the volatility, the more difficult it is to anticipate and act in advance.

Forecasts in general are critical elements in the supply chain as it would be nearly impossible to operate efficiently without them (either provided by external partners or estimated internally). Forecast sharing essentially acts as information flow from downstream customers to upstream suppliers, and they provide a basis for each party to act accordingly. If the forecast is accurate, it informs the supplier exactly how much to plan for and produce and at what period of time, leading to a lean and cost-efficient supply chain.

However, in real life nothing is certain. Therefore forecast by definition has an inherent uncertainty. In spite of that, when managed and shared correctly, forecasting is still an important method to communicate and provide directional guidance to the vendors.

Forecasts typically change with time as more information is surfaced and also imposing more challenges. For example, the hospital might have communicated a 6 month demand forecast in March that they would need to buy 100 devices per month from October, the beginning of flu season. However, 3 months later, due to some natural causes a new strain of flu virus started spreading. So the hospital revised their forecast and asked for 100 flu detection instruments from August through September and increased the October through December forecast to 200 instruments per month. This posed two challenges. First, recall that the lead time is 3 months, the newly added demand in August and September cannot be met unless the supplier figures out how to reduce lead time either by adding more shifts or diverting the subcomponents from other devices in production to prioritize the manufacturing of the flu detectors. The second challenge the supplier faced was to ramp-up capacity to deliver 200 per month from October which was just at the required lead time of 3 months. However, in real life when such an event happens, a supplier might not be able to meet the forecast and might end up delivering in partial quantities like 30 in July, 75 in August, and then finally would catch up and produce 100 in September. The supplier would have less flexibility within lead time and might not be able to fully accommodate forecast changes.

15.4 OPPORTUNITIES

In this chapter we will touch on various supply chain activities and showcase the underlying opportunities for data analytics (Alicke et al. 2016). Prior to any procurement activity the first step is to identify a vendor who will be supplying the item in demand. This step is often referred to as vendor selection.

Let's consider an example, where a hospital had decided to buy a new type of flu detection instrument. There could be many manufacturers who manufactured the required type of testing instruments. Therefore the first step for the hospital was to identify one or more vendors that could supply the necessary instrument. The second issue for the hospital was determining the logistics and delivery timeline. Where and when should the manufacturer send these instruments? Where did they manufacture it and were there any regulatory or import restrictions or tax implications? Due to sudden increase in demand and reduced lead time, would these instruments require faster shipment such as overnight or air shipment instead of slower and ground shipment? Would there be any final assembly or installation required at the hospital. How would they be serviced if some of them breaks down after arrival?

The volatility in forecast introduces risk to the continuity of business. Shortages and price fluctuation are some examples of risks which need to be managed as part of supply chain risk mitigation. For example, the actual order placed by the hospital for July could be 90 and the supplier could only supply 30, therefore there will be a shortage of 60. The supplier could also increase price as it has to pay for overtime, expedited shipping. So the price could change.

Managing inventory is also an important aspect where there are a lot of opportunities for applying data analytics and optimization. The sudden increase in demand for flu detectors left the hospital without the required number of machines and caused the supplier to lose potential revenue opportunities in July and August. In July the demand was there for 90 instruments and the vendor lost the opportunity to supply 60 units as they could supply only 30 instruments. Either party could have mitigated the challenge by maintaining an inventory of finished products. Inventory could be maintained at many levels. The hospital could have had inventory of these devices in the hospital itself for quick deployment or could have contractually made the supplier to maintain a certain level of stock at a nearby warehouse. Subsequently the vendor could have maintained an additional stock of all the components to assemble them quickly to meet such demand surges also. There are so many different types of inventory strategies to support various kinds of business models that it makes sense to use analytics to determine which inventory method is optimal for the specific business model.

15.5 VENDOR SELECTION USING TCO MODEL

The first analytical modeling we will discuss is total cost of ownership (TCO) (Linthicum 2016) and will apply it for vendor selection. Vendor selection is an important aspect of the supply chain and we will explore how we can leverage data analytics as part of the decision-making process. As all vendors were not the same, how did the hospital choose the right flu detection instrument supplier out of a pool of suppliers. The most common aspect about selection criteria was often the cost of the instruments. However, pricing was only one element and there should be many other elements to consider.

For example, let's say the instrument needed a battery to operate. So even multiple vendors could have the same price but the life of the battery could be different for different vendors. For one vendor the battery life could be 3 years and for another it could be 7. If one battery died sooner than the other, obviously, there would be an impact on the preference. Even though the pricings were to be the same, the capability was therefore not the same. Next element that the hospital might care about would be manufacturing capability. One vendor was very flexible and had a shorter lead time of say 4 months and the other might have a lead time of 6 months. The former was better equipped to react with forecast changes than the latter. Similarly, one manufacturer might be able to produce up to a 1000 units per month and the other could do a 100 per month. However, the first one may ask for a monthly minimum quantity that had to be purchased whereas the latter had no such restriction. Gathering all these data would actually help the hospital determine which one was better. So the first step was to collect data points on these elements before doing the analysis.

In order to build an analyzable data set, a supplier manager needs to get a set of answers from the supplier. First question is about capacity: how many units can the supplier manufacture in a month or how many units in a year. We need the information to assess the capacity of a vendor. Why is it important? Once the vendor is selected, the demand is going to be stochastic. That means the demand will fluctuate which will constantly keep changing the forecast. The actual demand will often come above or below the forecast. If the end product has a lot of demand there will be upside. The ability to satisfy the upside demand is therefore a requirement and a selection criteria as some vendors will be capable of meeting the demand and some may not be able to handle the fluctuation.

This is something which can be simulated. A comparative model could be built which captures the inventory, production capacity of the multiple vendors. The model should be able to demonstrate that by switching vendors which particular vendor will have a direct impact on the revenue or cost. As one would be able to fulfill in spite of a higher degree of forecast and demand fluctuations and others may not. The one that is unable to fulfill will result in lost revenue opportunity (Tayur et al 1999).

The models can take into account both quantitative and qualitative information. For example, capacity and lead time are quantitative data. However, quality and reliability are a bit harder to quantify. In the industry some of the typical measures are mean time to failure (MTTF) and mean time to repair (MTTR) (Gomes 2020). Similarly there is also break-fix or RMA (return merchandise authorization). If something breaks down and if it is under the manufacturer's warranty then the manufacturer will issue an RMA to replace the broken part. These metrics would actually help turn quality and reliability into quantitative data. MTTR is the time element which provides data on how long does the manufacturer takes to fix a broken part. The RMA process also provides insight into the count of broken parts which is also quantitative data.

Weaving it all together is the harder part. For example, a business which is already running and has a supplier decides to switch suppliers. The key is to quantify the downstream elements and their impact on revenue and cost via simulation. For example, if the capacity is different at the new supplier, that is going to change the lead time and therefore going to have a direct impact on the amount of inventory needed to manage fluctuating demands. While a higher capacity may be preferable from a demand fluctuation perspective, a longer lead time will call for a larger inventory which will probably drive up the inventory holding cost. So the difference in lead time will be modeled through inventory value and holding cost.

There are also more qualitative data sets. For example, the difference in accuracy between two vendors' products. If a medical instrument has accuracy issues, then what is the cost of a missed diagnostic for a hospital? If it ends up in a legal court, it can mean a ton of legal cost. All of these needs to be weaved together in a simulation to find the optimal vendor.

However, it is not always possible to build an end-to-end simulation model because this is quite hard to build due to interdependencies among functional teams within a company, and second often different data elements are owned by different teams. Visibility to actual data is often difficult to get as they may be in disparate systems owned by teams which could be working in silos. For example, the cost data could be in the financial or supply chain teams' systems but the repair data or RMA data could be in a different system managed by the hospital operations team. One option is to make simulations based on models which are based on actual data, such as PO information and reasonable approximations on the other parts. Although difficult, working toward a centralized end-to-end model is the only way to ensure one is optimizing globally.

To better illustrate the vendor selection problem, let us consider purchasing pacemakers as an example. It is a medical device that is surgically inserted inside a patient's body to help control abnormal heart rhythms. One of the key components is the battery which keeps it functioning. There are two competing manufacturers A and B, each with their own strength. Suppose the hospital's supply chain manager can only select one of these two manufacturers to supply the pacemakers, how should they pick? To start, they will need to understand the critical elements that will impact their decision. Based on their domain knowledge, they identified several key elements: demand for the pacemakers, unit cost, supply lead time, product reliability (MTTF), battery life, and heart rate detection accuracy. Note that some of these elements are product related and some are supply chain related, each with their own comparable specifications. It would be trivial if one vendor is significantly better than the other in all areas; however, this is rarely the case. TCO is a useful tool when making decisions that involve multiple criteria; the basic concept is to monetize all elements to dollars. The process could start with getting a quote, which includes the elements mentioned earlier. It is also important to get a sense of prioritization based on which element is more influential than the other. Then we deep dive into each of the elements and quantify the dollar impact. The end goal is to make a comparison between the vendors based on TCO.

Demand for the pacemaker is a critical element and can be difficult to forecast as it may depend on seasonality, product features, and many other factors. Not to complicate the problem, let us assume that the demand is independent of the product features and pricing and simply a function of time so that we could focus on how TCO assembles everything together. If demand is not independent of the product features and pricing, then extensive market research is required to understand the relationship among those elements; the goal is to model demand as a function of these variables. Now back to our problem, we first forecast demand as random variables, for example, in the upcoming weeks (could be days or months depending on the level of granularity of the historical data), what are the volumes needed in each of the weeks? Ideally, we should not only provide point estimates but a range or distribution for each of the estimates. For example, the forecast of pacemakers needed for next week is 200 (point estimate), the following week is between 100 and 300 (range), or even better, the forecasted pacemakers need for next week follows a Poisson distribution with a rate of 200 units per week (random variable). We provide more information about uncertainty as we move from point estimate to range to modeling forecast

TABLE 15.1

Week-over-Week Pacemaker Demand with Variance

Week	Mean Demand	Variance
W10	200	30
W11	150	20
W12	170	30
W13	300	50
W14	220	25
W15	180	18

as a random variable. To do so we can use historical data and apply various forecasting methods (seasonality decomposition, auto-regressive integrated moving average (ARIMA), SARIMA, neural network-based methods, etc.). Example of a forecast is shown in Table 15.1.

Thus far, we have quantified the uncertainty in demand. When demand is higher than expected we could encounter shortages and encounter excess inventory when demand is lower than expected. These imbalances have costs, normally defined as overage and underage costs. Overage costs would be the cost of held inventory, the risk of inventory getting expired, etc. Underage costs would be the cost when a patient requires a pacemaker and there is a shortage. In practice, shortage costs are much harder to estimate but is a cost that has been overlooked. A shorter lead time means the supplier could adjust to demand uncertainty faster leading to lower costs (both underage and overage). This also implies that the cost of lead time and demand uncertainty comes hand in hand (when demand is deterministic then lead time becomes irrelevant).

Similarly we model product reliability (MTTF), battery life, and heart rate detection accuracy as random variables as well. To summarize, how frequent this event occurs, the variance of frequency and the cost of an event happening are the inputs for the simulation model. For example, product A on average fails after 25 years of usage, with a variance of 2 years, and each time a failure occurs before 20 years the hospital needs to pay for the surgery and the material costs. If the failure happens after 20 years, then the patient pays for the surgery and the hospital pays for the new pacemaker.

Now that we have all the elements, we are ready to build a simulation model that simulates the real world via data-driven estimated parameters. Each simulation trial represents a realization from the random variable, we run the simulation for many times and then show the distribution of total costs. Since we have a time element in the simulation model, for each week, we have the distribution of TCO, then how do we compare the simulation results between supplier A and supplier B? A common approach is to take the expected TCO for each week and apply the net present value (NPV) to aggregate the expected TCOs into one number.

In addition the simulation results could be also used as a negotiation tool. A supplier manager may get insight on which element vendor A needs to adjust to be competitive. For example, if vendor A can bring down the price anywhere below $4500 then that will offset the holding cost of vendor B and be competitive in spite of having higher cost. So these simulations also provide insight to the supplier manager on negotiating different facets of the selection criteria.

15.6 VENDOR SELECTION USING HYPOTHESIS TESTING

Every analytical model has its challenges in terms of having the right data to yield accurate results. Sometimes it is not feasible for companies to obtain end-to-end parameters and build an accurate regression model for TCO analysis. Another approach for vendor selection is using hypothesis testing (Frost 2020). Assume a company is unable to get all the information as mentioned above but still wants to create a comparison between two vendors to find out who is better. Just on the basis of the quoted price one can say vendor B offering $9 unit price is better than vendor A with $10 unit price. However, how does the customer factor in the performance difference between the two vendors?

One way of measuring performance is to measure yield or analyse failure metrics. One of the commonly used failure metrics is MTTF. Assume that the hospital does not want to repair a machine once it fails, then MTTF is the right metric as it measures the average amount a nonrepairable instrument operates before it fails.

For example, let's say a hospital had bought two batches of devices from two vendors on a trial basis. From vendor A they had bought 1000 units and from vendor B they bought 200 units. Assume they had run all 1200 machines from both vendors for about a month. At the end of the month they had to reorder more instruments, now the hospital had to decide which one to order. So the hospital collected the month end status of all the machines and found out of 1000 units from vendor A, 5 were broken. On the other hand for vendor B, 3 out of 200 were broken. Now, the failures could be random. If the hospital bought another 1000 of A and 200 of B and again ran for a month it could have got a different quantity of broken instruments for the two vendors. So, how would the hospital conclude whether A was better than B in terms of reliability and order correctly? This was a case where two different quantities have to be compared and a method called T test could be used.

T test is statistical inference technique which checks if the averages between two different samples are similar. In this case, the T test would start with a hypothesis that both vendors had the same reliability. If the test resulted in a strong rejection of the hypothesis then it would conclusively indicate that they were not similar.

Statistical processes being based on probability also carries a certain amount of risk associated with the conclusion. In this case, there could be a risk where A was incorrectly chosen over B or A was not chosen in spite of A being better than B. However, there are also methods to minimize the possibilities of such errors, for example, collecting more data or more data over a longer period of time. The main advantage of this approach is it does not require a broad data set like the previous example of TCO analysis.

15.7 LOGISTICS CHALLENGE SOLVED THROUGH NETWORK OPTIMIZATION AND LINEAR PROGRAMING

Let's consider a case where a manufacturer was deciding where to set up a factory. Let us say China was chosen to be the manufacturing location. Next question would be where were the parts suppliers located? Were they in Europe or Americas? One could create a graph model by connecting the upstream supplier to the factory. The material could be converging to a node which could be a vendor-managed inventory (VMI) hub location. From the node it connected to the manufacturing location and eventually connected to the customers.

Creating the network topology design would be a complex process as there could be many possible combinations of different supply sources, different routes, etc. One approach would be to build a cost model of how much it cost to flow one unit from point A to B. Let us assume that one of the raw materials was magnets which needed to be sourced and there were two locations from where it could be sourced – US and China. So there could be two lines (arcs) for magnets going from supplier to the factory. One would be US to China and the second would be within China. So the cost model would indicate a different amount of cost for shipping one unit in each branch. However, the magnets could be shipped by air, sea, or train (for within China). So the cost would vary as it was dependent on the route. Now imagine the complexity when there were multiple parts and each with multiple manufacturing locations, converging to the factory ultimately. The complexity would increase if a second factory location was added in Mexico. It would require determining the amount of magnets to be taken from the US and China for each factory which would hinge on cost of goods, transportation cost, etc.

Location of customer would add additional branches in the network topology. The customer could be in Africa or could be in North America and that might call for putting warehouses close to the customers. This was how the logistics graph could be formulated and analyzed for optimal flow. For each possible goods movement there would be an associated cost incurred. One method of optimizing the cost is to use a method called linear programming (Shapiro 2009, Hillier and Lieberman 2010). Through this approach one could minimize the whole logistics cost so that the flow from one to the other times the cost of that flow. You know that's the total cost and your objective is to find a flow which has the best cost.

For example, say there were ten vendors, three different factories, and five warehouses to store finished goods to support global customers. Each supplier could ship components to any factory and any factory could ship finished goods to any inventory. So each cell in column 1 could have 3 arrows (directed arcs) pointing to each cell in the second column. In other words each arrow is a shipping lane. For instance, the magnet supplier in the US could ship magnets to factories in China, Mexico, and the UK. Similarly there would be five arrows from each factory to five of the warehouses, meaning factories in Mexico could ship finished products to warehouses in Canada, US, China, UK, and Brazil. However, each shipping lane or arrow would have its own cost. The end goal of linear programming model or network organization was to recommend for each vendor what was their target factory for shipping supply. It would also recommend from which factory how much finished goods should be shipped to which warehouses to best serve the end customers. This approach could be used for determining the optimum location for suppliers, factories, warehouses, distribution centers, etc.

When there are many choices in terms of flow of goods, through a complex network of suppliers, factories, warehouses, and customer locations, each path could be represented through a variable such as cost or time. If cost is

TABLE 15.2

Suppliers with Factory Locations and Warehouses

Supply	Factory	Warehouse
1. Magnet supplier (US)	1. Factory CN	1. Warehouse Canada
2. Magnet supplier (CN)		2. Warehouse US
3. Coil supplier (Russia)	2. Factory MX	3. Warehouse CN
4. Mechanical supplier (CN)		4. Warehouse UK
5. Package supplier (MX)	3. Factory UK	5. Warehouse US

chosen, then linear programming can be used to compare the costs of various paths and recommend the path of least cost. Also it could be used to find out what is the optimal path to serve the customer in the best possible way, or optimal quantities of inventory to keep, or where to locate factories and warehouses, suppliers, etc. (Table 15.2).

15.8 RISK MANAGEMENT THROUGH PROBABILITY SIMULATION

In the previous section we discussed how to optimize a complex network from suppliers to factories to warehouses and ultimately to the customers. In this graph view, each branch of the network connected a material source to its destination which could be, for example, a supplier from the US supplying magnet to China or a factory in Mexico shipping finished goods to Canada. Risk arises in the supply chain from many sources, such as geopolitical influences, natural disaster, and technology transition. However, a simple way to visualize risk would be to think as if a branch between a pair of source and destination is no longer available. The optimization variable such as cost or time, based on which the model was tuned, is no longer valid. This will require removal of a branch or multiple branches. If the factory goes off-line, all the branches connecting raw materials to the factory and factory to the warehouses get disrupted.

Due to the pandemic, factories in many parts of the world had to stop production. This led to serious disruption of supply of both raw materials and finished goods. However, the customers still need to be served. The shutdown may cause an increase in logistics cost and an increase in transportation time which will add to the production lead time increase. At the same time other factories have to increase production capacity. So net net, all parts of the network need to react to this change.

An earthquake is a frequently disruptive event where a supplier may have to shut down an entire factory and recalibrate all production machineries before resuming production. When a vendor goes off-line, its entire production capacity also disappears. Similar situation can happen if a supplier decides to shut down business or shut down production of a nonprofitable product which could be a critical raw material for a factory which was buying it. Assessing these scenarios and mitigating the challenges coming out of these situations is essentially risk mitigation (Heckmann 2016).

Quantitatively this risk could be assessed using probability distribution. If a supplier stops production for any reason, the lead time is going to increase. If the lead time increases, the inventory model would change, additional amount of inventory will be required to be carried. Or if this is a temporary shortage, a certain amount of time will be required to back fill all the demand. If the lead time is used as a probability distribution, it can help answer a lot of questions. Let's say the customer has given a forecast which articulates potential demand over specific time periods. Let's say the forecast calls for 10 units of demand in the next 15 days. The supplier can review its lead time and use the lead time probability distribution data and respond on how much finished goods can be supplied in what period of time with a quantifiable amount of certainty. It can commit to fulfilling the 10 units of demand in the next 15 days with a high probability. The lead time probability distribution would allow the factory to answer how many could be ordered or how long and how soon should it be ordered so that it could be fulfilled in a stipulated amount of time. Say the customer has an urgent need for 10 counts of the finished goods. Using probability, the factory can say what is the likelihood that it can fulfill the order. The factory could say something like that they can fulfill the demand with 95% confidence but if the customer desires 99% confidence then it could fulfill only 6.

Risk management simulation can be done to address different aspects of the supply chain which can be impacted by disruptions. For example, the original problem we discussed, an earthquake which puts a factory offline, will impact cost and drive it higher. Simulating that can lead to need for more locations or moving out from certain locations. Similarly impact of disruption on lead time or impact of quality on risk could be measured through this approach. The end goal is to measure the impact with a probabilistic confidence level.

15.9 INVENTORY MANAGEMENT MODELS

Managing inventory is one of the essentials of supply chain management. There is a lot of inventory, in every layer between the raw materials to finished state in varied stages of transformation. For example if we drill down into a supply chain of a company manufacturing audio systems. The company would be maintaining an inventory of finished audio systems in its inventory. It would also maintain an inventory of speakers in its factory to build more audio systems. In the same vein, the speaker suppliers will maintain an inventory of finished speakers on the outbound side and an inventory of magnets as one of the components on the factory side. Similarly the magnet vendor would carry inventory of finished magnets on the outbound and inbound inventory of raw alloys and metals.

The reason we need inventory is to deal with uncertainty of demand in every stage. For example, in every stage of the example given above there are forecasts. A customer will provide a forecast to the audio system manufacturer, who in return will aggregate demand from all customers and produce a forecast to the factory. The factory will provide a forecast to the upstream speaker manufacturer. The process will repeat up to the raw alloys and metal supplier. All stakeholders will plan according to the forecast to deal with the actual order. When the actual orders come from each stakeholder to its upstream supplier, the orders are rarely in line with the forecast and typically they are higher or lower than what has been forecasted. Suppliers at each stage deal with the uncertainties with additional inventories and there is a cost associated with it. The cost arises from buying the raw material inventory, making the quantity of finished products, and then holding the amount of inventory.

The problem which every supplier in each stage of the supply chain faces is how to minimize the cost by balancing between the overage cost and underage cost. Overage cost arises from excess of unsold inventory. Underage cost comes from the loss of profit because of lack of inventory. While optimal inventory management is a science, it's important to understand that it depends on the business model. There are well-established inventory optimization models and the one which fits the right business needs should be chosen. One such model is the Minimization model.

Let us consider two different market segments and discuss what is appropriate for each. First let us consider a newspaper company which publishes newspapers daily and tries to figure out how many copies to print each night. The cost of printing each copy is $1. The newspaper generates revenue from sources like readership, advertisements, classifieds, etc. Average readership ranges from 7000 to 10,000. So these are the variables that could be considered to figure out how many copies should be printed. At first glance, the right answer could be an average between 7000 and 10,000 or a maximum of 10,000. However, let us consider a few additional pieces of information. First, the revenue is $1M daily and it increases by $1M for every 10,000 new readership. Second, the cost of printing each copy is $1. In this case the opportunity cost is $1M and the cost of printing 10,000 more is only $10,000. Calculating the inventory cost is straightforward as the inventory loses all its value as soon as the next day's news gets printed. This would incentivize the newspaper company to print copies greater than 10,000 instead of printing less to minimize overage cost. This is the well-known problem in inventory theory called the 'newsvendor model' (Stevenson 2009).

Next, let us consider a coffee chain which sells coffee beans. The problem they are trying to solve is what level of inventory should be on hand at the stores to satisfy its customers coming in to buy coffee beans? Normally this translates to an optimal policy called Q-R policy(Hadley & Whitin 2012). Assume on an average the store sells 300 pounds of coffee in 30 days and it takes 10 days of shipping time to get any amount of replenishment. So on an average the store sells 10 pounds of coffee per day. In an idealistic world the company should order for replenishment when the inventory level dips to 100 pounds assuming it will get depleted in 10 days and in exactly 10 days replenishment will arrive. So an ideal scenario would be the company has 300 pounds of coffee beans on day 1 and whenever the stock depletes to 100, which is considered as a reorder point, replenishment order is placed and new stock is expected to arrive before stock depletes to zero. Of course in the real world the daily sale of beans varies and does not stay at 10 pounds per day. In that case the store has to figure out the right reorder point which is higher than 300 pounds but also has to minimize the cost of losing revenue, minimizing the cost of excess inventory. So the minimization model tries to find the lowest amount of total cost. The formulation of the problem will depend on the business model. For newspapers it's perishable as the value goes to zero the next day, for coffee beans its nonperishable. However, the end goal is to minimize the total cost.

15.10 MACHINE LEARNING-BASED FORECASTING

In the world of supply chain, forecasting is perhaps the most critical area where an enormous amount of attention is given from computational resources point of view. There are many techniques for developing forecasts and various types of analysis (IBM 2020). Let us take a situation where an audio speaker company was purchasing speakers from a speaker supplier. Every month the company sent a rolling 12-month forecast and also placed orders for speakers to the supplier. In other words, say in March of 2020 it sent a monthly forecast spanning from April 2020 to March 2021 and placed the order for April 2020. In April 2020 it would repeat the process and sent the forecast for May 2020 to April 2021 and placed order for May 2020. During this process the forecast of say March 2021 would change 12 times as each month the forecast was refreshed and finally an actual amount of order will be placed in February 2021. The changes in forecast of any particular month as it changed every time the forecast refreshed could be captured as a set of data. The series of monthly forecast amounts for different months indexed over time is called a time series of the forecast. Similarly the data of the actual orders placed indexed over time would be called a time series of the actuals.

There are a lot of conventional ways of developing a forecast model using historic time series data. This approach leverages past data over a certain period of time and produces a set of data for a span of time in the future. One of the conventional ways of developing a forecasting model is called the seasonality decomposition model. In this model it is possible to decompose or separate out the effects of trends, seasonality, cyclical behaviors, and noise in the historical data. Once the data are decomposed into these four categories, the decomposed data could be used to infer what could happen in future. Another model known as ARIMA model is useful in finding patterns and relationships between two or three different factors. So these are traditional approaches to forecasting using statistics. The basic downside for these models lies on the fact that they rely on historical data.

For example, the historic information of past 36 months of actual quantities of magnets ordered could be modeled to predict or forecast the next 12 months of demand. However, if we know that an automobile company is going to buy a lot of speakers for its new line of automobiles next month, obviously, that is going to have an impact on the 12-month forecast. Unfortunately, ARIMA model or the seasonality decomposition model (Hyndman and Athanasopoulos 2018) cannot take this information into account as these models take only data points from the past and therefore a future event is not a data point which these models take into consideration.

This limitation has been removed with the new machine learning-based forecasting models. Even the most basic neural network-based machine learning models can take multiple input, not just the ones that have happened before but also inputs like a future event. With more advanced machine learning models such as the recurrent neural network, the forecasting accuracy could be better than traditional conventional statistical forecasting models.

The higher accuracy is due to both flexibility and ability to consider a large number of inputs. Traditional models have specific formats and formulas which restrict its ability to ingest any kind of data. It can take in only those data which are needed specific to the formula in the chosen forecasting model. So these models take in only specific sets of data to produce the desired outputs. Machine learning is more flexible. In deep learning recurrent neural nets, the data provided do not need to have a specific form. Any number of data sets are fed in and the machine finds the best way to approximate it. That makes machine learning powerful, as it is more flexible. For example, when applying a linear regression model, we are limited by the model formulation that the response variable is a linear combination of the explanatory variables. In neural nets, similar assumptions are not necessary, instead inputs are provided to discover an approximate relationship. While the machine starts its learning process, it can come up with a linear or a curve model. That's one of its flexibility.

Machine learning also has its challenges. One of which is that it's harder to interpret because a lot of the machine learning models are like black box models. It takes in a bunch of input and produces an output. It's accurate, can take a lot of different data, accuracy is high but hard to interpret. Traditional statistical approaches are easier to explain as it is easier to identify the key contributing variable or reason. Another challenge is the amount of data needed to run a more complex model is significantly greater; in the context of the supply chain, as most data come from business transactions, there are not that many to begin with. As a result, traditional statistical approaches are often found to be more suitable.

15.11 CONCLUDING REMARKS

Supply chain analytics is an emerging field where new innovations in machine learning and artificial intelligence are being applied to solve supply chain problems and to monitor supply chain health to the minutest details. However, for a supply chain leader who or whose team is the primary user of the analytical tools, it's important to have a high level of understanding to adopt, implement, and validate the outputs. The outputs from these mathematically intense tools have to ultimately make sense and add value beyond proving the obvious.

Supply chain tasks such as vendor selection criteria, vendor performance monitoring, and risk management can be significantly automated and defensible if they are driven by actual data. Without getting lost in the minutia of mathematical algorithms, a supply chain manager needs to have some broad understanding of how the analytical models are developed to validate and rely on the results. Inventory management and forecasting are two areas where the supply chain world has been pouring in a lot of analytics resources. For a supply chain leader it is critical to use their business acumen to understand the capabilities and potentials of these methods and keep their expectations in check specially when adopting a new forecasting or inventory management system.

ABOUT THE CONTRIBUTOR

Tony Chen is a data scientist at Facebook whose forte is turning complex problems into mathematical models. He is an expert in machine learning and statistical modeling techniques which he uses to tackle supply chain problems for IT infrastructure. Tony is an expert in stochastic modeling and other statistical methods to measure supply chain performances and do predictive analysis. Tony has a doctoral degree in Operations Management from the University of Washington.

REFERENCES

Alicke, K., Glatzel, C., Karlsson, P., et al. 2016. Big data and the supply chain: The big-supply-chain analytics landscape (Part 1). https://www.mckinsey.com/business-functions/operations/our-insights/big-data-and-the-supply-chain-the-big-supply-chain-analytics-landscape-part-1 (Accessed December 26, 2020).

Chen, C. 2019. Got bad supply chain data? Here's what you need to know for data cleansing. LeanDNA Blog. https://www.leandna.com/blog/supply-chain-data-cleansing/ (Accessed December 26, 2020).

Frost, J. 2020. *Hypothesis Testing: An Intuitive Guide for Making Data Driven Decisions*. State College, PA: Statistics by Jim Publishing.

Gomes, P. 2020. MTTR and MTBF, what are they and what are their differences? OP Services. https://www.opservices.com/mttr-and-mtbf/ (Accessed December 26, 2020).

Hadley, G.F. and Whitin, T.M. 2012. *Analysis of Inventory Systems*. Prentice Hall International Series In Management And Quantitative Methods Series. Upper Saddle River, NJ: Prentice Hall.

Heckmann, I. 2016. *Towards Supply Chain Risk Analytics: Fundamentals, Simulation, Optimization*. Wiesbaden: Springer Gabler.

Hillier, F.S. and Lieberman, G.J. 2010. *Introduction to Operations Research*: 9th Edition. New York: McGraw-Hill Higher Education.

Hyndman, R., and Athanasopoulos, G. 2018. *Forecasting: Principles and Practice*. Melbourne: OTexts. https://otexts.com/fpp2/ (Accessed December 26, 2020).

IBM 2020. What is supply chain analytics? https://www.ibm.com/supply-chain/supply-chain-analytics (Accessed December 26, 2020).

Linthicum, D. 2016. Cloud computing's elusive total cost of ownership. https://www.cloudtp.com/doppler/cloud-computings-elusive-total-cost-ownership/ (Accessed December 24, 2020).

Robertson, P. 2021. *Supply Chain Analytics: Using Data to Optimize Supply Chain Processes*. Oxon: Routledge

Shapiro, J. 2009. *Modeling the Supply Chain*. Boston, MA: Cengage Learning.

Stevenson, W.J. 2009. *Operations Management*: 10th Edition, page 581. New York: Mc-GrawHill Education.

Tayur, S., Ganeshan, R., and Magazine, M. 1999. *Quantitative Models for Supply Chain Management*. New York: Springer.

16

Longevity Challenges in Avionics

Ashok Das

CONTENTS

16.1 Objective 332
16.2 Overview of Avionics Industry...332
16.3 Macro Trends Impacting the Industry..333
    16.3.1 Distinction from Other Industries 333
    16.3.2 Long Life Cycle 334
    16.3.3 Obsolescence...335
    16.3.4 Long Tail...335
    16.3.5 Quality and Reliability...336
16.4 Supply Chain 337
    16.4.1 Products...337
    16.4.2 Regulatory Control...337
    16.4.3 Demand 338
    16.4.4 Reactive Changes..339
    16.4.5 Spares and Support..339
    16.4.6 Relationship.. 341
    16.4.7 Liability 341
16.5 Case Study 343
    16.5.1 Cross-Functional Collaboration.. 343
16.6 Individual Performance.. 344
16.7 Organizational Leadership... 345
16.8 Concluding Remarks... 346
About the Contributor 347
References.. 347

DOI: 10.4324/9780429273155-19

16.1 OBJECTIVE

This chapter will focus on the aviation industry, particularly on the avionics supply chain. Avionics is a global market which requires an extreme level of reliability and calls for a supply chain that can sustain products for long product life cycles. It is also highly regulated where both the suppliers and products come to the market through a host of certification processes.

16.2 OVERVIEW OF AVIONICS INDUSTRY

The avionics industry is a part of the larger aviation industry, also known as aviation systems. There are many segments in the aviation industry as it includes aerospace and defense. The larger segments are airframes, engines, etc. The aviation systems are also a substantial piece of business, approximately valued as a 200-billion dollar industry. What comes under the aviation systems is avionics and other components of aviation systems such as aircraft electrical management systems, mechanical and hydraulic systems, and landing gears. The worldwide avionics market is currently valued at approximately $70 billion and was growing at a 4.8% compound annual growth rate prior to the Covid pandemic. Covid has really reset the market in a significant way. It was targeted to grow to close to $90 billion by 2024 but due to the significant drop in air travel due to the pandemic, the growth will be readjusted (Markets and Markets 2020).

The avionics systems are essentially the brain of the aircraft. It includes aircraft controls, flight management systems (FMS), various display systems and computing network systems (CNS). This segment also includes health monitoring systems which collect millions of data points from subsystems and sensors which are responsible for successful operation of the aircraft. Monitoring this data continuously is also part of avionics. Software plays a key role in all of these systems or platforms. These platforms go into various kinds of commercial, military, and business aircrafts and helicopters. As each aircraft is an expensive piece of equipment, the customers try to get as long a life as possible from these machines. To support the long service lives of the aircrafts, there is a lot of demand for upgrading avionics. While there is a segment of the avionics business dedicated to newer aircrafts, there is an equally large segment dedicated to aftermarket upgrading. It is quite common for customers with aircrafts which may be 10 or 15 years old to ask for new avionics for the aircraft to bring in new capabilities in fighter or commercial aircrafts.

16.3 MACRO TRENDS IMPACTING THE INDUSTRY

Avionics has become a growing industry with a global footprint. The primary market segment is in North America followed by Europe and Asia Pacific. Overall the relative market rankings have stayed the same over a long period of time primarily due to high barriers of entry. North America has maintained its lead primarily due to the high amount of investment in research and development over the years. Similar to the pharmaceutical industries, the initial research and development cost is very high to develop the necessary expertise. In addition to this, the industry needs a strong university system to continuously produce skilled professionals with technical degrees to support growth. All of these factors make it difficult for new countries to quickly acquire the necessary capabilities and catch up and disrupt the avionics industry. Though, avionics companies as such are still dominantly North America and Europe based, from the customer base standpoint, the bulk of the growth is coming from Asia Pacific and other parts of the world (Mordor Intelligence 2020).

Some of the major Avionics producers are Safran and Thales from France and Honeywell, Curtiss-Wright, and Collins Aerospace in the US. Most of the avionics market leaders have been in business over a very long period of time and have been supplying avionics products to the aircraft manufacturers for decades. These companies have been maintaining long-term relationships and have closely developed their products to support generations of aircrafts manufactured all over the globe.

16.3.1 Distinction from Other Industries

What distinguishes avionics from commercial electronic products are certain very stringent requirements. First and foremost is safety. For any item which is part of an aircraft that may be going on a mission or carrying passengers, safety is the number one requirement. The next important requirement is weight. In order to be fuel efficient and cost efficient, being lightweight is critical for any avionics product. The next important factor is being rugged and dependable. Depending on the type of plane the avionics system will go into, a certain level of elemental stress is expected on the aircraft. Stress could be in the form of mechanical vibration when the aircraft passes through a storm or any other turbulent weather systems. The avionics has to perform with the same level of reliability and efficiency under normal circumstances when the plane could be flying on autopilot to an extreme situation where the aircraft is hit by rough weather. For example, if an aircraft control system is compromised in some shape or form, it can have a disastrous effect on the plane and its passengers. Recently, for a certain passenger aircraft, there was a software/system glitch which led to the plane malfunctioning severely. This led to worldwide grounding of this particular type of passenger aircraft. So, level of safety, reliability, weight, and ruggedness are key factors of the avionics industry which distinguishes it from normal commercial electronics.

16.3.2 Long Life Cycle

As the aircrafts are expected to have a long service life, so do the avionics inside the aircrafts. Avionics platforms continue to be in service over a long period of time and require maintenance and upgrades. From large commercial jets to fighter aircrafts, all aircrafts go through multiple civil or governmental owners, and during the process may upgrade the avionics several times. There are fighter jets which may have gone through three generations of avionics upgrades and could be still in active duty. Each piece of equipment costs a lot of money so every customer and end user wants to utilize the piece of equipment for as long as it can. So old avionics design lingers on in active duty.

For example, an aircraft sold in 2010 was installed with an avionics system made of electronics components which were available before 2010. Let's say it had an electronic circuit board card with a processor. Since then the processor had reached its end of life and was no longer manufactured after 2013. Assuming the aircraft has been in operation and the card malfunctions in 2021, the first option would be to upgrade it with a newer version of the card that is available now, provided the newer card is backward compatible in form, fit, and function. If that is not the case then the next option could be to replace with a card manufactured before 2013 if available in the spares inventory. A more expensive but sometimes unavoidable option could be to replace the larger avionics system if the card is irreplaceable. The complexity increases when the same card needs to be replaced for an entire fleet of aircrafts or the fleet is more than 20 years old. This type of demand from older generations of aircrafts is common in avionics (Richter and Walther 2017).

16.3.3 Obsolescence

The number one issue in the avionics industry is obsolescence of components because of old designs (FAA 2015). For example, avionics uses a lot of semiconductor-based electronics which are also used by devices in personal electronics, health care, computers, etc. However, the demand for these types of components coming from the avionics industry is much smaller in comparison to those industries. Consequently, the customers or the verticals such as smart phones, IT hardware, automotives, health care devices, etc., who consume most of the supply, end up controlling the supply space and dictating the supplier behavior. These industries also change their designs at a much faster pace. For example, leading smartphone manufacturers often come up with new generations of products every 6 months to a year. As the smartphone design evolves, the underlying components also transition at the same speed and bulk of the supply of the components switch to the next generation of the technology to support the newer design.

The avionics industry, however, does not change their platforms' designs so rapidly and that creates a challenge in the continuity of supply. The component manufacturers switch their production to support the bigger consumers and are not always able to support the demand arising from the avionics industry. As a result, the avionics manufacturers have to rely on components which are about to or already obsolete. An avionics supply chain manager may still have to procure those components and have to deal with obsolescence management to keep multiple generations of avionics components in production.

16.3.4 Long Tail

The second biggest challenge in the avionics supply chain arises from the need to sustain a long tail of products. As aircrafts can easily have 30 years of service life, a specific generation of commercial jet planes could get sold to different carriers multiple times during the course of their long service life and would be flying in different parts of the world. However, in order to sustain their service life, all the aircrafts will need a corresponding specific generation of spare parts. Some carriers may decide to upgrade to a subsequent generation and some carriers may not. Bottom line, there has to be spare parts from the same specific generation to keep the planes in the air. Therefore, an avionics manufacturer has to maintain inventory of a long tail of products across multiple generations. These spares get installed infrequently as it depends on when matching generations of platforms come for repair. This makes the avionics spares business a very high mix but low-volume business where the manufacturer has to maintain a lot of spares spanning over multiple generations against infrequent demand.

On one hand, avionics manufacturers have to deal with the cost of managing small volumes of a lot of spares, but on the other hand demand for older generation parts produces greater margins. For example, when a relatively newer generation of aircrafts are launched, it will require spares in a greater volume as the fleet in active service is growing. So the manufacturer may carry a larger inventory to support the growth platforms but will also carry tens of hundreds of parts for the smaller demand platforms. As the platform ages and demand slows down, the manufacturer at some point will end the life of the system and stop making it. It will also become difficult to manufacture as the subcomponents used will also reach the end of their lives and be no longer available in the market. Eventually, the only source of the spares would be whatever extra or spare inventory, the original manufacturer decides to build and stocks before it stops producing them. Customers would pay more for the spares as their availability becomes limited. It is also cheaper to replace a subcomponent with a spare than replacing the entire system with a newer one. So the spares start to fetch a higher margin for the avionics manufacturer as the platform ages. This prevents the manufacturer from getting rid of old spares inventory and has to carry it. (Satair 2020)

16.3.5 Quality and Reliability

The avionics industry has a lot of parts which can be manufactured by subcontractors who also manufacture products for general consumer electronics. However, there are often special electronic or mechanical components which call for specialized manufacturing skill sets from the contract manufacturer. The contract manufacturers need to be able to mass produce specialized electronic circuit boards which have to meet much higher safety specifications or military specifications. Often contract manufacturers need to have a separate aerospace division to meet the manufacturing needs for such specialized parts. Apart from the complexity of the electronic circuitry, there are higher levels of mechanical requirements and standards of metal works for cabinetry and other parts that are needed. The products have to be more rugged and be able to operate reliably in a moving aircraft under extreme temperature and other environmental conditions. These often required special types of material or metals going through special processes, which may require special welding and bonding technology which only specialized sheet metal companies can produce.

16.4 SUPPLY CHAIN

16.4.1 Products

In avionics there are different types of platforms such as flight management systems (FMS), air traffic services flight notification (AFN) systems, computing & network systems, and health monitoring systems. Most of these are essentially electronics-based systems. These are electronics circuit boards which are put together inside a metal cabinet using connectors and wiring harnessing to connect the circuitries inside. These products go through stringent assembly processes followed by extended testing for ruggedness where the platforms are operated with varying environmental conditions. The platforms go through testing under extreme vibration and shock for a certain period of time to ensure uninterrupted performance. However, in order to meet these test standards, both electrical and mechanical parts are designed in a specialized way so they can withstand the ruggedizing process.

For example, an existing fleet of passenger aircraft may go through an upgrade of its Wi-Fi system to take advantage of 5G which is the next generation of telecommunication technology (Avionics International 2020). As technology improves, existing FMS and navigation systems all need to be upgraded to keep the planes flying as both the planes and airports and aviation in the sky advances through technology.

16.4.2 Regulatory Control

The avionics industry is managed by regulations to ensure the necessary controls for mass producing extremely reliable platforms as well as to ensure security and export and import restrictions. The manufacturers, subcontractors, and component or subassembly manufacturers all need to have certain regulatory certifications to manufacture avionics platforms and components (Tip 2018). Some of the most basic certifications they need to have are AS9100 and AS9145 (SimpleQue 2020). AS9100 demonstrates that the manufacturer meets the basic requirements of quality, safety, etc. AS9145 puts additional standards on quality, approval controls, information security, export and trade regulations, etc. In order to meet these standards the manufacturers need to effectively demonstrate their abilities to put controls and manage their processes. So, there are two basic platforms in avionics: civil platforms and military platforms. In civil platforms there is a little more freedom with respect to procurement in terms of nongovernmental pressure or regulations other than Federal Aviation Administration regulations. The chosen suppliers have to have at minimum AS9100 and AS9145 certifications. Certain cases may require suppliers to be National Aerospace and Defense Contractor certified which is again an elevated level of certification. Apart from compliance guidelines, there are process flow downs for each major aircraft manufacturer which are the manufacturer's own technical and quality requirements. The suppliers for different avionics companies in turn need to adhere to them to stay as a viable supplier. For defense or aerospace, the list of suppliers becomes even smaller as there are more stringent requirements for export control certification. The suppliers need to adhere to these in order to be allowed to supply for government and military contracts and may need special certification issued by a specific governmental organization like the state department.

Due to all these tight restrictions, the supply base is pretty stable and mature and very few new suppliers are jumping into the space. It is a restricted supply base which also creates a capacity challenge especially when the airline market goes through a boom. Supply chain continuity becomes a challenge when the demand in the aircraft for the avionics platform increases as the capacity to increase production is relatively limited.

16.4.3 Demand

Overall there is a certain element of long cycle demand in the avionics industry. Typically the avionics manufacturers sign agreements with the aircraft manufacturers to produce a platform. These agreements run anywhere from 5 to 20 years. These agreements are done based on a certain level of year-over-year commitment or projection of demand by the aircraft manufacturer over the span of the agreement. The pricing of the platform takes into account this demand. When the span is short, say 4–5 years, the pricing structure could be firm and fixed or could be determined annually. On the other hand when the contract is much longer term like 20 years, then there could be stipulations that will be built in the contract which will dictate price changes over that period of time. However, outside of these contract periods there could be demand fluctuations due to changes in market dynamics. These market fluctuations also impact supply and hence price. For example, prior to the pandemic, the demand fluctuation was more on the increasing side. The demand was trending positively and customers' orders were coming mostly higher than the projections or commitments over the contract periods. However, the grounding of the worldwide fleet of a certain short-haul aircraft by aviation authorities globally swung the fluctuation to the negative side. The current pandemic also made a deep impact on the aviation and avionics industry as it created a massive demand fluctuation.

16.4.4 Reactive Changes

The type of events mentioned above often reshape the aircraft industry and its supply chain, as these events unearth issues which are often centered around passenger safety. Subsequently these result in strong repercussions on engineering design changes in all aspects of an aircraft including its avionics platforms. The avionics systems also get modified and enhanced to address the new passenger safety issues. For example, the recent event of grounding the worldwide fleet of a particular short-haul passenger aircraft has led to a large number of changes in aviation regulations. The way aircrafts are certified have undergone a massive change. That has added new demand on the control system of the aircraft and the avionics where there are new software and new systems being introduced. One of the biggest changes that has happened in recent times is hardening against cyberattacks. The control systems need to be able to protect themselves from being hacked. Protecting against malicious software has driven the cybersecurity requirements and has introduced a new set of design challenges that the avionics industry has to address.

16.4.5 Spares and Support

Earlier we have discussed the importance of spares in the avionics industry due to the long service life of aircrafts. Aircraft owners spend a significant amount of money in manufacture, repair, and overhaul (MRO) spending (Cooper et al. 2018). While excess and obsolescence is a challenge for any after-sales spares business, avionics is no exception; it is also plagued by unpredictable demand. Predicting the demand is a major challenge.

A portion of the customer base may prefer to get covered by a contractual agreement (CSA). Say a customer has ten aircrafts and puts a company under a CSA for 10 years, then it becomes very predictable for the manufacturer to set up a pool of spares based on the count of aircrafts and the contract duration. However, there is a large segment of customers for after-market spares, where demand just pops up. An avionics manufacturer has to deal with such drop shift demand quite often. So the overall demand is a combination of some long-term CSA-generated demand, as well as a large amount of drop ship demand. This makes the task of predicting or forecasting the demand of certain components quite challenging. In this era of inventory optimization, suppliers cannot keep on piling up components hoping or expecting that some spare order will show up within a predicted amount of time frame. So, on one hand the manufacturers need to tightly control inventory but at the same time manufacturers don't want to let go of this regularly and in some cases high margin spare parts marketplace. Manufacturers address it through a mixed approach based on historical usage data and criticality of parts. Based on the age of the parts and the age of the design, manufacturers may decide to store a fixed number of years' worth of finished goods supply on hand. The quantity could be determined based on historical consumption. In addition, the manufacturer may also procure and store additional raw material or subcomponents which may become or have already become obsolete and leave the rest of the subcomponents to be procured at normal lead time if they are relatively easily available in the market.

One of the biggest challenges in avionics spares and after-market service business is on-time delivery. As the manufacturers may not have the visibility of the demand especially with drop ship demand, it ends up impacting the manufacturer's ability to do on-time delivery. This then adversely affects its business relationships with large customers, who reward or punish a supplier based on the supplier's on-time delivery performance.

In addition to the manufacturers maintaining a pool of spares, there is also a broker network for spares. There is a group of electronics distributors who specialize in this area. Their business model is to source high demand obsolete parts and procure and store them and then sell them with a markup. The manufacturers often source really old and obsolete spares from this category of brokers and resell it to the end customer with a markup. Though there is a limited number of brokers, counterfeit electronics parts are a threat in the relatively open market. So it's very important to utilize a trusted broker or a certified broker to procure spares made by the original equipment manufacturer.

16.4.6 Relationship

Manufacturing contracts with upstream avionic components or parts suppliers usually span about 5 years from when the manufacturer has to produce and supply parts or subcomponents. Five years is roughly the amount of time for a platform to stay in production and get sold by the avionics company, and the upstream component suppliers are expected to produce their products to support it. However, there is a category of parts which are sole sourced. These are controlled parts which are either specially designed by the avionics platform manufacturer as part of their unique design or it could be dictated by or based on the downstream end customer's specifications. The downstream end customer could be the aircraft manufacturer who is going to buy the avionics platform or could be the airlines or authorities who own a fleet of aircrafts with the specific avionics needs. When a supplier is making such a sole-source part which is customized to requirements, the avionics company does not have any other sources but the specific sole-source supplier to procure these parts. So a dependency gets formed for continuity of supply.

Due to the long life of the avionics platforms (Aersale 2019), a supplier manager for the avionics platform company has to put long-term sole-source contracts to mitigate these risks. The sole-source suppliers are managed on a long-term basis which is dictated by contracts which can span 15–20 years. This binds the supplier to manufacture or support parts beyond the usual 5 years. These contracts between the avionics platform manufacturer and its upstream suppliers often mirror the contract between the avionics platform manufacturer and its downstream customer like the airframe. These contracts protect the customer both from price increase and continuity of supply. This also offers price and investment protection to the suppliers. Without this kind of contract, the suppliers may increase the price as they are aware that their customers have nowhere else to go and designing or finding an alternative takes time. However, in most cases, the suppliers are offered a 5–6 year projection or a volume of business by the avionics platform manufacturers. This is used as the basis for price negotiation. The negotiated price also allows the platform manufacturer to negotiate their own price with their downstream customer.

16.4.7 Liability

Similar to the pharmaceutical or medical industry, the airline industry has an enormous amount of responsibility toward its customers. This translates to taking responsibility and carrying liability for accidents and losses. When any such unfortunate event happens, an end-to-end root cause analysis is done to identify the responsible authorities who owned and operated the aircraft. The root cause finding extends to the manufacturer of the aircraft and ultimately to the platform inside the aircraft which has malfunctioned inside the aircraft. The final objectives is to pin point to the specific hardware or software component whose malfunctioning led to the incident. One of the key elements that an avionics supplier manager has to negotiate is liability with their suppliers. The amount of liability cascades down in the ecosystem. The cost or amount of liability is shared. For example, the component supplier who manufactures a part which will go inside an avionics platform will try to limit its liabilities. One of the biggest costs in the aviation industry is the liability that can come out of failure of a product that has been supplied into an aircraft which may be substantial in certain cases. Avionics companies themselves have to carry a certain level of a liability which itself could be in hundreds of millions of dollars. However, the supply base usually tries to restrict that liability limit so it is not possible for the avionics platform manufacturers to pass on the entire liability to its supply base. The suppliers would argue that they are not responsible for the workings of the entire system and interaction with other components even though the failure of the entire system may be pinpointed to a specific component. So the liability is not shared equitably. It is more of a much lower, more restricted liability that they share with the avionics manufacturer. A supplier manager has to negotiate with the supply base to minimize the exposure to the higher liability limit.

Similar gaps exist in warranty cost and support. As the cost of a component is significantly lower than the cost of the entire platform, it will be incredibly hard for a supplier manager to negotiate a contract where a supplier will accept responsibility disproportionate to the cost of what they are supplying. The supplier will always try to make things very specific to the part that they have provided and protect their exposure from being forced to take broader responsibility and or provide larger monetary compensation based on the entire system that has been built on the specific part. For example, a failure of an air speed sensor may cause an airplane's control system to misread wind speed. However, the platform could have had additional sensors for redundancy or self-testing to detect fault in the sensor or other alternate checks and measures. Avionics manufacturers are also responsible for providing the warranty for the entire platform. Component manufacturers usually restrict their cost only to replacement of what they have provided. This is one of the key negotiation nuances which a supplier manager has to perform with their suppliers and show how much cost validity can be transferred. While the supplier managers will always take as much as they can, nine out of ten times, it is more limited than what the avionics manufacturer has to commit to its customers.

16.5 CASE STUDY

This is an example where cost is managed through cross-functional collaboration. This is quite common as the avionics industry thrives on innovation. However, innovation often comes at a price which puts the entire business and sustenance of innovation in jeopardy. This is an example where a financial challenge is solved by engineering and supply chain working together.

16.5.1 Cross-Functional Collaboration

For an upcoming aircraft, several new avionics systems needed to be completely designed from scratch. The avionics industry was collaborating with the aircraft manufacturer to develop new power generation, and management system and remote data control, among other avionics systems. During the course of the project, one of the biggest challenges the avionics company started to realize was runaway development cost. As cost kept piling up while the systems were being designed, the avionics company realized prior to getting into the final phase of the development that unit cost will be prohibitively high.

Now the challenge was how to make the engagement economically viable. The challenge was given to a cross functional group. The objective was to identify what needed to be done to regain profitability in the near term instead of looking at a very distant future demand that would turn into profitability in the long term. And that is where cross-functional collaboration started as the teams came together and looked at the issue from various functional angles to identify what things could be done and also identify what are those elements in the design that need to be cut down to size. While engineering looked at tweaking design, the supply chain team reviewed the cost of labor and transport and identified the expensive manufacturing practices and where it was taking more labor hours. They looked at how to reduce those costs and make the production more efficient. The sourcing side of the supply chain looked at the savings that could be obtained by moving the manufacturing from high overhead regions to equally capable but lower overhead regions.

Based on months of analysis and financial modeling, a final optimized model emerged. The team jointly explored if it was possible for manufacturing to be transferred in its entirety to another lower cost manufacturing region which could make the project viable. If the project was to be moved to the best cost country where there was already existing aviation expertise and manufacturing capacity at a reduced labor hours, the project could become profitable within a reasonable time frame. In addition the supply chain ecosystem was developed and parts procurement was localized as much as possible. It was possible to do the entire manufacturing and testing and shipping directly from there to the aircraft manufacturer. This brought about a radical transformation in the cost structure and it further turned the entire projection in the long haul of 20–25 years to be more profitable compared to what it looked like in the recent past. This was a big display of the power of cross-functional team work and cross-functional mindsets where the engineering and supply chain teams played their respective roles and finance helped to drive the optimal decision.

16.6 INDIVIDUAL PERFORMANCE

For a newcomer in avionics or an individual contributor, most of the expected leadership traits are very similar to what is expected in any supply chain vertical. However, there are some specific nuances to the avionics industry. Avionics is a very regulated industry, which has products with very long life cycles and iconic aircrafts which have been flying for years. In addition, a lot of customers are governmental agencies or military agencies which have their own requirements and stringent requirements. Any individual who wants to make it in avionics supply chain as a career needs to be ready to face a lot of regulatory authorities and needs to grow awareness about a lot of regulatory requirements. One would really need to be well versed with all the nuances of business regulation, customer regulation, and understand the nuances and complexities which limit procurement options. The supplier manager needs to understand and be able to successfully manage business continuity within the constraints of who can be brought in as a supplier, who can be chosen as a manufacturing partner, who can be chosen as revenue sharing partner, etc. This knowledge is critical for every supply chain professional coming to this area and they should be ready to be able to grasp that promptly. For example, a sourcing manager coming in from another industry will find sourcing in avionics is quite different. They may take months or even a year to understand the difference and to understand the technical complexity, regulatory requirements when it comes to military procurement, and other compliance requirements that they need to follow. There are volumes to read and understand, rules and regulations like Federal Acquisition Rules (FAR) that need to be put to use in order to be able to interview suppliers. Learning these will be required to ramp-up as a supply chain professional in the avionics industry.

The other trait that is needed to be successful is to be able to develop a certain level of patience. Things don't happen quickly in a regulated industry because there are so many regulations and restrictive business practices with many levels of approvals. One needs to have patience. At the same time any individual who wants to succeed needs to demonstrate passion and energy. Perhaps the best way to describe this is to borrow from the concept of Four E's and one P (Welch 2005). A supply chain professional needs to have "Energy" and needs to maintain their energy and the stamina and their passion to get work done in spite of all the restrictive practices that are around them. They need to be able to "Energize". They need to make sure that people they are working with get some level of that energy and enthusiasm that they have. They need to have "Edge" where they can make decisions without fear, and then the most important element is "Execution". Ultimately it boils down to how as a leader the supply chain professional is executing and how they are closing a task that is under the purview. Those four E's are very important. Outside of that curiosity to learn about the industry, products, challenges, and ability to look around corners, a sense of boundaryless behavior is very important. These are key for not just the avionics industry but everywhere. One should be able to demonstrate leadership and influence without ownership over cross-functional team boundaries. These are some of the critical skills that leaders look at from an individual contributor coming into the industry or who wants to move up in the business.

16.7 ORGANIZATIONAL LEADERSHIP

The four key objectives for a leader are delivery, cost, cash, and people. Those are four pillars that leaders themselves and each of their leaders should be focusing on. Delivery is the most important key performance index (KPI) and on-time delivery with the big customers is most critical. The customers often use on-time delivery as the most important criteria in terms of their decision-making for their next orders. Similarly the leaders need to hold their supplier to the same level of accountability and grade them based on delivery performance as well as quality and cost performance. So as a supply chain professional being able to maintain on-time delivery is perhaps the number one KPI. The second, of course, is cost and how as a leader one keeps the overall cost of supply chain manageable. This is a constant struggle as costs trend up due to obsolescence and special requirements in the avionics industry and often the customers are not ready to accept that when it comes to price. So, how does a leader make their organization help the business through supply chain? Keeping cost low, producing cost savings by negotiating lower cost while maintaining delivery performance is critical. Third is liquidity. Cash on hand or liquidity is always king, so one of the biggest problems that companies face today is maintaining a strong cash position and a supply chain plays a direct role in this process. A leader has to help the business hold on to the cash as long as possible, through better on-hand inventory management and leveraging vendor-managed inventory programs, maximizing payment terms and using it as a leverage. Finally the most important aspect is being able to manage people and create a culture of success. Being able to build a cohesive and resilient team which can manage through the cyclical aviation business and manage delivery, cost, quality, and people themselves is ultimately the key to success.

16.8 CONCLUDING REMARKS

One of the most critical challenges in the avionics supply chain is to hold on the right quantity of spare parts for its customers. As planes get older, the subcomponents needed to build the avionics systems also get older or be at end of life. That means the replacement system cannot be built due to lack of one more critical component. Often they need to be replaced with new parts which would require a redesign and a requalification. Passing down cost to the customers is often not possible which puts the avionics supply chain in the critical path to find the optimal solution.

The avionics supply chain is often shaped by regulatory requirements change or from lessons learnt from critical events. This also puts liability risk on components and is a topic that is managed by the supply chain leader. Designing avionics systems requires integration of a lot of complex technologies. A supplier manager can optimize cost and create values through cross-functional collaborative leadership and also by listening to the customer and figuring out the right solution.

ABOUT THE CONTRIBUTOR

Ashok Das is a supply chain leader in the aviation industry. He has been in leadership roles for over 20 years in a Fortune 500 company, with the most recent 6 years in the aviation industry, where he leads sourcing for aviation systems business. His area of expertise spans across sourcing strategy development, sourcing operations and procurement. Ashok has a GE Master's in Supply Chain Management from the University of Michigan, an Executive MBA from IIM, Bangalore and is an engineering graduate from the Indian Institute of Technology.

REFERENCES

Aersale. 2019. Aircraft Life Cycle Management: A Breakdown of Your Aircraft Life Cycle. https://www.aersale.com/media-center/aircraft-life-cycle-management (Accessed December 27, 2020).

Avionics International. June/July 2020. What Will 5G Bring to the Aviation Industry? http://interactive.aviationtoday.com/avionicsmagazine/june-july-2020/electronic-flight-bag-applications-advance-including-those-with-nasas-tasar/ (Accessed December 26, 2020).

Cooper, T., Reagan, I., and Porter, C. et al. 2018. Global Fleet & MRO Market Forecast Commentary 2019–2029. Oliver Wyman. https://www.oliverwyman.com/our-expertise/insights/2019/jan/global-fleet-mro-market-forecast-commentary-2019-2029.html (Accessed December 26, 2020).

FAA (Federal Aviation Administration). Final Report November 2015. DOT/FAA/TC-15/33. Obsolescence and Lifecycle Management for Avionics. http://www.tc.faa.gov/its/worldpac/techrpt/tc15-33.pdf (Accessed December 26, 2020).

Markets and Markets. 2020. Avionics Market by End User (OEM, Aftermarket), System (FMS, CNS, Health Monitoring, Electrical & Emergency and Software), Platform (Commercial, Military, Business Jets & General Aviation, Helicopters), and Region - Global Forecast to 2024. https://www.marketsandmarkets.com/Market-Reports/commercial-avionic-system-market-138098845.html (Accessed December 26, 2020).

Mordor Intelligence. 2020. Commercial Aircraft Avionics Market - Growth, Trends, and Forecasts (2020–2025). https://www.mordorintelligence.com/industry-reports/commercial-aircraft-avionic-systems-market. (Accessed December 26, 2020).

Richter, K., and Walther, J. 2017. *Supply Chain Integration Challenges in Commercial Aerospace: A Comprehensive Perspective on the Aviation Value Chain.* Cham: Springer.

Satair. 2020. The top 10 risks the aviation industry is facing. https://blog.satair.com/ten-risk-in-aviation-industry (Accessed December 26, 2020).

SimpleQue. 2020. AS9145 – APQP and PPAP for Aerospace. https://www.simpleque.com/as9145-apqp-and-ppap-for-aerospace/ (Last accessed December 26, 2020).

Tip Technologies. 2018. The Importance of Quality in the Aviation, Space and Defense Industry. https://www.tiptech.com/blog/articles/the-importance-of-quality-in-the-aviation-space-and-defense-industry/ (Accessed December 27, 2020).

Welch, J., and Welch, S. 2005. *Winning.* London: Thorsons.

17

Bridging Innovation and Governance in Biotech

David Passmore

CONTENTS

17.1 Objective 349
17.2 Overview of Biopharmaceutical Industry... 350
17.3 Macro Trends Impacting the Industry... 350
    17.3.1 Drug Development Timeline... 352
    17.3.2 Leaders of the Industry.. 352
17.4 Supply Chain 353
    17.4.1 Contract with CDMO 353
    17.4.2 Choosing a CDMO Partner 355
    17.4.3 Protecting IP 355
    17.4.4 Managing the Relationship.. 356
17.5 Case Studies.. 356
    17.5.1 Changes Brought in by Pandemic....................................... 356
17.6 Individual Performance... 358
17.7 Concluding Remarks... 358
About the Contributor.. 358
References... 359

17.1 OBJECTIVE

This chapter will focus on the biopharmaceutical industry and its supply chain. The primary product of this industry is biological pharmaceutical compounds. Pharmaceutical products are typically produced by synthesis of chemicals. In comparison, biopharmaceutical products are medicines that are manufactured from living organisms such as bacteria, yeast, and mammalian cells. This chapter will focus on the supply chain and manufacturing aspects of biopharmaceutical medicines.

DOI: 10.4324/9780429273155-20

17.2 OVERVIEW OF BIOPHARMACEUTICAL INDUSTRY

The biopharma industry is dominated by large pharmaceutical companies who have internal manufacturing capabilities and their own supply chains and service providers. There is a large number of contract development and manufacturing organizations (CDMO) as well as contract research organizations (CRO) that provide services and supply to the biopharma industry. Traditional pharmaceuticals consist of small molecules which means the pharmaceutical drugs are made of organic compounds with low molecular weight. In the last 10 or 20 years, there has been a transition in the industry toward medicines which are produced from living organisms and may contain compounds derived from human, animal, yeast, and other microorganisms. Typically, these are different kinds of proteins which are generally expressed by cells, so essentially this can be described as cell-based protein pharmaceuticals. Now, as a result the pharmaceutical industry comprises both small molecule synthetic drugs and biologics (FDA 2018) which are generally protein-based products. The manufacturers of the latter are referred to as biopharmaceuticals.

CDMO are entities which provide services to the biopharma industry and those services include development and manufacturing of the biologic drugs. There are many CDMOs worldwide with global manufacturing capacities (Tyson 2020).

The main biologic markets (Globe Newswire 2020) are the largest economic markets such as China (Research and Markets 2019), Japan, the European Union, and the United States. Biopharma companies are the branded companies that sell end products, and CDMOs typically supply or manufacture for the biopharmaceutical companies. This industry comprises several very large biopharma companies that have their own supply chains and manufacture the majority of their own drugs; however, they may still go to CDMOs for extra capacity. Additionally, there are many small- and medium-sized biotech companies that cannot, have not, or would not invest in internal manufacturing capabilities and depend on CDMOs for manufacturing of their products.

17.3 MACRO TRENDS IMPACTING THE INDUSTRY

Supply chain for a biopharma company primarily involves managing and dealing with the CDMOs. This is especially applicable for the mid- and small-sized pharmaceuticals who primarily drive their production through the CDMOs (Beutin and Schmidt 2019). For the small- and mid-sized biotech companies, their primary concern is time, as time is more important than money. In a typical scenario they would raise capital from venture capital investment firms. At an early stage the companies will create a compound with the potential to become a drug and this phase is known as the discovery phase. Step-by-step they will transition their discovery to prospective drugs and then into actual drugs. In order to reach the final state they need approval from regulatory authorities such as the Food and Drug Administration (FDA). The whole process and the entire market is a highly regulated industry (Handfield 2012).

From the get go there is a tremendous amount of time pressure which requires the biopharma companies to move quickly from discovery research, meaning identifying early candidates and selecting lead candidates which would then go to CDMOs for good laboratory practice (GLP) process development and good manufacturing process (GMP) production. Biopharma companies would contract with CDMOs once they have a lead compound, while still in discovery phase but having a compound which is sufficiently stable for manufacturing CDMOs in large quantities. Again, the main pressure is about time and the goal is to move quickly from discovery into clinical development. In order to do that, biopharma companies need to work with a CDMO to develop a GLP production process. GLP (Slomiany 2009) is a little bit of a misnomer. It doesn't mean that in discovery there is good and bad laboratory practice. It is all GLP, but GLP is a technical term which relates to the level of oversights and regulation and quality control and the necessary documentation that is put in place to control production quality. It's a quality control in documentation of the manufacturing QC and using standard practices and methods that have been approved and proven to be highly reproducible to ensure product consistency. As part of manufacturing a drug which will eventually move from nonclinical development to clinical testing in humans, a GLP process is required to produce GMP-grade drug product, and that's what a CDMO will do initially for the biopharma. They will take the compounds to the next step from discovery which are researched under non-GLP conditions to GLP/GMP production.

Biotechs will execute contracts with CDMO organizations to manufacture their lead compound under GLP conditions so that it can be submitted to FDA for approval to do clinical trials. Once they receive approval from the FDA to do the clinical trials, a biopharma company will contract the CDMO to manufacture large quantities of the product which will be used for the next stage of studies. It's very much like prototype development where those different stages are highly regulated and the CDMOs have the knowledge, process, capacity, equipment, and resources in place to manufacture a biologic which is a significantly big investment in itself. CDMOs play a critical role in the clinical development process, and CGMP production is critical during post-approval manufacturing of products.

17.3.1 Drug Development Timeline

One of the key aspects of drug development is to understand the timeline. Though there is a lot of pressure to move quickly but within the regulated guidelines, some of that journey is going to be driven simply by the underlying science which has to continue until the drug compound is proven to work as intended. Nobody can necessarily control this part as it depends on the drug's mechanism of action and intended uses. Failure to establish clinical proof of concept (POC) is definitely one of the risks in this industry. Typically, biotech companies will spend 2 to more than 4 years in discovery to come up with lead compounds that will enter preclinical development with a CDMO. It will take a CDMO anywhere from 14 to 20 months to develop a GLP process and produce enough GMP material that would then go to clinical trials. And then once the actual clinical trial starts, it can take many years to move through and demonstrate the efficacy of the product in clinical trials. The clinical development process can take 5–10 years and could be a very long and expensive process, and during that entire length of time the CDMO has to manufacture the needed material. Biopharma will continue to need GMP-grade drug compounds to provide for human clinical trials which is costly especially over a long period of time.

17.3.2 Leaders of the Industry

Biopharma industry leaders are determined by who has the top selling biologic products. One of the top selling drugs currently is a tumor necrosis factor (TNF) inhibitor (Humira) produced and sold by AbbVie. Humira had sales over $19B USD in 2019 (Blankenship 2020). TNF is a part of the body's inflammatory response. Humira is indicated for rheumatoid arthritis and other inflammatory autoimmune diseases. So, it's the number one selling drug and has been for many years. Other top selling drugs are used for treatment of cancer. One example is Opdivo produced by Bristol-Myers Squibb which had sales of $8B USD in 2019. These are the two main indications for biologics: cancer and inflammatory or autoimmune diseases. So, biologics have been growing in number of sales over the last 5 or 10 years as cancer therapeutics and the related trends increase.

17.4 SUPPLY CHAIN

Products and Processes

As a supply chain person within a biotech or mid-sized biopharma company that has or is going to create a manufacturing contract with a CDMO, it is important that they understand the products. It goes back to understanding the basic science behind the product. However, while understanding the underlying science is important, that is not the primary purpose of a supply chain person. The most important thing for them is to make sure that the drug is made correctly; meaning that it is approved by the regulators and therefore the role is much more about understanding the process and deliverables of the CDMO.

A supply chain person needs to know about the manufacturing process, how it's done right, and the potential problems to avoid. The key skills would be understanding standard manufacturing processes, understanding the challenges like regulatory issues, process issues, the process science, and then secondarily the sort of the research behind the product. In addition to the process science, a supply chain person needs to detail nuances of manufacturing, what are the various standard practices and regulatory steps, and what's important within regulations and for obtaining FDA approval of the manufacturing processes. It is important to understand various manufacturing processes as there are different production cell lines and different processes have different features, benefits, and price points.

17.4.1 Contract with CDMO

Selection and contracting are two key steps to a biopharma company's journey with a CDMO. Starting from research discovery of lead candidate compounds and then transitioning them into GLP/GMP products is an extensive contracting process. A strong experience in handling such types of contracts is highly desired as these contracts may take 3–6 months to negotiate. One of the key decisions that a biopharma firm will make during this process is its choice of cell lines. Cell lines refer to the propagation of cellular culture which is used to manufacture the biologic. The CDMOs either have access to or have developed their own proprietary cell line. The cell lines have different features with respect to productivity. For example, higher versus lower productivity cell lines determine the grams of product per liter. It's advantageous to have higher productivity but that will also typically cost more because those cell lines are proprietary with the CDMOs or the licensee. The CDMO might itself be licensed and then sublicensed to the biotech company but the better cell lines will cost more in reality. So, it's a trade-off between cost and time (Sison 2019).

Hypothetically, if a biopharma has all the money in the world and is really focused on time it can choose a high productivity cell line even if it costs more. That will also impact the company's decisions around process development and all of these are negotiated in the contract with a CDMO. The process chosen will also determine recovery and yield. For instance, a biopharma firm can opt for a standard process which may not have the highest yield. Yield implies the percent of product that is recovered at the end. If this is low, in order to move faster, one has to produce more and that may cost more but will save time. On the other hand, the alternative would be to take more time developing an optimal process that has higher productivity, but it takes a longer time to achieve this. So, those are some of the typical trade-offs in the supply chain and the supply chain person negotiating the contract with the CDMO needs to understand these trade-offs (Downey 2016).

Once a CDMO is chosen a biopharma company is fairly locked with that firm. Of course it can always exit a contract, but that just means it has to start over again with another CDMO and go through a lengthy negotiation process again and start all over which is why it's rarely done. So, it's an important choice and there's huge switching costs. Therefore, a supplier manager has to get it right from the beginning.

A supply chain manager has to manage the long-term relationship between the biopharma and the CDMO during the drug development. As the biopharma goes from R&D and then through initial phases of FDA approval and human clinical trials, for a couple of years the firm is fairly well locked in with its initial choice of CDMO. Once it gets into clinical trials and if it looks like the product has safety and efficacy and it's time to scale up and increase the number of patients in the trial, it will need more amounts of materials to be manufactured. There are certain inflection points where a biopharma could afford to switch processes that were developed at one CDMO over to another one. That can happen, but even that's costly and has some associated risks. Because the process may not transfer from one to another smoothly, there are very high switching costs and hence the contract with the CDMO is a foundational piece in the whole equation.

17.4.2 Choosing a CDMO Partner

CDMOs typically bid for new contracts within biopharma and it is very competitive because if a contract is successful, there is a long runway. The global biologics CDMO world is a fairly fragmented market so there are many providers with about three to six industry leaders in the industry. In 2020 it was about a $9 billion industry. It is an expanding market which is growing at approximately 11% compound annual growth (Reportlinker 2020). The biopharma market is growing, which is thus driving the CDMO industry growth. The number of compounds which are in discovery and clinical trials, many of which require CDMO manufacturing contracts, is also growing.

There are high barriers of entry for the new CDMOs as the investment requirement is high. Investment requirements to establish the facilities and all the resources and people are significant. Hundreds of millions of dollars are needed just to go in and set up a proper facility. There is a clear and significant barrier of entry. However, as the market is big and growing at a rapid rate, there are a lot of companies and hundreds of CDMOs that service that $40 billion yearly market.

17.4.3 Protecting IP

Protecting intellectual property (IP) is critical in this industry. Typically the IP is really around the identity of a drug. In case of a biologic, it will be the identity of the protein or the protein sequence for example. It is a concern particularly in the early R&D stage or maybe before a biopharma firm obtains its patents. So, the main IP protection is to patent each of these biologic drugs which the biotech company owns. This is generally valid as long as the companies are operating in a country which is part of the international patent system. Some countries are not part of the international patent system and so in early discovery phases, a lot of biopharma will not really want to do business with companies in such countries, because they would be concerned about IP protection, which in this case it would be the biologic sequence. However, by the time it goes to CDMO, there isn't much to steal because by that time the biopharma firm typically owns the product and patents. The patents are probably issued before the contract is completed. The patents are even published and so actually it becomes public knowledge. However, by the time the product is with a CDMO, the biopharma company has a contract which safeguards them from IP contamination, infringement, or any other risk (Haehl 2017).

17.4.4 Managing the Relationship

Cost of the product plays a key role in managing relationships. They need to be managed as depending on the processes. Typically a supplier manager has to go and inspect the process and the people and get first-hand knowledge and experience with the site and confidence that the site is set up correctly and that processes are being run correctly. The supplier manager is ultimately responsible to provide feedback and ensure the processes meet the service level agreements and are on schedule. Another key expectation is delivery against quantity required to support trials and production. The CDMO has to be accountable for preparing and stocking all the materials that go into the process and on-time delivery (Ohms 2020).

17.5 CASE STUDIES

This is an example of how two of the key factors for biopharma, i.e. time and regulation, got accelerated when it came to solving a challenge. This is an example where the time pressure is not just felt by the manufacturer for faster time to market but also for regulators and customers. This posed a unique opportunity to the regulatory process and practice and may in fact shape the future of regulation and regulatory time periods which are so intrinsic to the industry.

17.5.1 Changes Brought in by Pandemic

This is a real life example that has come up due to the COVID pandemic. The pandemic is so urgent and so impactful on the world, it has impacted the entire pharmaceutical industry including its supply chain. The biotech and biopharma companies and regulators such as the FDA and similar regulatory authorities around the globe were compelled to get together and start exploring on what can be done differently to speed things up much faster than usual to create a drug to abate the pandemic.

For example, one of the therapeutics that is being developed for COVID therapy is a type of biologic compound called therapeutic antibodies. It is a passive immunization, different from a vaccine, as a vaccine is something that's injected into a person and causes the body to develop antibodies which then protect from the actual virus. Vaccines take longer to develop but what can be done faster is identifying protective antibodies and manufacturing those and using them as passive therapeutics. Injecting those antibodies into people will provide 3–4 weeks of temporary protection while the antibodies last in the body.

That is one strategy which is being pursued, and several of those are in clinical trials now. So, the companies and their supply chain managers and regulators got together and explored what are the minimum steps that absolutely must be followed and how the processes can be modified to get these therapeutics into clinical development as quickly as possible. A lot of the solution is around process and will call for the manufacturing process to be modified and a lot of the preclinical development steps to be shortened. A supply chain manager working for this kind of product has to have full understanding to deliver products which are created through these new steps while keeping focus on safety as they would still need to prove the therapeutics are safe for clinical development. That'll happen in the upcoming 4–6 months. So, the industry has changed what would normally take 2 or 3 years to get it done to 4–6 months.

Depending on the expected success of this, the industry may review and consider whether in future other drugs can get approved through a similar accelerated pipeline. If a COVID vaccine can be done from the lessons learnt, what prevents the process from being extended and followed all the time? There's some risk and in this case, the target is a virus and so it's something external to the human body versus a target that's native to the body like a cancer cell. Therefore there are fewer safety concerns when targeting something like a virus, but at least for the former category of target, the process could become accelerated if the vaccine is successful. This would impact everybody in the drug development process including the supply chain. So, even in this disruptive time the companies and regulators are able to speed up the process significantly and potentially revolutionize development times (FDA 2020).

For certain new cases like the Pandemic where global regulatory bodies have significantly shortened the approval process based on the urgency to move quickly and where one can be relatively assured of the safety, the drug development process will change in future (Dutton 2020). In this case it has been applied for antibody therapeutics and has moved quickly through the phases of the drug development process. As the antibody therapeutic demonstrates efficacy through the testing period it will ultimately become an approved therapeutic against the pandemic.

17.6 INDIVIDUAL PERFORMANCE

The employment prospects are quite good in the biopharma supply chain. Finding the right people is one of the key challenges for these industries. Therefore, the career prospects are very promising for people that want to work in this industry. For somebody who is not in or coming into the industry, understanding the regulations and the process is very important. At the entry level, typically this industry will hire someone who may not have any experience but has a graduate-level degree or equivalent background in discovery research. The entry-level folks are expected to gain in-depth knowledge of the regulations and the manufacturing processes and their challenges. They need to understand what a correct process will look like and whether the FDA is likely to approve. They also need to learn how to deal with situations where a drug development process is put on hold and understand what has failed in the manufacturing process or what has not been done to specifications which held the clinical development. They need to understand the risks in the process and whether it could be fixed or need to start over.

17.7 CONCLUDING REMARKS

The biologic market is driven by a large global demand as it offers some of the most advanced therapies for cancer, autoimmune, and other conditions. The biotech companies use cutting edge tools and processes to develop these drugs which are produced from living organisms or contain components of living organisms. While the drug brand owner develops the drug, the actual behind the scene manufacturing is often done by CDMOs. The relationship management is critical for the timely release of the drug into the market as the CDMOs are ultimately responsible for adhering to the GLP/GMP and other manufacturing-related processes which are instrumental in getting approval and maintaining manufacturing consistency of the drug.

ABOUT THE CONTRIBUTOR

David Passmore is an accomplished biotechnology professional who has been an active member in the team developing a new drug Nivolumab (Opdivo®), used for treating certain types of cancer. David has led numerous business development and drug development processes for antibody-based therapies for cancer and immune-related diseases. As a business development leader, David has worked extensively with CDMOs and in the discovery and early development of biologic drugs. David has an MBA from Pepperdine University.

REFERENCES

Beutin, N., and Schmidt, H. 2019. Current trends and strategic options in the pharma CDMO market. PWC. https://www.pwc.de/de/gesundheitswesen-und-pharma/studie-pharma-cdmo-market.pdf (Accessed December 26 2020).

Blankenship, K. 2020. The Top 20 Drugs by Global Sales in 2019. Questex LLC. https://www.fiercepharma.com/special-report/top-20-drugs-by-global-sales-2019 (Accessed January 2021).

Downey, W. 2016. Biopharmaceutical Contract Manufacturing Contract Negotiations. Contract Pharma. https://www.contractpharma.com/issues/2016-04-01/view_features/biopharmaceutical-contract-manufacturing-contract-negotiations/ (Accessed December 26 2020).

Dutton, G. 2020. How COVID-19 is Changing FDA Approval and Clearance Processes. Biospace. https://www.biospace.com/article/how-covid-19-is-changing-fda-approval-and-clearance-processes/ (Accessed December 26 2020).

FDA 2020. Coronavirus Treatment Acceleration Program (CTAP). https://www.fda.gov/drugs/coronavirus-covid-19-drugs/coronavirus-treatment-acceleration-program-ctap. (Accessed December 26 2020).

FDA. 2018. What Are "Biologics" Questions and Answers. https://www.fda.gov/about-fda/center-biologics-evaluation-and-research-cber/what-are-biologics-questions-and-answers (Accessed December 26, 2020).

Globe Newswire. 2020. Global Biologics Industry. https://www.globenewswire.com/news-release/2020/12/02/2138318/0/en/Global-Biologics-Industry.html (Accessed December 26 2020).

Haehl, K. 2017. Do You Make These 3 Mistakes in Your CDMO Contract Negotiations? https://www.outsourcedpharma.com/doc/do-you-make-these-mistakes-in-your-cdmo-contract-negotiations-0001 (Accessed December 26 2020).

Handfield, R. 2012. *Biopharmaceutical Supply Chains: Distribution, Regulatory, Systems and Structural Changes Ahead.* Boca Raton, FL: CRC/Taylor& Francis.

Ohms, C. 2020. The 4 Most Important Elements of a Successful Sponsor-CDMO Relationship. Outsourced Pharma. https://www.outsourcedpharma.com/doc/the-most-important-elements-of-a-successful-sponsor-cdmo-relationship-0001 (Accessed December 26 2020).

Research and Markets. October 2019. Biologics Outsourcing Industry Outlook to 2023- The Fastest-Growing Regional Market is China. https://www.globenewswire.com/news-release/2019/12/17/1961396/0/en/Biologics-Outsourcing-Industry-Outlook-to-2023-The-Fastest-Growing-Regional-Market-is-China.html (Accessed December 26 2020).

Reportlinker. 2020. Biologics Contract Development and Manufacturing Organization (CDMO) Market - Growth, Trends, Forecasts (2020 - 2025). https://www.reportlinker.com/p05986883/Biologics-Contract-Development-and-Manufacturing-Organization-CDMO-Market-Growth-Trends-Forecasts.html?utm_source=PRN (Accessed June 1, 2021).

Sison, R. 2019. CDMO Selection: 3 Questions to Save Time and Reduce Stress When Reviewing an MSA. Outsourced Pharma. https://www.outsourcedpharma.com/doc/cdmo-selection-questions-to-save-time-and-reduce-stress-when-reviewing-an-msa-0001 (Accessed December 26 2020).

Slomiany, M. 2009. *The Indispensable Guide to Good Laboratory Practice (GLP)*. USA: Pinchurst Press.

Tyson, T. 2020. The Future of the CDMO Market: The Art of the Possible, Pharma's Almanac. https://www.pharmasalmanac.com/articles/the-future-of-the-cdmo-market-the-art-of-the-possible (Accessed December 26 2020).

18

Global Trendsetting in Apparel

Anupama Kapoor

CONTENTS

18.1 Objective 361
18.2 Overview of the Apparel Industry 362
18.3 The Brands and the Manufacturers 363
18.4 Macro Trends Impacting the Industry Landscape 364
18.5 Supply Chain Management.. 364
    18.5.1 Shift in Perspective... 364
    18.5.2 Product.. 365
    18.5.3 Logistics 367
    18.5.4 Supply Base... 368
    18.5.5 Trade Regulations... 369
18.6 Case Study 371
    18.6.1 Dealing with Volatility 371
18.7 Individual Performance...372
18.8 Organizational Leadership...374
18.9 Concluding Remarks...375
About the Contributor...376
References...376

18.1 OBJECTIVE

In this chapter we will focus on the apparel industry and discuss the nuances of its supply chain. Some of the unique nuances are the sources of volatility and unpredictability which can stem from changes in fashion trends or preferences or from shifts in trade relationships between nations. This is also an industry which is fragmented between local brands which compete with global brands and the supplier managers have to always think on their feet and trust their judgement and organization's sense of fashion instead of

DOI: 10.4324/9780429273155-21

heavily relying on data analytics and expecting the business to be repeatable, which makes this vertical somewhat unique.

18.2 OVERVIEW OF THE APPAREL INDUSTRY

The apparel vertical extends the trends in the fashion industry into a design concept for products for consumers which shifts seasonally (Wang 2019). That concept is then translated into actual physical products and goes all the way from development of the product to sourcing to actual manufacturing and then the production, shipping, logistics, arrival, and going to the marketplace via distribution to retail outlets. It involves determining the pricing, executing marketing plans, and then finally creating revenue through sales to the consumer. The apparel vertical covers all generations and fashions. It consists of all types of garments from pants to shorts to dresses to socks and all such categories.

The whole process from design to production to sales can be referred to as the end-to-end cycle. The length of the cycle time varies by company and by product. There are corporations who are efficient and have shortened the cycle to increase their speed to market and it could be as short as 6 months. However, typically these cycles average anywhere from 12 to 15 months.

Typically, the design concepts are developed in-house by taking trend directions and conceptual ideas for what a brand wants the season's offering to be and converting those ideas into an actual design. The designed product is then translated by a product development organization which then works with suppliers to mass manufacture. For example, the process will involve taking a design into a prototype, which is then reviewed to include the models for fittings analyzed for silhouette, styling, color, etc. Simultaneously the pricing is determined based on material, complexity of manufacturing, and checked for viability to actually hitting the brand's margin and price point. The products need to make sense and developers need to think about whether a design really translates to price value for the consumer? Is there going to be any risks to production? Will people across generations like it? Is the manufacturing scalable for mass production or will it have major challenges due to fallouts in manufacturing? What could be some of the issues in terms of manufacturing that could change based on capability or capacity to handle technical issues.

The supply chain team has to analyze all of these risks and has to make sure that it is translatable when it comes to production. Only after doing that kind of scrutiny, the product gets released for manufacturing and then gets handed to the downstream supplier management teams. Then the next step in the supply chain process is to create the purchase orders and the actual production starts. Simultaneously the product is passed off to the marketing, sales and merchandising teams to determine how to get the products faster into the hands of the customers. The path could be through wholesale partners or through the brand's own retail stores or through e-commerce platforms. During the process of brand merchandising, sales and marketing team decides on how they want to showcase it to the consumer, how the point of sales marketing would be done, etc. They will determine the price-value equation for the consumer, if there is going to be any special messaging to the consumer, whether they are playing up to a sustainability perspective, how they connect with the consumer, etc. This is how the end-to-end process is planned.

18.3 THE BRANDS AND THE MANUFACTURERS

Apparel is a massive industry. The global market is around 1.5 trillion dollars (Shahbandeh 2020). The largest markets are the US, European Union (EU), and China. The Asia Pacific region has one of the highest densities of population and is the fastest growing market. The global apparel market like most other industries has been critically impacted by the pandemic which is likely to change this growth trend.

All over the world there are additional local brands that compete with larger global brands. The US market itself is more than a hundred billion dollar industry, but it's a fragmented market. In the US, there are retailing giants such as Walmart, Target, and Amazon. Almost all of them have both retail store or e-commerce presence. The e-commerce stores push significant volume and they sell both national brands as well as private labels. There are other big players who are major department stores such as Macy's, Kohls, Nordstrom, and Bloomingdales. There are also world-renowned brands such as Levis, Wrangler, PVH (owner of popular brands such as Tommy Hilfiger and Calvin Klein), Abercrombie, American Eagle, and Nike. There are also brands who make specialized apparels such as Activewear or outdoor sports gear North Face, and Patagonia. The market also has a lot of boutique brands that have loyal followings or local consumers. There are also a lot of international brands such as H&M, Zara, and Uniqlo that participate in the US market and also sell into many countries, and so this creates a giant and fragmented market just in the US alone.

18.4 MACRO TRENDS IMPACTING THE INDUSTRY LANDSCAPE

Apparel is one of the most volatile industries (Ahmed et al. 2020) and one of the reasons is because in spite of its size, it's not an essential service. Its volatility stems from its complete dependency on the customer who is actually buying the product, what they are looking for, whether they feel like they need new clothes, etc.

On the other hand there is the trend that determines what product actually sells or does not sell and that is the influence held by some of the big fashion houses. There used to be a lot of trend direction coming from Europe at one point in time from the fashion shows out of Paris and Milan which used to drive trends. Now the market is shifting a little bit and Europe is no longer the only major driver. A lot of trend direction is coming from Asia as well. So, Japan, Korea, and China are driving a lot of the trend because they have a lot of buying power. There are a lot of macro forces playing into the industry as well which includes the current economic situation, the change in the supply chain dynamics in terms of availability of raw materials that go into manufacturing the product, the trade and the tariffs decisions that big countries and government are making, improvement in logistics and shipping and changing locations of manufacturing hubs, etc. As those macroeconomic forces tend to change, they end up continuously reshaping the supply chain landscape and influences how apparel is manufactured and how it finally gets to the consumer.

18.5 SUPPLY CHAIN MANAGEMENT

18.5.1 Shift in Perspective

For the apparel industry, the focus in the supply chain used to be solely on manufacturing and cost. Supply chain teams from branded manufacturers used to focus primarily on making sure that the manufacturing was set up properly for business continuity and the objective was to ensure that they were able to manufacture the product without interruption and were able to get the raw material continuously. The overarching goal was to manufacture at the cheapest cost location and be able to supply the goods on time when they were needed (Ha-Brookshire, 2017).

Now, the dynamics are so volatile that the focus of the supply chain has shifted. Cost obviously is still important but executives and supply chains are more focused on de-risking the supply chain. This is because there is so much forecast volatility which has been caused by change in market dynamics. There has been a seismic shift from the older generation to the millennials into those in Generation Z who have very different buying habits from older generations. Previous generations were slightly more predictable, and they were mainly focused on what the product was and their personal preferences versus the younger generations.

The younger generations are more self-aware and there is a lot of influencing that happens from their peer groups on what is happening in general and all over the world. They are conscious about taking into account aspects like sustainability and saving the planet, movements against discrimination or transparency around where certain products are manufactured and how they're manufactured. It's becoming more and more important for the supply chain to focus not just on the manufacturing and cost but also on what other macro factors could potentially impact, either the demand or the actual manufacturing of where and how they're manufacturing. This has been a big shift for the apparel supply chain in the last few years. The manufacturers are still learning in terms of how to react to these shifts and changes which makes the job of a supply chain even more difficult (Kochar, 2020).

18.5.2 Product

Successful supplier managers really need to understand their products. Understanding the product requires a supplier manager to learn what goes into manufacturing of the products like raw materials, raw material cost, raw material availability, raw material risk, and then into the actual manufacturing. It requires developing knowledge on various facets of manufacturing, for example, if there are specific techniques that are needed to manufacture a product, any technology that is needed, the actual manufacturing of the product, and testing of the product to test for ensuring the materials that are going into the products are within the guideline of the industry. Of course, expecting to understand the whole cost aspect of manufacturing is equally important for a supply chain manager. They have to ensure that the cost of making the product is still within the target margin requirements in accordance with the financial goals of the company. From there on it goes to the actual management of the supply chain from where the manufacturing is going to happen and what would be the lead time of manufacturing to delivery at retail outlets and customer's doors.

A supply chain manager has to determine the location of manufacturing and optimize in accordance to the proximity of raw material sources and where the target consumers are. Both have to make sense from a cost and logistics point of view (Fernie and Grant 2019). After decisions are made on what product is to be manufactured, where, and how, the next step is to decide how much needs to be manufactured.

Dealing with the volatility in real time is also an aspect of understanding the product. This requires developing a business acumen to assess how much the product is going to sell, who the intended end consumer is, and making sure that the manufacturing supply chain is flexible. There has to be enough flexibility, given the long lead time for retail manufacturing and products so if there was to be a shift in the demand closer to when the product hits the market place, the supply chain has to be able to flex up or down in whatever way is needed.

As an example, if the first lot of products hits the market place and the consumer starts buying and there is some kind of feedback either about the product design or quality, there needs to be some amount of flexibility in the supply chain so that the product could be quickly modified based on the feedback to what the consumer wants it to be.

In addition to being able to adjust the products design or messaging based on early feedback, the other requirement is ability to manufacture more quickly. For example, a product was launched based on a forecast sales volume of 100,000 units for a particular season. If it sells out faster than expected, production needs to be ramped and additional quantities need to be made within the season. The supply chain needs to have that flexibility within the raw material procurement, manufacturing, and logistics and that is something that needs to be thought out and planned while creating the products.

As an example let us consider what goes on in manufacturing a pair of jeans. Seventy percent of raw material that goes into manufacturing of Jeans is the actual denim material. Apart from the denim material there are a lot of other things that go into manufacturing the product. This includes the button, the thread that is used to sew the garment, the fusible material to give it the body, etc. Then the jeans go through a lot of processes such as laundry for washing, so there are a lot of chemicals that are needed in the actual processing of the garment after it's sown. From there the pair of jeans go into the packaging where the labels or the hang tag gets attached. In some cases some retailers may ask for stickers to put on jeans so the consumer can see what the size is. Then the product goes into packaging where it may require polybags and then it goes into a carton. These are some of the steps in manufacturing before it goes into the hands of logistics who delivers the jeans to the retail store.

18.5.3 Logistics

One of the major characteristics of the apparel industry is the volatility of demand. One way of handling the volatility is dividing the market into two segments: high fashion and basic requirements. The volatility is often more visible in the high fashion segment. For more basic products like T-shirts or jeans or casual pants, demand is a little more predictable. For some of these clothing articles in the basic segment, the forecast is mainly for replenishment of products. For instance straight cut jeans, which consumers buy day in and out usually get replenished twice a year. That is relatively easier for the manufacturers to forecast. But then when it gets into a bit more of the volatile segments, there are a couple different ways of tackling it. One is deciding on from where to source the product. For example, if the decision is made to source from an Asian country like Bangladesh for selling in the US market, the lead time between when the actual product is manufactured and shipped out to the retail floor could be as long as anywhere from 4 to 6 months. That is a pretty long lead time and so in order to be a little bit more agile and flexible in terms of responding to shifts in demand, a supply chain manager could decide to source an item closer to where the product is intended to be sold.

So, for example, the product could be manufactured in Mexico or in Dominican Republic or in Columbia which is closer to the US. In addition if the product is a really niche product which can be sold at a high price and profit margin, one could even make it within the US though the cost of manufacturing is typically higher in the US but it has advantages too in terms of logistics. So one way of injecting flexibility is by nearshoring it or manufacturing closer to home (Fernie and Grant 2019).

Another way of shrinking the lead time of getting products from wherever they are manufactured to the market place is by actually staging some of the products. Staging can be done for the raw materials so the manufacturers have access to additional amounts of raw materials on hand to be flexible. This cuts down the procurement time by 30–60 days by holding the raw material at the manufacturing location. In certain cases, the products are staged in a finished goods state. Rather than shipping everything at once and holding in the distribution center or at the retailer, the finished product could be staged with the manufacturer also and shipped as needed directly. So, that's the other thing that the supply chain is starting to do so that it can replenish closer and faster in response to the volatility in demand.

18.5.4 Supply Base

The supply base has been shifting pretty dramatically over the course of the last 10 years. Ten years back, the entire supply base was very fragmented. There were a few really big manufacturers which were large corporations which had global footprints and massive factories that were all over the globe. In addition there were a lot of smaller manufacturers who could keep their cost low as they were small and kept their overheads low. There used to be hundreds and thousands of small manufacturers in countries such as Bangladesh and China. That was the way of sourcing for a lot of companies who were looking for quick production, cheap cost, and fast fashion.

Over the course of the last 10 years, a couple of very unfortunate incidents have reshaped the supply base to become a lot more streamlined (Banerjee Saxena 2020). The incidents exposed the appalling manufacturing conditions which raised global awareness of the actual manufacturing environment and inadequate infrastructure. The negative image caused consumers to boycott brands and reject products with such negative images. Regulations and corporate social responsibility teams and watchdog organizations have since then significantly changed the supply base by putting safe guards.

Along with increased awareness, there are a lot of nongovernmental organizations (NGO) that are now becoming very active about policing the manufacturers and really questioning where the products are being manufactured. So, most companies and most brands are now becoming extremely careful about who they're sourcing from as they thrive on having a positive image. This has created a pretty dramatic rationalization of manufacturing. A lot of smaller manufacturers now have to demonstrate how they are shoring up the entire manufacturing infrastructure according to building guidelines, workers safety, and workers treatment and also demonstrate how raw materials are procured and overall processes are followed in an ethical and socially responsible way.

Apparel companies have adopted stringent laws about how they select or source manufacturing. They are pushing the supply base to understand what are the key terms of engagement guidelines and what are the minimum requirements for anybody to engage and manufacture products for their brands. There are also third-party agencies that are doing this kind of certification of suppliers now, and some of the big brands have internal compliance departments as well which certify any supplier that has either been brought into the supply chain or is currently manufacturing products. Over the last 10 years, there has been a lot of streamlining of the supply base. Some of which definitely has driven costs associated with using smaller manufacturers so the bigger corporations with larger factories and global footprints have obviously started winning a bigger share of the global market. They are being able to meet the standards that the brands are looking for. They have the necessary financial strength and ability to invest in more automated ways of working which call for more sustainable manufacturing practices, whether it's water recycling or making sure the building safety is shored up, installing cameras to make sure there is no ill treatment of workers on the floor, or setting up worker well-being program which could be providing them with medical facilities, education for their kids, and education in terms of basic health and hygiene, fire safety, etc. (Banerjee Saxena 2020).

Consequently the larger suppliers are winning the source base as they have the means to establish the processes and maintain it through audits, installation of cameras on the retail floors and other checks and balances. Most companies would rather engage with bigger manufacturing companies and have longer term relationships so that they significantly de-risk their supply chains and not have to worry every season about who they are going to manufacture with. The larger factories are being able to provide assurance of supply and the brands do not worry whether the supplier would be able to meet their requirements adequately and would the suppliers survive through economic instability such as a fire or natural disaster or a pandemic as has been going on in the last 6 months with Covid.

The pandemic has resulted in the shutdown of a lot of manufacturing capacity. A lot of the small manufacturers are without business and have not been able to survive. For corporations it becomes really important to have a stable and reliable source base rather than just focusing on cost and then quick replenishment which sometimes could prove to be too risky and so they have shifted their focus to sustainable sourcing (Muthu and Gardetti 2020).

18.5.5 Trade Regulations

Trade regulations play a significant role connecting apparel manufacturers to the consumers (Kunz et al 2016). Most of the apparel production in the world happens offshore from the country of consumption. For example, a lot of manufacturing happens in countries such as China, India, Vietnam, and Cambodia. The mass production has moved far away from the US or European Union. The production has moved away from the countries where a majority of the products are finally sold purely for cost and scalability. This has, however, created a major dependence on trade regulations and dynamics between the country of production and the country of consumption. This has a big impact on the supply chain and the entire apparel industry because the dynamics are very volatile and they even change from season to season. It sometimes changes with change in government. It has become imperative for companies to have strong governmental affairs and legal department to anticipate and manage the changing trade relationships. This is something which was a lot less impactful and a lot less risky in the past than it is today.

As an example, assume there were bilateral trade treaties between the US and Sub-Saharan African countries which allowed the countries to enjoy special trade agreements for apparel. The clothes manufactured in these countries could be shipped duty free to the US. These relationships change for whatever reason if something happens. Hypothetically speaking, if something was to happen in Nigeria or Kenya then the US might decide to take the preferential trade agreement away. If the US did not agree with some of the practices of one of the countries, say Kenya, it might then pass orders which would impact all products coming out of Kenya. Suddenly by those sanctions or changes in trade practices, instead of being shipped duty free, a 15%–40% duty and taxes could be imposed. A duty of such proportion would wipe out the entire cost margin of the products. Then it becomes a huge financial burden to the manufacturer and the brands as one or both parties have to share the cost of the tariff for anything that was manufactured out of Kenya coming into the US.

The cost cannot be passed to the consumers because the consumer is not ready to pay the extra amount and will simply choose another brand who is manufacturing out of a different country. Most of the time either the apparel manufacturer that is manufacturing it or the brand that's selling it have to absorb the additional cost. It impacts both the product and the financial profitability for both the manufacturing side as well as the brand side. Of course continuing in such a situation is obviously not financially viable in the long term. So, in such situations a supplier manager has to figure out how to reduce cost and move to a different sourcing country immediately. This is an important aspect of the apparel supply chain where the trade regulations constantly change and the supply chain needs to track changes and de-risk the supply chain.

18.6 CASE STUDY

This is an example of volatility in the apparel industry. Volatility happens both at sourcing end as well at the consumer end and mitigating risk is a constant challenge in this vertical. This example demonstrates how the ecosystem reacted to the pandemic and managed business continuity. This is an example where the supply chain had to be informative, innovative and had to react in real time to manage around market volatility.

18.6.1 Dealing with Volatility

Like every other global industry, the pandemic has also massively impacted the apparel industry. When the pandemic started, the industry was completely unprepared for what this would do at both ends of the supply chains. On the demand side, the shutting down of all stores caused demand to come to a screeching halt. Everything that was in the pipeline, i.e. in manufacturing, in shipping or even on the floor of the stores was basically stuck as nothing was selling. On the other hand, even the manufacturing source base came to a screeching halt as this was a global phenomenon and factories all over the world had to shut down because of Covid.

In a way everything came to a grinding halt for a little bit. But then the market dynamics started to shift toward online platforms as delivery was considered an essential service. In addition there were a few retailers such as Walmart and Target that were open mainly as providers of essential services and groceries. The apparel industry started seeing some pretty significant increase in demand for even apparel from these essential service and grocery stores which it had never seen in the past. The neighborhood providers for essential services and groceries also became the go to place for apparel consumers as they didn't have the ability to go anywhere else and shop for Athletic wear, Jeans, T-shirts, casual dresses, etc. So on the demand side, there was a sudden spike in demand for apparel coming out of online stores and grocery & essential outlets. So, apparel brands had to figure out the challenge of how to cater to this new demand while the manufacturing had already shut down. Apparel brands had to really identify what were the manufacturing locations that were open and still able to service their needs.

So, for example, China was the first one to shut down in December and January. But as the year progressed to late spring, China had already managed to curb the spread of the pandemic. Factories actually started opening up because they had gone through the peak and then they figured out how to deal with Covid and handle it and cases came down. So, manufacturing in China opened up again. That was an opportunity for a lot of manufacturers and suppliers to shift production in China instead of other countries where they would typically manufacture. There were also cases manufacturers moved production to Mexico, particularly for the US market, as there were still certain companies that were open or willing to manufacture products. They had put in pretty significant measures of how to manage Covid infection within the manufacturing area and were operating at 50% capacity.

Both the apparel brands and their manufacturers figured out how to change the whole manufacturing setup in the factories with social distancing, setting up protocols for testing within the facility, etc. Similarly the supply chain teams tracked local governments' restrictions and willing manufacturers. They figured out the whole manufacturing piece of where to source the product and where this kind of surge was happening as manufacturing for the most part was still shut down. The apparel companies worked with customers and clients to figure out if they were able to use existing inventory to service some of the needs. The existing inventory could have been originally targeted for high fashion and intended for higher end retailers. But as retailers were mostly shut down, the brands had to explore if they could redirect that material to essential stores mentioned above. Re-ticketing had to be done as these items were intended for a different price point and now had to be sold at massive discounts. However, sometimes it's better to sell something than not to sell anything at all.

So, those are some of the difficult decisions that a lot of brands had to take into consideration because they on one hand faced the prospect of zero revenue and on the other hand have demand coming in and had to find the products in the pipeline to service it.

18.7 INDIVIDUAL PERFORMANCE

For an individual contributor or somebody who wants to get into the supply chain in the apparel industry, the first requirement would be to develop an understanding of the products. Basic product knowledge is really important because there are so many nuances that go into manufacturing of apparel products which are very specific and very different from manufacturing in other industries. For example, for a machine or a computer, the design is pretty standardized in a product vertical and one can dictate when that product needs to change or should change. However, for apparel, there is no control over what the product should be or could be or will be because it is dictated by what the consumer needs, wants, and trends and shifts in the industry happen on a real-time basis.

As volatility is an integral part of this industry, an understanding of products is very important because they change so much and so frequently and have such a big impact on where to manufacture, how to manufacture, and what to manufacture. A supply manager will need to know what goes into the product in order to know whether they can change it or cannot change it and what impact it will have on the cost or actual lead time if they are to change the product dynamically. Understanding of the product and its time-line are two of the most important aspects.

Most manufacturing companies and brands in the retail industry have a go to market calendar. This means measuring and knowing how long it takes from the moment of conceptualization and design all the way to the product reaching the market place, which is key because there are so many different variables in the schedules and deliverables in the end-to-end process.

In addition, there are so many different people that touch the decision-making process in the end-to-end supply chain. Understanding and navigating this complex decision-making process and knowing who drives which part of the supply chain is very critical. There are so many moving parts from design, to product development, to forecasting, to understanding financials, to merchandising, to marketing, to laying out the logistics, to the actual manufacturing. Apparel has a lot of complexities.

A supply chain manager can work in any of the above aspects of the apparel supply chain. So making a selection of the specific area they want to work in needs to be well thought out. A supplier manager can start at one particular area and focus on one aspect of the supply chain but eventually need to develop understanding of the end-to-end supply chain in order to become a manager or an executive. Developing an in-depth understanding of each aspect of the end to end process is super critical because without that, it is going to be hard to manage the whole process.

In summary, in order to be successful as an individual contributor it is critical to develop understanding of the products, cycle, timeline, nuances of different aspects of manufacturing, shipping, and selling to the consumer. The ability to know when to be flexible and change because it's a very dynamic industry is also equally important as apparel is not an industry which has set processes, templates, or set ways of working which can be replicated season after season.

Sometimes 90% of things will change in a season from its previous season. It could be either because of product changes or manufacturing location changes from the previous season or the technique of manufacturing changes or the need for the manufacturing practices is different. Sometimes it could be caused by changes in trade regulations or changes in logistics and shipping lanes.

There is always something that is shifting in the entire manufacturing process. So, being able to learn continously, move and modify, be flexible, and also figure out how to adapt while still keeping the integrity of the product is very critical. It is essential for the supply managers to be able to think outside the box and think on one's feet to make quick decisions. Every day they have to deal with some kind of new situation so it's quite typical for a supplier manager to have the stamina to solve a different kind of fire every day that needs to be put out as something has changed and is not working the way it's supposed to be.

18.8 ORGANIZATIONAL LEADERSHIP

From a leadership perspective, the successful leaders in this industry are mostly the leaders who grew up through the ranks. As there are so many nuances to apparel and the way the apparel industry works, without knowing the ins and outs of the working, manufacturing, supply chain, and changing dynamics in the industry, it's very difficult for a person to come in and lead and make decisions quickly. There are plenty of instances where leaders, coming in from other industries, often find that the apparel supply chain to be behind in terms of automation compared to other industries. They may question why the apparel industry is behind in driving decisions based on data and analytics as compared to others. While that makes sense from an outsider's perspective, the reason why analytics has not proliferated as much in the industry is because of volatility. In order to apply data analytics, the systems require a certain amount of data that spans a few years or longer period of time to produce some kind of trends for providing directional guidance (Wong et al. 2013). The apparel industry data changes constantly making it challenging to produce reliable data for applying similar levels of analytics. Moreover, because of the volatility, the trend directions may even be less reliable because apparel is widely related to what a consumer wants at the moment.

So, it's really hard to use some of that automation and tools that could be used in a brick and mortar industry or in finance, and they cannot be easily implemented in the apparel industry. So, from an executive perspective, one has to find the right balance. In this day and age of technology, modern tools and data analysis technologies have to be used. A strong leader is expected to be forward thinking and can drive adoption of new technologies. However, at the end of day they need to find the balance of how much to drive based on their understanding of the constantly changing nature of the business and how much of it can be analyzed and automated.

This aspect of leadership is super important and that's something lot of industries outside technology are going through right now because we are in this digital age where everybody does not necessarily have the right tools to go digital and this industry is slow in adoption of modern technologies like 3D imaging technology, or new manufacturing technology to really automate the industry the way that is needed. There is a lot of work that needs to be done in terms of innovation and demonstrate foresight when it comes to the executive level. A leader has to be flexible, be able to think ahead, and solve problems on their feet.

A leader in the industry is also expected to be able to manage a large organization. Apparel is a people-heavy industry, from manufacturers to retailers to the people who work in stores to packers, etc. There is not a lot of stuff which can just be automated and done through machines. A lot of the processes are very heavy in terms of actual individual people. So, obviously the ability to manage large people-heavy organizations is something that the leadership in this industry must be able to demonstrate. At any point in time, a leader may need to be managing a thousand people organization as part of the supply chain. That is something the leaders of this industry have to be able to handle.

18.9 CONCLUDING REMARKS

The apparel industry is one of the largest industries in the world with over a trillion dollars. The market is highly fragmented. The global brands compete with the local brands. There has been a shift in the apparel supply chain as the Gen Z and millennials are more aware about from where their apparels are sourced and how they are being sourced and those factors influence their purchasing decisions. As a result of this the apparel industry has been shifting more toward larger manufacturers who have deeper pockets to ensure right processes are being followed while sourcing for apparel manufacturing.

Overall the apparel industry is very dynamic and new epicenters of fashion trends are emerging in Asia. For a supply chain leader, it's very important to understand the end-to-end processes as there are often shifts due to trade regulations, fashion trend changes, logistics challenges, seasonality, etc., which a supplier manager needs to adjust for while endeavoring to meet the seasonal fashion window. Understanding of these factors is also key for determining profitability and making a particular line of apparels viable.

ABOUT THE CONTRIBUTOR

Anupama Kapoor is a VP of Global Product Development & Sourcing at Levi Strauss. She has extensive knowledge of the global apparel industry and enjoys a unique and expert perspective of it through more than two decades of changing trends in fashion. She has set up numerous supply chain operations and qualified sources in Asia, Europe, Africa and Latin America and has streamlined the supply chain processes for Levi Strauss. Anupama is a graduate from the National Institute of Fashion Technology, India.

REFERENCES

Ahmed, I., Berg, A., Balchandani, A. et al. 2020. The State of Fashion 2020. McKinsey. https://www.mckinsey.com/~/media/mckinsey/industries/retail/our%20insights/the%20state%20of%20fashion%202020%20navigating%20uncertainty/the-state-of-fashion-2020-final.pdf (Accessed December 27, 2020).

Banerjee Saxena, S. 2020. *Labor, Global Supply Chains, and the Garment Industry in South Asia: Bangladesh after Rana Plaza*. Oxon: Routledge.

Fernie, J., and Grant, D. 2019. *Fashion Logistics: Insights into the Fashion Retail Supply Chain*. London: Kogan Page.

Ha-Brookshire, J. 2017. *Global Sourcing in the Textile and Apparel Industry*. New York: Bloomsbury.

Kochar, S. 2020. Top 9 Manufacturing Trends Shaping The Future of Fashion. Techpacker. https://techpacker.com/blog/manufacturing/5-top-trends-in-fashion-manufacturing-technology/ (Accessed December 27, 2020).

Kunz, G., Karpova, E., and Garner, M. 2016. *Going Global: The Textile and Apparel Industry*. New York: Bloomsbury.

Muthu, S, and Gardetti, M. 2020. *Sustainability in the Textile and Apparel Industries: Sourcing Synthetic and Novel Alternative Raw Materials*. Cham: Springer.

Shahbandeh, M. 2020. Global Apparel Market - Statistics & Facts. Statista. https://www.statista.com/topics/5091/apparel-market-worldwide/ (Accessed December 26, 2020).

Wang, Y. 2019. *Fashion Supply Chain and Logistics Management*. Oxon: Routledge.

Wong, C., Guo, Z, and Leung, S. 2013. *Optimizing Decision Making in the Apparel Supply Chain Using Artificial Intelligence (AI): From Production to Retail*. Oxford: Woodhead Publishing.

19

Bootstrapping Supply Chain in a Startup

John Jacobson

CONTENTS

19.1 Objective 378
19.2 Early Phase: Setting Up the Foundation 378
    19.2.1 Stabilize...378
    19.2.2 Undoing and Redoing...379
    19.2.3 Laying the Foundation 380
    19.2.4 Identify Key Hires 382
19.3 Launch: Rubber Meets the Road 382
    19.3.1 Launch Blockers... 383
    19.3.2 Lessons Learnt: Evolve the Supply Base 384
    19.3.3 Strengths and Weakness of Partners 385
19.4 Supply Chain Challenges... 386
    19.4.1 Repositioning Resources 386
    19.4.2 Multisourcing and Slicing the Business.............................. 387
    19.4.3 Relationship Management 389
19.5 Supply Chain Systems 390
    19.5.1 Prioritizing the Internal Pain Points 390
    19.5.2 ERP Systems...391
19.6 Case Studies... 392
    19.6.1 Dealing with Hyper Growth.. 392
19.7 Culture 393
    19.7.1 Dealing with Failures.. 394
19.8 Concluding Remarks...395
About the Contributor.. 396
References.. 396

DOI: 10.4324/9780429273155-22

19.1 OBJECTIVE

The focus of this chapter is on how to scale a supply chain organization from startup to maturity. As an early-stage startup starts ramping-up capacity to mass produce its products, the supply chain organization within the startup also starts growing. As this journey continues, it faces many kinds of challenges ranging from bringing up its supply base, building a team of people, forging relationships with its suppliers, implementing new IT systems, etc. All of these activities happen while the organization continues to design and develop new products and faces myriads of supply, quality- and product-related issues. The challenges evolve as the company moves from early phase to more mature stage and so do the teams, processes and tools. In addition to dealing with a plethora of challenges, bringing up a supply chain from scratch takes a lot of personal commitment and sacrifice. Startups have a different environment, and in order to thrive and grow it is very important to have culture and value alignment between the supply chain leaders and the organization.

19.2 EARLY PHASE: SETTING UP THE FOUNDATION

19.2.1 Stabilize

Typically a startup does not have a supply chain team or a dedicated person from day one. As it starts developing its first product, the leadership team realizes the need for a dedicated supply chain team. The initial evolution can happen in many ways. Supply chain activity starts with members from the product development team starting to do sourcing activities for components needed to build prototypes (Cohen 2015). Similarly, the finance team may get involved in paying for the goods and services and in the process dedicates a buyer and planner to issue purchase orders (PO) and pay invoices. In the same vein the head of engineering or CEO or head of operations reviews key contracts or takes external legal help to put contracts with suppliers in place. Quite often the members of the product development team involved in sourcing or the members of the finance team doing the actual procurement are involved in an ad hoc manner and have other tasks too. As the involvement grows, the supply chain team gets formed and it is often recognized by hiring a dedicated sourcing person or a buyer/planner. These folks may still report to the engineering or manufacturing or financial suborganization. Hiring a leader or a manager or head of manufacturing often marks the founding of a supply chain team.

Quite often the newly hired leader walks into a chaotic situation and needs to stabilize it. As the early members are executing supply chain functions in an ad hoc manner, they tend to have more of a task orientation, as they have to jump from fire fight of the day to the next. Surviving the day becomes the priority. A supply chain leader, coming in, is expected to stabilize the situation. They are expected to coalesce the activity owners into a team with certain dedicated functions and scope and with directional guidance, without dropping the balls already in the air. This stabilization process has three notable aspects which includes taking stock of the current situation, building an appropriate foundation and identifying key hires (DiResta et al. 2015).

19.2.2 Undoing and Redoing

Quite often the early supply chain decisions are influenced heavily by how well a supplier supports the engineering team, and how responsive they are to design changes and prototype deliveries. Other considerations (such as financial stability, quality control, cost competitiveness, ability to scale, etc.,) may not have been fully taken into account during the supplier selection phase. A startup typically has limited amounts of resources and is funded by the investors on a milestone basis. Completion of product development is typically a key milestone. Finding customer and mass scale production comes in the next phase as without the former spending time and resources on the latter gets lower prioritization. This is why often the decisions made in choosing a component or supplier hinges mainly on speed and not really optimized for cost, ensuring quality for high-volume production. The leaders who make decisions in the very early stage could be very engineering savvy but not necessarily supply chain savvy. Also sometimes decisions are made based on personal contacts of decision makers. For example, an engineering leader requests a contract manufacturer to build the product based on past working experience. Now the environment, volume of business, and working relationship between the previous organization and the contract manufacturer could be quite different from the current situation. The past expereicne could be based on a much larger scale where it was a stable high-volume-based business relationship and got appropriate attention from the supplier. The present scenario could be completely opposite. This may not necessarily be right fit in the new situation where the startup could be cash strapped and have a very limited demand.

A supply chain leader coming into the role may have to undo and redo some of these decisions, which may include component supplier selection and manufacturing partner selection. Making these changes can be painful, but failing to make these necessary changes to the supply chain can lead to even more difficulties down the road, when the business may not be able to scale or control quality, and the financial implications can be harder for a startup to handle. Deciding what changes to make and when, should be driven by data, based on sound analysis of the risks, and be aligned across the company's leadership team. If changes are to be made, they need to be done in a way so that risk of disruptions and delays are minimized. Sometimes the challenge is simpler where a supply chain leader comes in and can build on existing supplier relationships. Improved cost structure, payment terms, shorter lead times, are examples of improvements that can be made through negotiation and relationship building with key suppliers. These situations can become win-win where supplier performance and alignment can improve without impacting existing sourcing relationships, product design decisions or development schedules. However, certain choices could be tough where the previous leaders may have signed contracts which are not really favorable from a long-term perspective. The supply chain leader has to assess and is expected to develop an exit plan from such relationships or when the startup has inadvertently locked itself in a situation. Depending on the situation it could call for a quick disengage and reengage with a new supplier or some other long-term plans such as introducing a competitor to the existing supplier and moving business over time. A supplier manager has to be careful because sometimes the early decision makers may have given personal commitments or have emotional ties with the decision they have made and may be reluctant to follow the supply chain leader's suggestions. Objectivity and a data-driven approach are keys to building credibility. Demonstrating some quick wins helps in gaining trust from the leadership.

19.2.3 Laying the Foundation

As an early-stage startup starts developing its products it begins laying the foundations for the future. Organizations such as product development, sales, marketing and operations start taking shape. Inside these organizations functional suborganizations such as supply chain, quality and IT starts getting defined and their leaders emerge. Running the business during high growth stages is like driving a train while throwing track in front of the train. You have to build infrastructure and process while you scale, and growth is limited by how fast you can build. At some point, everything breaks – the key is knowing what will break first, and making sure you can address the issues. Hope is not a plan, so take a hard look at what is working, what is not working, and build a realistic plan. Along with the internal metamorphosis, the supply chain leader also starts transforming the external relationships with the supply base. One of the key steps is to understand the critical relationships and understand if those relationships are aligned with the goals and objectives of the startup.

There is no such thing as free lunch, a startup also has to pay for the goods and services it procures. Yet, successful startups create something beyond the transactional exchanges with its supply base. It creates a chemistry of mutual trust and belief on the prospect of long-term shared interest. Often a startup approaching a large-scale manufacturing partner gets disenchanted due to lack of interest from the name brand manufacturer. It could be because a high-volume player is simply not scaled to give high attention to a very low demand customer. A startup is a learning organization as it scales from an idea into a product. During the course of development to production, it learns what works and what does not from various perspectives such as engineering, manufacturing and customer expectation. And a major outcome of that process is to be able to adjust and change on the fly. This could mean the product has to change at the last moment from how it was manufactured with many other corrections. A supply chain leader is expected to have an eye for recognizing these needs and build relationships with partners who are also open to learning and iterating. For example, consider a supplier trying to mass produce a speaker which has traditionally required hand-crafted precision for quality, and the challenge would be how does it translate that level of quality in a mass manufacturing production line. This will require not only learning how to design the speakers in a mass manufacturable way but also investing in test equipment and changing the assembly process to meet the quality requirements. Both sides have to be eager to learn together and develop the new manufacturing process together.

It's a supply chain leader's responsibility to envision the direction of the future of the product development and manufacturing state and forge the right relationships. Startups have an infectious amount of positivity on its potential for innovation and success. Successful startups often garner sponsorship from its supply base where suppliers offer preferential discounts for goods in service in spite of low volume. Suppliers may also do additional investment in terms of offering resources in the form of additional technical or business support. Even putting the best engineer in a project with the startup is an investment from the supplier. The suppliers do that for their vested interest which could arise from their trust on the startup's leadership team, venture capital investor's clout, prospect of innovation and potential growth. While there could be a plethora of reasons from the suppliers, recognizing the right relationships and balancing between creating an advantage and not getting into a locked position which would be a disadvantage is very important for the startup. Counselling the leadership team to bet on the right suppliers and then maintaining ability to influence with the external peers is what supply chain leadership is all about at this stage.

19.2.4 Identify Key Hires

Hiring the lynch pin positions is undoubtedly the most important task at the early stage. The supply chain leader has to put together a team of people who can put together the right processes and right functional teams in place to cover sourcing, planning, logistics, fulfillment, spares, etc. Getting resources in startups is always challenging as the CEO has to constantly balance between demonstrating profitability and need for investment. In other words resources are precious in the supply chain and the supply chain leader has to be ready to get the position filled quickly but with the right type of candidates.

Startups typically lack tools and processes to maintain hyper focus on product development. The early recruits of the leader have to be well aligned and can deliver in a borderless and ambiguous environment. The candidates' cultural fit with the organization is the key along with ability to deliver, lead and recruit to build the team further. These lynchpin roles are typically for candidates with a dominant area of expertise such as forecasting and demand planning, logistics and commodity management. At the same time the candidates should have broad experience to get out of their core and seamlessly contribute adjacent functional areas.

19.3 LAUNCH: RUBBER MEETS THE ROAD

The supply chain organization gets really tested when the startup gets ready for the final release of its first product. After months or years of development, the prototypes have gone through phases of improvement and delivered a manufacturable product which has met all the testing and qualification criteria. Now it's the time to mass manufacture and turn on general availability and generate revenue. For a startup coming to this moment of truth could be tumultuous. It may find all kinds of issues which it may not have considered or planned for while being only focused on the development of the product. A contract manufacturer may have been able to produce prototypes during the development phases. However, due to small quantities of prototypes needed, the challenges which come with large-volume production start surfacing for the first time as and when the startup starts to rampup. These are blockers which can make and break a startup and need to be solved immediately. This could be a moment to realize who in the supply base is not really aligned and incapable of moving on and needs to be replaced for scale.

19.3.1 Launch Blockers

A startup was about to launch its first product when its leadership team realized they had a problem in hand. The supplier of the choice of the enclosure which was one of the critical mechanical parts started failing to produce and had quality issues. To successfully launch the product and meet customer demand, the launch volume needed to be 3000 units per month. In spite of committing to this volume, the supplier struggled to produce even a 1000 per month.

The mechanical enclosure turned out to be an extremely difficult part to manufacture in mass quantities. Part of it was due to the molding material that was used to build it. It was a type of epoxy material called bulk molding compound which is made of fiberglass, epoxy and wood chips. The material is formed using a compression-molding process to get its final shape. But the surface of the molded parts turned out to be very difficult to paint and achieve the desired surface finish. The supplier underestimated the difficulty of scaling the painting process and they ended up promising a lot more than what they could actually deliver. The company lacked the expertise on the sourcing team to adequately identify and mitigate the risks associated with using this material ahead of time.

The supplier had enough capacity to mold the enclosures, but it did not have enough capacity to paint them and support the launch commitments. The supplier was without any option and was reluctant to accept that they had a problem. These kinds of trying moments were not atypical in a startup where a single supplier can push it to a make or break point. The supplier dug in and refused to do anything except make more commitments to improve that were not likely to happen in the timeframe necessary. This was one of the situations where the supply chain leader had to make a tough call. In this situation it was communicating to the supplier to either find another supplier with proper painting capability and capacity, or else one would be chosen for them. The situation had turned into a standoff. There was a real possibility that the supplier would not cooperate and could even decide to quit the business altogether.

In this case the conflict with the supplier had to be resolved by coming to an understanding which would provide a short-term solution to unblock the launch. The urgency of the launch window and the limited cooperation from the supplier made the situation more tense. The supply chain leader ultimately convinced the supplier to sell unfinished enclosures to another supplier who could paint them and supplement the overall supply capability.

19.3.2 Lessons Learnt: Evolve the Supply Base

Two key lessons were learnt during the launch. First, the compression molding for the bulk material was more difficult to handle than anyone expected. The company lacked the expertise to source and develop this process, and was too reliant on a supplier who had a reputation for over-committing and under-delivering. When the company ran into trouble at launch, it had no resources with the right expertise that it could deploy in China to help solve the problems. Secondly, the supplier was too prideful and stubborn to put the needs of their customer above their own ego. This kind of supplier should never have been selected as a single source for this key component. On the first issue, the startup went on to hire additional resources that had the expertise and could be deployed on-site at the supplier factory as needed. On the second point, the start-up developed and executed a plan to qualify a second supply chain to mold and paint this enclosure. Eventually both suppliers got over all the quality and manufacturing issues associated with the material and painting requirements and ramped-up to a much higher monthly production volume. Again lessons learnt were surviving the short term with a creative solution that put together a set of partnerships both in the immediate term and longer term that enabled the company to scale the product volume. The approach created a lower risk path for the original vendor the time and space to improve because the additional supplier could support the launch with the extra paint capacity. In parallel, a longer term initiative that created a completely separate and redundant supply chain for the enclosure was put in place that enabled the product to continue to scale reliably and smoothly.

19.3.3 Strengths and Weakness of Partners

Launches and production ramps are probably the most challenging events because all of the supplier capabilities, tooling, assumptions, and plans get pressure tested. It becomes immediately evident, which suppliers are ready to scale and which are not. If the sourcing strategy is sound, supplier selection was done well, and the new product introduction process was robust, ramps tend to go smoothly. If any one of these areas was weak, those weaknesses can get amplified during a ramp. It is very important to have a strong supplier management and technical presence that can be on-site to support the most technically challenging manufacturing processes and/or weakest suppliers. Organizationally, this means creating global teams with the right technical and geographical presence to fully support the critical points in the supply chain.

In order to develop a good strategy that delivers both strong engagement during the new product development phase, and also the capability to drive cost efficiencies at scale, different suppliers are selected for different purposes. During the development phase, the startup may choose to engage a supplier who is really strong at NPI, for example, really good in supporting development of a product which is a high-touch process to support iterations between the startup and the supplier. The startup engages with it and the partner supports the launch. In parallel, a second supplier is selected that maybe doesn't have the capability to support the product development phase, but is really good at efficiently manufacturing parts that have designs that are mature. By building a sourcing strategy around two partners with complementary skills, you can achieve the best of both worlds: great support during product development and the ability to split the business in such a way as to leverage a lower cost supplier that can support larger scale.

In the contract manufacturing and component supplier world, there are partners who have very strong in NPI capabilities, have teams that communicate well in English, and can scale reliably. These companies are the right kind of partner to engage while developing the design, making lots of changes, and developing the manufacturing process. Once the design and manufacturing processes are fully established, and the scale of the business supports it, a second supplier can be added that has particular strengths in manufacturing but without the cost structure of a large support team., Choosing a second partner that may not have as many engineering resources, or fewer people that can interface with an English speaking development team, but has a strong manufacturing capability and a lower cost structure is a great way to fully achieve time to market, cost, quality and delivery objectives.

19.4 SUPPLY CHAIN CHALLENGES

In this section we will discuss more issues which may come up as the startup matures. As the startup grows, it gets to a state where it may have a portfolio of products in various stages of lifecycle and revenue starts ramping-up as well as the customer base and volume of business gorws. From a supply chain lens, during this phase the supplier list starts growing, second sources get added and risk mitigation strategies are put in place for continuity of business. As the supply chain scales, it also demands internal investment in tools and processes to keep the production machinery manageable. The teams, processes and tools all of them would need investment in this phase.

19.4.1 Repositioning Resources

Bringing up a supply chain remotely is very difficult. This is valid for not just during the early phases of design or supply chain bring up but all through the life cycle of a product. When building a supply chain in a different geographic region, there is a point in time when a startup must have boots on the ground to do the things that need to be done. If a design engineer in the US wants the contract manufacturer located in China to do a modification, communicating that via email or conference call will of course produce the needed results, but may cost several iterations. However, having people on the ground who will sit with engineers at the factory floor and getting them to do what the engineering at home wants will produce results at a different performance level. It will be much faster and with better accuracy and efficiency. A startup will definitely need quality engineers that can basically camp out at a supplier to oversee delivery of exactly what is needed and can lead a root cause corrective action process in real time. A supply chain leader will definitely need sourcing managers who can manage the commercial relationships with suppliers on the ground especially during an NPI phase because there are just so many things that come up. Being able to be in the meetings every day with the supplier and managing the product from the ground up is really important.

In the previous example where the startup had challenges with painting of the compression-molded mechanical part, it also discovered another issue stemming from geographical distance with the manufacturer. The startup had established an operations team in a particular far eastern country assuming the contract manufacturing would happen in that country. However, the contract manufacturer decided to move out from that location to another country. This event left behind an operations team who would support manufacturing but was not in a geographic position to support the business. As the startup started realizing that they were having major problems and not getting the mechanical product painted correctly, the operations team tried to remotely solve the issues and faced a lot of difficulty inspite of being in the same time zone and speaking the same language. This was a really important transition that needed to be managed and created a need for building an operations team in close proximity to the most challenging manufacturing processes and highest risk suppliers. Though the existing operations team had some of the necessary skills and knowledge, the velocity and urgency needed to solve the painting challenge made it obvious that the startup would not be able to scale due to the geographical distance between the existing operations team and the manufacturer. It ended up needing to build an entirely new operations team in the specific manufacturing region.

For launching any kind of product which may be technically complex or targeted for a premier consumer market having stringent cosmetic requirements or has extremely precise and low tolerance margin, achieving that by exchanging written specifications and contracts will yield the necessary result is not a pragmatic expectation. Qualifying a supplier based on presentations or initial factory visit may end up in a situation where a supplier over promises and under delivers. Even the supplier could be producing products for competitors which are extremely high quality, but believing that process would automatically be replicated could be far-fetched. Achieving perfection requires supervision, learning and an expectation that the higher the requirements, the higher the cost.

A supply chain leader sets this expectation and gets a budget for building an operations team colocated with suppliers or fights for travel budgets to send people at suppliers' factories for extended periods of time. Quite often the desired level of perfection does not exist in the supply base and cannot be achieved by escalations with the suppliers. It will only push the supplier into a frozen state where they do not know what to do or will not invest as the cost will increase making the endeavor unprofitable. A startup can only get past this by providing the expertise pretty much on the ground and wherever it is needed. Building up this capability is really a big part of scaling and success as well.

19.4.2 Multisourcing and Slicing the Business

Quite often, the standard approach in supply chain management is to do multisourcing as a part of natural progression for risk mitigation and business continuity. However, there are nuances which should be considered before taking this step specially with critical manufacturing partners and suppliers. A key factor to consider is whether the business is big enough to split into multiple suppliers. A contract manufacturer or key component supplier will be investing its precious resources, floor space and equipment capacity into the startup's business. It is important that the growth rate and size of that business opportunity are taken into consideration before any decision to split the business across a second (or multiple) suppliers. Adding a second source reduces the volume opportunity for the original partner, and this can also potentially reduces ththe strategic value of the business to either supplier.

A supply chain leader really needs to be careful about making these decisions. It can end up being in a situation where a supplier who is totally invested in the startup's products and management and is putting all their best resources and has given great support and then once a decision to split is made, it fundamentally changes the equation. At that point the supplier realizes the business isn't great anymore for them. And all of a sudden that introduces instability into how they think about the business, how they plan for it, how they set resources internally. And those can become difficult and challenging problems. Sometimes these challenges are not clear at the executive level for quick adjustments. For example, a supplier may move its top engineers to a different project. The people on day-to-day engagement with the supplier may realize that the key engineer is missing but may not be able to connect to the reason behind it and may not escalate. The repercussions of the decision to split business as taken by the startup's executive may eventually manifest as a product schedule slips at a later point but at that time the damage would be done. Such situations could be also detectable more easily if there are people on the ground working closely with the suppliers, otherwise one may never notice until all of a sudden things just start happening and nobody really knows why.

From a startup's perspective, there will be a point where it has to identify the threshold for splitting the business. It's basically a function of growth and how it can be managed. However, the counter to it is if a customer is below a certain dollar value or a certain percentage (e.g. 10%) of the suppliers business it may get challenged with just being marginalized. It's a bit of a balancing act as the startup also has to ensure that the current partner is growing and the business is developing for them in a healthy way. Also another perspective is to understand the sources of opportunity. A contract manufacturer primarily brings the cost of labor for integration which does not change by switching manufacturing partners. The cost savings may come more easily from the underlying material. Going after the components by second sourcing or going one level down or deeper into the supply chain probably drives more cost efficiency than just the manufacturing partner would.

19.4.3 Relationship Management

A big part of the scaling factor is also building the right relationships with the suppliers. Building a personal relationship is also a critical task for a supply chain leader, especially for a startup with limited volume and resources. Visiting the executives of critical supply base members is probably an extremely important task for a supply chain leader. As the startup would be making big investments into these partners, it's imperative for the supply chain leaders to know the key partners and get to meet them. It's worth mentioning the importance of that investment and having facetime. Especially doing business in Asia, it's like a personal thing. A leader could get so much done on handshakes in Asia as handshakes are probably more meaningful than some fancy contract. A supply chain leader can get critical business partners to pay attention to their business. It makes a huge difference if a supply chain leader takes the time to share the startup's business plan with a supplier's leadership team, provides updates on the health of the business, gets to know each other and knows something about their families as opposed to being just some business card in the suppliers' drawer. All these touch points end up giving a lot of options when things start going south, especially when demand does not materialize at the expected forecasted level.

In situations where the actual demand falls far short of forecast and the startup needs to cut production significantly in a quarter, a supply chain leader should be able to pick up the phone and have a conversation based on the close relationship. One can get a lot of help that way and get a lot of understanding that way versus just emailing a new spreadsheet or a new forecast. For example, a situation may arise when a startup may need the contract manufacturer to purchase more material for getting a more favorable price or meet a commitment that was made to the component vendors. This may mean the contract manufacturer has to carry more material in its books than normal. In such cases when a startup needs its manufacturing partner to pull in additional material or invest more resources, a supply chain leader should be able to leverage their relationships and make that phone call and get that kind of support. These relationships are often tested when the startup attempts to scale as the investment has to come first. If a leader does not develop those relationships, especially in Asia, the startup is not likely going to be successful.

19.5 SUPPLY CHAIN SYSTEMS

An important part of scaling up a startup is doing investment to build systems or tools and processes. While developing systems may require a lot of knowledge in information technology, data sciences, etc. but as a sponsor the role of a supply chain leader is to recognize when it is time to get a systems team together to implement supply chain solutions. The leader should be able to articulate, get funding and mobilize resources at the right time because there is a point where the organization has to have these systems come together, otherwise it cannot build a staffing plan, a sourcing plan, a product roadmap and a growth plan. If the fundamentals are overlooked, then the startup may start failing to take orders, to manage inventory and to fulfill the product shipment to customers, and it has to be all integrated in the approach.

19.5.1 Prioritizing the Internal Pain Points

A lot of startups get to a point of time where they really start struggling with their systems of record. It becomes painfully difficult to get visibility, to get the data that are needed to see what is happening. It becomes hard to keep inventory straight if a startup does not have great systems and processes around supply and inventory management. And all of these end up being things that affect really the nuts and bolts of the company. How long it takes to close the books at the end of the month is a function of how clean a startup can manage its transactions during the month and reconcile everything and close it by the end of the month. If it's out of control, the reconciliatory process can take 2 weeks to close a month. And that's obviously really challenging and difficult for finance teams who are trying to report to investors on a timely basis.

A supply chain leader who may not care about the nitty gritty of the technology that needs to be implemented, however, needs to be able to build a team to deliver and monitor its progress through milestones. As an executive, the supply chain leader is the final owner and is accountable for putting the right controls in place. They are also responsible for managing and delivering to the business.

One way of getting a budget and implementation resources is to leverage the growth plan. For example, a narrative could be as follows: the next year target is to reach $200 million in revenue. This year the revenue has been $100 million. In order to double the revenue, the first step is to develop an understanding of where the weaknesses are and what the barriers are from being able to grow at the target rate. That helps in identifying the solutions which will move the startup forward. Implementation of robust data management tools and processes often fall in this category which will help the startup to move forward and achieve its success metrics. No startup CEO likes to spend money on the ERP (enterprise resource planning) systems. It's not fun as it's not a new product or it's not a new marketing campaign and it definitely doesn't grow the top line.

So, these are investments that are hard to make because in some cases, a leader who has past experience using the tools may not have necessary concrete data to make his case. In order to obtain the necessary budget and support, a leader has to articulate the ROI (return on investment) and what success will look like if the investment is done. Its equally important to articulate the risk or the problem that may occur if the investment is not made. Often when there is no shared experience of a disruptive event, it is sometimes challenging to convince the leadership team for additional budget. It is almost like before walking across a rickety old bridge, predicting one may make it across or one may not as one of the boards may give away anytime. Convincing with what if scenarios require persistence and is always a challenge for expensive and extended ERP implementation. For small companies it often boils down to the leadership in making the call to implement and making the investment and committing to completing it. A lot of companies end up doing a partial ERP solution that does not get fully implemented.

19.5.2 ERP Systems

ERP systems usually have different modules. For example, there could be a financial module which is used for managing earnings, expenses, etc., and then there could be a manufacturing module to keep track of inventory, components, etc. Startups are frugal and usually the early implementations stems from the need to manage the finances. Then as operations start scaling, the need for more modules of the ERP system arises. This also creates a challenge as it requires significant resources for implementation. A supply chain leader often gets questioned whether the system in place is the right system to scale or not. A lot of times during the course of this evaluation the leadership team realizes that the current tool may not scale. Or the leadership may conclude that the previous ERP system was a wrong choice and fixing it is almost worse than starting from scratch. In either case a new challenge arises as the leaders have to figure out how to run the business today while putting in a plan in place to replace the existing infrastructure over time.

Often this may mean building a completely separate instance of the same or a new ERP system, setting it up and running it in a test environment and then finally cutting over to the new system at some point in time. These are huge projects that demand tons of resources both in terms of people across functionally and cash. Sometimes to control resources and budget the leader pushes off from making the decision to implement. Kicking the can down the road does not make the decision-making any easier later because there is never any natural window which just opens up for implementation. So often some tough decisions need to be made by the supply chain leaders along the way while balancing it with other priorities. Another unavoidable challenge arises from doing ERP implementation in an opportunistic but isolated manner as availability of budget ultimately dictates the scope of any implementation. So, instead of doing implementations in an integrated manner, inventory could be running in one set of ERP tools and purchasing and order management in another or they could be running on different instances of the same tool. The implementations should be done with an aligned view on future growth scenarios.

19.6 CASE STUDIES

This is an example where a startup ended up innovating in the manufacturing process along with suppliers. The supplier faced a lot of difficulty during this learning period but came out as a winner at the end. For the organization this arduous process required patience but ultimately came up with a competitive advantage of having a cutting edge supply base.

19.6.1 Dealing with Hyper Growth

A startup needed to mass manufacture a very high-quality speaker. The speaker industry had not really changed in the last 40 years and high quality was always achieved through hand crafting. In a way the manufacturing was done based on human experts who developed ears through years of practice.

Of course this manual approach to quality and precision limited the production capacity of the audio speaker supplier. To take this level of precision to mass production, manufacturing in much large quantities was the primary challenge. The supply base only offered the traditional audio expert based suppliers who had never sent high-quality speakers in very large volumes to contract manufacturers for final assembly into the entertainment systems. It was a completely different world as instead of using a pair of human ears to discern between pass and rejection in quality testing, an array of characterization instruments were to be leveraged. These would be used to test each speaker uniformly but stringently as incoming quality inspection at the contract manufacturer. This resulted in a high amount of failures in incoming quality inspection. The assembler rejected and returned the speakers back to the suppliers at a rate that the suppliers had never experienced before.

The startup ended up significantly influencing how speakers are manufactured with the supplier even though they had been in business manufacturing speakers that way for 30 years. There was definitely a new way to build the speakers that the startup had to impose on the supplier.

Key to all this process from a supply chain perspective was to find suitable suppliers that were capable of learning, interested in learning and interested in understanding the differences that the startup's business would mean to them. The startup's supply chain leader played a pivotal role in garnering executive support from these suppliers and to get them to come along and to look at challenge with a fresh perspective. The speaker manufacturers understood the startup's motivation and business model and their underlying growth opportunity for the speaker business. The suppliers were really miserable and unhappy for a period of time as the learning curve was awfully painful. But once the startup's engineering and supply chain team got them through it, they did continue and appreciated the effort as in the end the opportunity made it worth their while. It also opened up a business opportunity that allowed them to take a step forward as a supplier, got better and also added revenue to their businesses.

19.7 CULTURE

The culture inside a startup is often created by the early leaders. The persona of the organization develops based on the leadership styles of the leaders and how they react to crisis. Leaders get recognized for how they behave during crises and not during good times. It evolves as the early employees get feedback from the leaders. They learn based on the type of actions that get rewarded and recognized by the leaders and do more of those as opposed to those actions which receive discouragements. The behaviors start permeating through the organization via the functional leaders. A lot of startups are more evolved as they consciously manage it via employee communications, frequent all hands, employee calibration and by looking for cultural fit during the recruitment process. As a startup begins from a concept, its fruition often heavily depends on its culture. Cultural fit is, however, a two-way street. While a startup looks for cultural fit while hiring, potential employees should also assess whether the specific startup and its culture suits them. So it's important to understand cultural fitness while picking up a startup both as an individual contributor and as a leader.

Startups, particularly in high tech areas, are demanding environments. By definition, resources are limited. The company's very survival is at stake, and this means that jobs seldom fit nicely into 40 hour work weeks., Joining a startup, basically means you are signing up to do whatever it takes to make the company successful. In return, the employee is positioning him or herself to benefit financially by sharing in the value that is being created. A prospective employee may choose a startup primarily based on its growth prospects in an emerging field or because of the innovations the startup is promising to deliver. One may not really give much consideration to the cultural aspects of the company and could end up making a bad decision by joining a company which may not have a supportive culture. Sometimes startups are just really a collective of very competitive, talented and smart people that have come together but have no idea how to work together. There could be all kinds of internal strife, jealousy and egos, which could produce a miserable experience for its employees.

Companies having a unique culture where a prospective employee would be able to thrive should be a critical criteria while choosing a startup. Founders or early leaders should be able to authentically portray their experience, their confidence and track record of building successful companies. A prospective employee should get a sense of how they would be able to contribute to that success and growth of the startup as well as learn a few new things during that journey. Beyond the monetary prospect, the startup should be able to pique the prospective employees interest in the field of the startup and mission and vision of its leaders and generate genuine interest to be part of the startup.

19.7.1 Dealing with Failures

A startup often offers tremendous growth opportunities which are not available in mature companies. Great company cultures are often based on honesty and ability to make mistakes and be able to talk frankly about what the problems are and what is not working while continuing with the journey as a team. A startup culture recognizes the fact that its stretched in all directions and things happen that are not planned. It is important for a startup to have an open environment where one can come in and present an issue or present an honest assessment of what's happening and what needs to be done about it.

One of the humbling things about working in a startup is that there are failures that one has to deal with. Anybody who does not fail and does not learn is not being honest with themselves. The most important lessons are always around things that do not go well and things where mistakes are committed or a bad decision is made. Those are the things that an emerging leader in a startup takes away and those are the things that make them better the next time. It's really important to focus on and not be afraid to deal with the fact that one needs to learn from their mistakes and you being humble enough to recognize that one makes mistakes all the time.

So when choosing a company, it really comes down to the quality of the people that will be working together. So while weighing in the economic prospect, it is very important to assess whether the executives are really leaders or just a group of talented people with egos and baggage as it could ultimately be a blocker to the employee's growth and the company's as well. Executives could be charismatic or may have compelling personal stories and may have secured a huge amount of venture capital investment but may still be immature or have no past experience or have no idea of how to lead a company. Based on how the leader negotiates or how the company engages during the recruitment process are often early indicators of these issues.

19.8 CONCLUDING REMARKS

As a startup goes through several milestones to be successful. It goes through a series of fundraising and ultimately generates enough revenue to become cash positive and may become a public company. At every step the supply chain organization also passes through milestones like first prototype completion, launch of first product and then its mass production. As a team the supply chain team also metamorphoses from a single person team to a structured organization with suborganization focused on forecasting, planning, sustaining, etc.

This journey is, however, not easy and there are a lot of challenges. A startup with a strong and resilient culture and who learns from its mistakes can innovate its way to succcess. When a startup grows, its partners also grow along and therefore the strengths of these partnerships are equally important for both sides. The relationship with suppliers and partners is a critical part of success in a startup supply chain. A successful startup is a rare combination of a great product, great execution and a great culture with a boat load of luck. However, they are always great opportunities to learn and develop leadership skills.

ABOUT THE CONTRIBUTOR

John Jacobson is a serial supply chain leader who builds and scales up billion-dollar global supply chain operations. He has led numerous leadership roles and built high-performing global supply chain and operations teams which he has integrated across cultural and geographic boundaries. John is pre-eminent in developing the modern outsourced electronics manufacturing supply chain with uncompromising quality and customer satisfaction. John graduated from Stanford with a Master's in Manufacturing Engineering.

REFERENCES

Cohen, A. 2015. *Prototype to Product: A Practical Guide for Getting to Market.* Sebastopol: O'Reilly Media.
DiResta, D., Forrest, B, and Vinyard, R. 2015. *The Hardware Startup: Building Your Product, Business, and Brand.* Sebastopol: O'Reilly Media.

Section 4

Case Studies in Supply Chain Management

20

Case Studies in Supply Chain Management

Asish Datta

Sanjib Sengupta

Debasis Bagchi
Pharmacy and Health Sciences

Sudip Gupta

CONTENTS

20.1   Objective 400
20.2   Benefits of Consignment Stocking System against Supply
      Disruptions 400
      20.2.1 Learnings 401
      20.2.2 Key Success Factors 402
20.3   Technological Disruption in Supply Chain 402
      20.3.1 Learnings 403
      20.3.2 Key Success Factors 404
20.4   Capital Cost Savings in Exchange of Increasing Inventory
      Holding Cost 404
      20.4.1 Learnings 404
      20.4.2 Key Success Factors 405
20.5   Vertical Integration 405
      20.5.1 Learnings 406
      20.5.2 Key Success Factors 406
20.6   Strategic Outsourcing of Manufacturing for Low Value-Added
      Products 406
      20.6.1 Learnings 407
      20.6.2 Key Success Factors 407

DOI: 10.4324/9780429273155-24

20.7   Collective Negotiation for Procurement Efficiency 407
      20.7.1 Learnings 408
      20.7.2 Key Success Factors 408
20.8   Unchecked Cost Reduction Causing Supply Chain Disruption 408
      20.8.1 Learnings 409
      20.8.2 Key Success Factors 409
20.9   Concluding Remarks 410
References 411

20.1 OBJECTIVE

The compilation in this chapter is a 'Practitioner's view'. It draws its matter as well as inspiration from first-hand experiences in dealing with and problem-solving in areas of supply chain management (SCM) (Bowersox et al. 2007, Giani et al. 2013, Myerson 2012, Shanton 2017). The underlying principles of the solutions are drawn from tested academic recommendations, as adapted to industry-specific situations. The content therefore seeks to blend principles with practice and should therefore provide holistic guidance.

SCM is at the same time is a complex as well as a simple subject. Complex because it is widely different for different industries and needs a great deal of understanding and clarity to be detailed down to its elements (Ghiani et al. 2013, Lucas 1989, Rushton et al. 2017). It is simple because once detailed, it also affords itself to simple, logical, and at times obvious solutions. It is not necessarily only concerned with significant investments and recommendations from large consulting organizations. While that may be required in cases, the best practitioners are the people who are a part of the supply chain daily and who know by study and experience how it can be made more efficient. SCM of course is an overly broad subject and touches every aspect of an organization's core business operation, from planning and procurement to production and distribution. It touches logistics, efficiencies, and economies at each stage (Bowersox et al. 2003, Ghiani et al. 2013, Rushton et al. 2017).

20.2 BENEFITS OF CONSIGNMENT STOCKING SYSTEM AGAINST SUPPLY DISRUPTIONS

The initiative: A leading chemical manufacturer decided to enter into long-term procurement contracts with a preferred list of vendors for certain key raw materials.

Historically, these raw materials always had lead-time challenges. There were always supply uncertainties owing to either issues at the vendors' end or due to logistical disruptions. Vendors for these materials were typically located at significant distances from the company's manufacturing facilities and therefore had long delivery lead times.

The rationale for these long-term procurement arrangements was founded on certain key commitments and arrangements required to be agreed upon by both parties. These when executed transparently would result in various benefits primarily from an SCM perspective.

Examples of some such commitments are as follows:

Customer commitments: (1) Assurance of 'off-takes' or commitment to purchase to a select list of vendors for a period of time (usually a financial year), with a low margin of tolerance, (2) Indicate a reasonable periodic consumption pattern to the vendors (duly considering reasonable estimates of spikes due to seasonality factors), with low margins of tolerance, (3) Make available adequate storage facilities at the factory site for the vendors to be able to store materials, (4) Consumed materials via 'call-ups', to be triggered frequently to meet production needs and with simultaneous intimation to vendor, etc.

Vendor commitments: (1) Make available adequate inventories at the customer's nominated facilities to cater to the 'call-ups', on a 'just-in-time' (JIT) basis. (The vendor would need to naturally consider his delivery lead times to be able to determine how much inventory would need to be maintained.) (2) Assume liability for preagreed penalties in case of supply disruptions – a deterrent to be exercised by buyer in cases of negligence. (3) Commercial terms and conditions (including credit period and payments for supplies) to be triggered based on 'call-ups' by the customer. (4) Inventory at customer facilities yet to be 'called up', would be for all practical purposes to be considered 'Vendor's Inventory'. Accordingly, all holding costs for the same would in theory be that of the vendor. In practice, this would be determined based on commercial negotiations and would lead to determination of the final procurement price. (5) Given the certainty generated by the longer-term commitment in this arrangement, the vendors would offer the best prices for his supplies (Ghiani et al. 2013, Myerson 2012).

20.2.1 Learnings

Consignment stocking system (CSS) is a commercially astute method for optimizing the cost of logistics in feedstock management by assuring committed JIT supplies with less than commensurate immediate resource commitments (Stanton 2017).

The following specific advantages could be derived from this arrangement

- Continuous availability of inventory on a 'JIT basis'.
- Optimal working capital exposure to operationalize this arrangement with minimal immediate resource commitment
- Benefit of best prices on a long-term basis
- A 'win-win situation' for both the vendor and the customer enabling certainty and efficiency of operations.

20.2.2 Key Success Factors

To ensure success of such an arrangement the following should be kept in mind

- The contracting parties should be willing to work as partners in true spirit. Transparency is the key.
- Commercial success will depend on detailing and accounting for all costs (particularly holding and financial costs) and then computing the net price benefits as compared to a normal market-oriented short-term commitment.
- Need for a proactive approach
- Dispute Resolution Mechanism (DRM) in place to quell potential for serious disagreements at an early stage and in a transparent manner.

20.3 TECHNOLOGICAL DISRUPTION IN SUPPLY CHAIN

The initiative: When computerized tinting machines were introduced it completely revolutionized the retail painting business. It enabled retailers to customize any paint as per customer's wish on the fly. Customers were no longer forced to consider a prefixed offer of shades. Retailers also did not need to manage many different SKUs (stock keeping units of paints, which were often not totally aligned to customer choice or requirement. However, retailers who didn't have the machine, their customers had to choose from the prefixed offerings.

What the new technology enabled was a much wider offer of shades that could be obtained by using a computerized color match of several tints and mixing the same with the 'base' to develop a much broader offer than what was prevailing in the standardized shade cards prevailing then, all of this was made possible at the point of sale which allowed the end customer to gain a first-hand experience of the end effect of the shade before buying.

In addition, greater flexibility was provided in terms of minimum quantities of offtake. In addition, the computerized machines also had the ability to provide real-time inventories with the dealer network, which would therefore enable replenishments on a proactive basis, without the lead time inbuilt in information collection by physical visits by salesmen to the dealers.

While the commercial potential (broader product offering, efficiencies in inventory management, enhanced customer satisfaction and therefore stickiness to brand, deeper market making, etc.) of this technology was immense, the key challenge was to convince dealers to invest in these expensive tinting machines at the point of sale. While the dealers did conceptualize the potential benefits well, they were hesitant to bear the risk of upfront investment before experiencing the benefits as the retail paint business was highly price sensitive and low-margin business.

Realizing the significant benefits of this technological advancement and to ensure competitive advantage by being early off the block in introducing this technology, a leading paint manufacturer decided to finance these machines. They purchased these machines in bulk for certain strategic dealers initially and for a more dispersed set in phased manner. By procuring in large quantities, the paint manufacturer was able to avail excellent prices for the machines. The machines were then installed at the dealer premises and therefore became available at the point of sale to the end customer. To provide support for this initiative, the company also provided for a pool of **technical support officers** the services of whom could be availed of by the dealers for problem-solving when required. The result was the creation of a dedicated dealer network as most of the dealers were hooked on to the company's products. Also, with real-time transmission of sales and inventory data, it was possible for the company to ensure accelerated replenishment of supplies, thereby bringing in added efficiency in its supply chain. The return on investment in the capital cost of the machines and resource costs was quickly recovered by savings in the areas stated.

20.3.1 Learnings

Seizing the potential benefits of technological disruptions and allocating capital innovatively in support goes a long way in expanding an accruing disproportionate benefit – in the areas of SCM in this case. This initiative has generated one of the best return on investment (ROI) for capital deployed in channel partner housed investments. Innovative expansion of product offering/options for the end customer, consequentially increasing channel inventory and improving competitive advantage, has paid rich dividends to the entire supply chain ecosystem (Lucas 1989, Shanton 2017).

20.3.2 Key Success Factors

Early recognition of opportunity for broader engagement with channel partners, transparency, risk taking, and a congenial working arrangement helps deepen as well as expand channels and brings greater efficiency in channel management.

20.4 CAPITAL COST SAVINGS IN EXCHANGE OF INCREASING INVENTORY HOLDING COST

The initiative: Governments in certain countries often allow buyers to store imported materials in customs-approved 'bonded warehouses'. What this arrangement entails is that in case of imports, the quantities of materials that are not going to be consumed immediately can be stored under 'Bond' with the customs authorities for predefined periods of time. Import duty payments on these materials are deferred till the material is required to be 'de-bonded' for consumption.

Besides the obvious working capital conservation by being able to defer payment of customs duties, the arrangement of a 'bonded warehouse' has the potential for securing significant procurement advantages for the buyer (as well as the seller) by enabling negotiation of better commercial terms (on price and payment) for strategic purchases in bulk.

The underlying rationale here is that the advantages of price that can be obtained on bulk purchase and the savings in finance costs on working capital savings (on deferred duties and ensuring supplies JIT from bond) significantly outweigh any incremental transaction or holding costs that might result from such an arrangement. Such an arrangement brings significant efficiencies in SCM. This route has been used by several companies with significant import content in their manufacture to maximize potential benefits from it.

20.4.1 Learnings

The clear advantages for the company are as follows:

1. Negotiation of more advantageous terms and conditions for the buyer, besides being a significant deal for the seller. This can be particularly advantageous for materials which have cyclicality in supplies/prices.

2. Enables operation of a JIT concept of material handling even for materials with long delivery deadlines.
3. Cash-out flows on import duties can be deferred, enabling alternate avenues of capital deployment amount.

20.4.2 Key Success Factors

Some key areas to focus on are as follows:

1. Shelf-life and storage conditions of the imported items to prevent wastage/losses.
2. Keeping a track on holding/handling costs to ensure that they do not offset price advantages.
3. Complete awareness of regulatory requirements in this regard and compliance with the same.

20.5 VERTICAL INTEGRATION

The initiative: Manufacturing of paint requires a diverse type of chemicals as feedstock. These would include items such as resins, phthalic anhydride, penta-erythritol, and acrylic emulsions to name a few. Some key attributes about supplies of these are as follows:

- Requirement for these is usually in bulk
- Supply sources are widely geographically dispersed
- Historically, these have been prone to both supply and price disruptions

Strategically therefore, to the extent possible, it is best to try and reduce dependence on external sources for these key materials as they have potential to cause significant disruptions in the supply chain in the event of supply uncertainties.

Certain key players in this industry have realized this and invested in manufacturing some of these through captive units in proximity of their paint manufacturing units. These units could either be through setting up subsidiaries or even as divisions of the same company.

Initiatives in backward/vertical integration have gone a long way in assuring required levels of efficiency to the supply chain while at the same time ensuring economizing in cost of supplies and ensuring quality.

20.5.1 Learnings

A strategy of backward integration, particularly in process-oriented industries (e.g. paint), not only makes for dedicated sources of production but also enables excellent management of the supply chain and therefore reduced dependence on an external ecosystem that could be unreliable at times. These backward or vertical integration initiatives could be structured either as separate subsidiaries or divisions of the main company, depending on the regulatory requirements/advantages.

20.5.2 Key Success Factors

Exercise strong governance as disruptions in a captive setup can significantly derail operations in the manufacturing unit

20.6 STRATEGIC OUTSOURCING OF MANUFACTURING FOR LOW VALUE-ADDED PRODUCTS

The initiative: As a part of offering a complete product portfolio to the market, large companies often resort to subcontracting the manufacture of low value-added products to a selected group of outsourced vendors and selling the products through the company's regular distribution channel.

A continuing offer of low-value products may be necessitated by one or more of the undernoted reasons and is therefore essential despite almost invariably not being margin accretive.

- Competitive pressures for a complete offering
- Complementary offerings to the main products
- Ensure cheaper cost of manufacture
- Niche manufacturing processes
- Derisking alternative sources of manufacture
- Distribution economies, etc.

Examples of outsourced manufacture can be seen across industries such as footwear, paint, foods, and various accessories.

For example, *colored cement paints* and *wall putty* are two widely used low-cost products. Former is used as a low-cost option for colored walls and floors and the latter is used for drywall preparation. They are used in high volume and are very low-value products when compared with the larger offering of paint products. Manufacturing of both cement paints and wall putty involves lots of pollution during the manufacturing process. Therefore, this needs to be managed within a dedicated and managed environment. Similarly, enamel paints, primers, and thinner too are low-value hazardous items.

There are stringent regulations governing the manufacture of such products. Rather than exposing the main value-added manufacturing facility to these restrictions, it is better to have the same done through a dedicated and controlled environment in an outsourced mode.

Outsourced manufacturers can provide a viable alternative in such cases at lower costs as they have lower labor and infrastructure costs and are willing to offer cost plus models to manage their own competitive pressures. Moreover, with almost zero sales and distribution costs as they market these products through the main company's distribution network, they are able to further reduce procurement costs for the main company.

20.6.1 Learnings

A strategy of outsourcing lower value-added usually commoditized offerings helps an organization focus all its attention to servicing the higher margin and more technologically advanced spaces in the market while retaining an expanded product offering to service every segment. This is usually achieved by creating a geographically spread out ecosystem of vendors from whom finished goods (branded with the buying organizations logo) are *bought in* and offered through the company's existing distribution network. As these products are manufactured under a strict governance framework of the main organization, adherence to required quality standards is ensured.

20.6.2 Key Success Factors

Need to exercise governance-related oversight periodically over outsourced partners to ensure adherence to norms.

20.7 COLLECTIVE NEGOTIATION FOR PROCUREMENT EFFICIENCY

The initiative: In many ways, collaboration in a commercial space can lead to adversely impacting competition, in the sense of it unfairly influencing commercial negotiations by 'ganging up'. There are ways, however, where a concept of collective negotiation by both sides (the buyers and the sellers) in a common virtual marketplace can bring advantages to both. Discrete and geographically dispersed buyers and sellers can pool in their supplies and requirements to create a sum of parts to negotiate on the entire pie and thereafter manage offtakes in a disaggregated manner, but at the best commercial terms. The industry-specific terminology for such an arrangement is economic price negotiation (EPN).

Such an arrangement has great benefits in case of non-commoditized niche items where supplies are scarcer and have complexity around international import regulations. For a paint manufacturer in Asia, a lot of raw materials such as pigments, colorants, and specialty polymers including rutile, host perm violet, carbon black FW200, Degussa black, araldite, emulsion, selected polymers, etc. needs to be imported. These imports are usually made from supply sources in Europe, the US, Japan, South Korea as well as certain other countries.

To derive the best procurement benefits from cost, freight, and dealing with governmental regulations, buyers and sellers collaborate across entities in the Asian, European, and African market. They have structured a negotiation and purchasing process where instead of importing their requirements individually, these buyers have decided to collaborate with their collective purchase requirements and approach suppliers with their aggregated demand to conduct an EPN. Suppliers are then required to provide them with a common quote together with other commercial terms and conditions on a long-term basis.

Ordering for these materials is then done by the buyers jointly, with specific individual delivery instructions.

20.7.1 Learnings

Collaboration helps buyers get economical prices for selected expensive and imported raw materials, which are regularly used by these companies. Similarly, sellers get the benefit of aggregated commitments on a long-term basis.

20.7.2 Key Success Factors

Transparent cross-country collaboration among competitors creates a win-win situation for each.

20.8 UNCHECKED COST REDUCTION CAUSING SUPPLY CHAIN DISRUPTION

The initiative: A chemical product manufacturing company was having a lot of disruptions in its supply chain. It noticed that its supplier performance had deteriorated. Disruptions had increased due to delivery delay and lack of communication from suppliers. Previously the manufacturer would get better communication on stock out parts from suppliers and get updates on expected time of arrival for new stock when inquired. Response time had also worsened from suppliers. Delivery times for shipped parts had also become unpredictable.

The company decided to investigate further and noticed that in some cases the long-term suppliers had been replaced by new partners who offered lower cost. In certain cases, the long-term partners had made adjustments to provide additional cost savings. Overall these changes were done to meet the cost goals provided by the manufacturer's supply chain managers. While doing the analysis it was found that the suppliers who made adjustments had accommodated by changing their service levels. In one case a supplier, in order to optimize its own cost, had moved from their own supporting warehouse to a very low-cost outsourced warehousing service who had no information on replenishment timeline for stock out parts. Second, some of the suppliers had also switched to a low-cost carrier who had a very small fleet and was often subjected to mechanical problems and delays due to weather conditions.

20.8.1 Learnings

Being viewed as a cost center, supply chain organizations often drive down the cost of goods and services procured to the possible minimum. For chemical products which are extremely commoditized, the supplier managers are constantly looking for cost savings opportunities to maintain profit margin. In order to drive for lowest cost, long-term supplier partners may get replaced by relatively untested new partners or they end up making subpar adjustments to meet the cost pressure.

20.8.2 Key Success Factors

'Should Cost' modeling is one way of benchmarking the cost target where the supply chain manager models the anticipated cost from its suppliers. Building a model based on understanding suppliers' cost, reasonable profit margin, internal sales, general administration cost, etc. a supply chain manager can put itself in the supplier's shoes to build a target cost. Relentless cost negotiation is harmful and if suppliers are trying to buy a business by giving abnormal price reduction that is a red flag. Supplier health monitoring is a critical responsibility of supply chain which should monitor the business margin and business growth of suppliers too. The cheapest product is not necessarily the most economical one. The cost goals may also need to be recalibrated with its internal cross-functional team because ultimately the dissatisfaction can reach to the end customers if the manufacturer is also being forced to compromise under cost pressure from its own customers.

20.9 CONCLUDING REMARKS

A total of seven strategic supply chain initiatives have been presented in this chapter to highlight the multiple facets of SCM and operational initiatives that have been put in play to obtain best benefits. While the solutions presented in the cases listed above are specific to situations and challenges in chemical industries, the underlying principles that help solve these are more generic and capable of adapting to other situations also.

This chapter is in alliance with all the previous chapters and should help to understand the vast possibilities of SCM from a more hands-on perspective. It is hoped that this chapter will assist the readers to further strategize and optimize the specific aspects of their own supply chains.

Asish Datta, B.Sc. (Hons.), B.Com. ACMA, has held multiple leadership roles in finance. He was a Sr. Vice President (Finance & Accounts) and CFO at Berger Paints and later served as the Head of Finance in Shalimar Paints. He had been an advisor for Jenson & Nicholson (J&N) and after that worked as the CEO of a reputed tour and travel company.

Sudip Gupta, MA, served as the divisional head of supply chain at Akzo Noble, Bangalore, and later as vice president of Sudarshan Cargo followed by head of supply chain for Nippon Paints, India. He also served as a senior general manager of Bangalore Management Association, Bangalore, where he is currently working as an independent consultant.

Sanjib Sengupta served as Senior Vice President, Administration & Corporate Affairs for INDAL/Hindalco, Kolkata, India. He served in various leadership capabilities at ARCELOR MITTAL, RPG Group – CESC Limited, Reliance Power and Brahmaputra Metallics Limited. Eventually, he started his own manufacturing company in Ranchi, Jharkhand, and currently settled in Kolkata, India.

REFERENCES

Bowersox, D., Closs, D., and Cooper, M. 2007. *Supply Chain Logistics Management*. New York: McGraw-Hill.

Ghiani, G., Laporte, G., and Musmanno, R. 2013. *Introduction to Logistics Systems Management*. West Sussex: John Wiley.

Lucas, R. 1989. Liberalization of Indian trade and industrial licensing: A disaggregated econometric model with simulations. *Journal of Developmental Economics*. 31(1): 141–175.

Myerson, P. 2012. *Lean Supply Chain and Logistics Management*. New York: McGraw-Hill.

Rushton, A., Baker, P., and Croucher, P. 2017. *The Handbook of Logistics and Distribution Management*. London: Kogan Page.

Stanton, D. 2017. *Supply Chain Management for Dummies*. Hoboken, NJ: John Wiley.

21

Dabbawala: A Hundred-Year-Old Six Sigma Logistics Concept

Debasis Bagchi

Asish Datta

Manashi Bagchi

Sourya Datta

CONTENTS

21.1   Introduction: Long before Logistics...413
21.2   Dabbawalas: The Past and Present 414
21.3   Organizational Strategic Focus 416
21.4   Process Excellence in Logistics...418
     21.4.1 Process Innovation for Operational Accuracy......................419
21.5   Creating a Brand in Logistics 421
21.6   Concluding Remarks.. 423
Disclaimer 423
References.. 424

21.1 INTRODUCTION: LONG BEFORE LOGISTICS

Supply chain management as a business practice is getting continuously popular around the globe (Branch 2010, Bartlett et al. 2007). One of the most vital issues is logistics or in other words timely and efficient delivery of goods at the right place and time, at the same time keeping the inventory at minimum (Christopher 2016, Liu 2012).

DOI: 10.4324/9780429273155-25

In the modern world, we take priority mail services like overnight delivery or assurance of delivery of packages within a stipulated amount of time from global postal and shipping organizations as something for granted. However, there is a massive amount of supply chain strategies and investment in a vast network of people, fleets of planes, trucks, and an endless amount of computers and other IT hardware and software which is keeping these logistics machines running. It's only very recently as the world has entered into the age of shared economy, door-to-door food delivery service has started. Due to the pandemic induced need for social distancing, the peer-to-peer delivery logistics network for food has expanded to grocery, medicine, and other delivery services. It has become a big enterprise and created job opportunities globally. However, this is also supported by massive innovation in data center infrastructure, mapping, and global positioning systems along with data communication and cellular technologies.

Long before the age of computers a peer-to-peer delivery logistics service started in the city of Mumbai. As a huge influx of migrant workers moved from different parts of India to Mumbai for jobs the "Dabbawala" was started. It started with about 100 people in the 1890s. The objective of the organization was to deliver home-cooked meals to the respective offices in and around Mumbai. The Dabbawalas used to collect and pack lunch boxes in their designated containers with designated color codes and deliver these lunch boxes to their respective offices within reasonable time and collect the empty lunch boxes in the afternoon and bring them back the empty lunch boxes to their residence. Their motto was on-time delivery of fresh food at a reasonable cost for low- to middle-income (Roncaglia 2013, Chakraborty and Hargude 2015) customers. This chapter provides an overview on the operation logistics of Dabbawala, which has been continuously growing in Greater Mumbai since then.

21.2 DABBAWALAS: THE PAST AND PRESENT

Dabbawalas (also termed as "tiffin wallas" or spelled as "Dubbawallas" or "Dabbawallahs") is a popular and growing lunch delivery service in Greater Mumbai (earlier known as Bombay) in India. In Hindi, Dabba means "box" and walas indicates "carriers" (Roncaglia 2013). In late nineteenth century a big influx of people from all over India and neighboring countries of varying communities migrated to Greater Mumbai in search of job and potential small business opportunities (Chakraborty and Hargude 2015). Because of the lack of enough potential job opportunities in villages, illiterate workers and farmers from rural areas moved to Mumbai in search of low-income job opportunities. Low-income communities hardly had enough money to buy regular lunches (Baindur Macário 2013). Inexpensive canteens, fast-food outlets, or cafeterias were available during that time, and the restaurants around were expensive and out of their reach (Mallik and Mukherjee 2007).

During that time out of sheer necessity, Mahadeo Havaji Bachche conceptualized the idea of delivering home-made fresh lunches to the low- and middle-income groups. He initially started a lunch delivery service with about a hundred men in 1890 in Mumbai. Mahadeo Havaji Bachche attempted to unionize the Dabbawalas in 1930. He registered a charitable trust known as "Nutan Mumbai Tiffin Box Suppliers Trust" in 1956. Subsequently in 1968, a commercial division of this trust was registered as Nutan Mumbai Tiffin Box Supplier's Association (Mallik and Mukherjee 2007).

The lunch boxes were initially collected from homes, in and around Mumbai, in late morning and delivered home-cooked fresh lunches to people at work using bicycles, handcarts and trains, and empty lunch boxes were returned in the afternoon (Matani et al. 2015). The original concept was that meals were freshly prepared at homes, food was hot, tasty and nutritious at a reasonable cost. Later, Dabbawalas were used by meal suppliers and cafeterias in Greater Mumbai, who used to pay them to ferry lunchboxes with ready-cooked meals from central kitchens to customers and back (Matani et al. 2015).

Both business and demand grew exponentially, and currently in Mumbai, a total of 5,000 Dabbawalas deliver lunches every day to more than 200,000 people and complete 400,000 transactions (Pathak 2010). An average Dabbawala charges Rs. 800–1200 (approximately USD $ 11.5–17) per lunch box per month, and the annual turnover of Dabbawala is Rs. 400–450 million (about USD $ 6–6.5 million). On average, every Dabbawala carries 35–40 lunch boxes and on average covers 40–70 km per day. Depending on the delivery distance, the cost varies from Rs. 800 to 1200 per lunch box per month. Additional charges are applied for additional delivery of drinking water as well as hot and cold beverages in thermos flasks. It is interesting to note that with this relatively low investment in manpower and low transportation cost, in part, due to the low cost of suburban rail services operated by Indian Railways and heavily subsidized by the Central Government, the return on investment (ROI) is quite lucrative. *New York Times* reported in 2007 that this highly specialized organization grows between 5% and 10% annually (Mi Dahlgaard Park and Näslund 2013, Percot 2005).

21.3 ORGANIZATIONAL STRATEGIC FOCUS

The Dabbawala organizational structure has been exhibited in Figure 21.1. This Dabbawala organization is called as "Mumbai Dabba Vahantuk Mandal" in their local language. As a supply chain organization it is very lean. The administrative side is extremely flat with the primary employee base focused on delivery and supervision of service quality. The primary operation is run by approximately 5000 Dabbawalas whose work is supervised by 635 supervisors reporting to 9 directors.

As a logistics service provider, the Dabbawala organization has a "customers first" culture along with frugality and relationship. It keeps both the capital investment and the operational costs at the lowest. All the Dabbawalas know their customers personally and are able to build a great rapport with them. Ultimate motto is not to deviate from their core competency and commitment but to maximize the service with existing infrastructure and not to be overdependent on modern technology. They focus on maintaining healthy relationships, staying emotionally united and fostering strong cooperation among colleagues, and competing to maintain market share. The organization penalizes employees for noncompliance (Fugate et al. 2010).

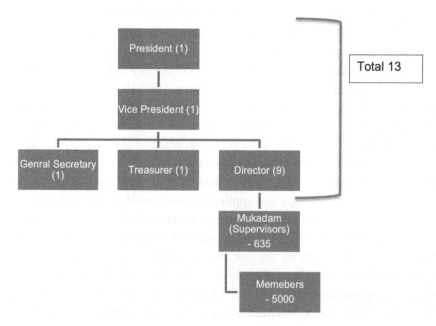

FIGURE 21.1
Hierarchical chart of the dabbawala organization.

Due to their strategic focus, Dabbawala's delivery system is almost error free. A survey reported that out of 6 million transactions, there may be just one or two errors. Based on a set of process assumptions, Dabbawala organization received a Six Sigma rating in a survey (which is defined as 3.4 defective errors per million) conducted by *Forbes Global Magazine* for outstanding service and efficiency (Percot 2005, Thomke 2012). Dabbawala is certified by International Organization for Standardization (ISO) and received an outstanding service organization award from National Institution for Quality and Reliability (NIQR). During July 2005, an unusual heavy monsoon totally collapsed the entire Mumbai city; however, in that catastrophic circumstances, Mumbai's Dabbawalas delivered prepared meals to their customers in their respective offices and returned the empty lunch boxes to their homes (Table 21.1).

TABLE 21.1

The SWOT (Strength, Weakness, Opportunities and Threat) Analysis of Dabbawala Team Operations

Strength	Weakness	Opportunities	Threat
• Teamwork • Strict discipline • Honesty and integrity • Customer satisfaction • Time management • Home-cooked fresh food provides nutrition and satisfaction • Cost-effective nutritious food for customers • Absolutely no preservative in home-cooked meal • Operational consistency • Commitment to service • Minimum inventory • Ownership and pride in the their service	• Totally dependent on Mumbai local train service • Limited job opportunities for the team members because of low educational standards	• Expanding network in other cities and suburbs • Open dabbawala franchise in other major cities • Collaborate with caterers to serve variety of service and meals-on-demand, such as (i) diet food, (ii) Continental foods, (iii) Chinese, Japanese, Korean or Thai Foods, (iv) restricted foods for fasting, (v) halal or Kosher foods	• Flexible time or shift workers/ work from home • Growth of readymade food supplies from other popular food delivery services, e.g. in India : Swiggy, Zomato and others • During any political situations, Dabbawalas may face rigorous time-consuming lunchbox and security checks, which will unnecessarily delay the entire process

21.4 PROCESS EXCELLENCE IN LOGISTICS

The delivery process has evolved over a long time and simplified to a level of operational excellence. As time of the essence, redundancies are kept to handle disruptions such as monsoon or public transport interruptions. The Dabbawalas have also invented an innovative and minimalist addressing schema so that destinations are visible, easy to read and manage. This is enhanced with an innovative yet simplistic and time-tested color coding system as the alphanumeric characters need to be read and actioned by delivery persons with varied levels of literacy.

Prepared lunch boxes pickup time starts between 9:30 and 10:30 AM depending on the location, while the empty lunch boxes are returned after 4:40 PM. The Dabbawala normally reports to the residence/central canteen area along with the designated marking container for lunch box pickup. Sorting of lunch boxes is done between stations at 11:20 AM and 12.30 PM and the delivery of the lunch boxes takes place between 12:30 PM and 2:30 PM, delivered in the ground floor reception/security area of each office. Thus, the people enjoy home-cooked fresh food during lunch. Collection of empty lunch boxes starts between 2:40 and 3:30 PM from the same location, i.e. ground floor reception or security, while the sorting of empty lunch boxes starts between 3:30 and 4:40 PM (Thomke 2012, Pathak 2010).

The overall duties are divided into three groups. The first group collects lunch boxes from residences/central canteens. The second group conducts sorting, collection and delivering the lunch boxes at specific stations, both lunch boxes with food as well as empty boxes. The third group collects the lunch boxes at delivery stations and distributes them to the designated offices and thus, repeat the process to return to the station within the specified timing mentioned. The ultimate mantra or focus of every Dabbawala is to be on time and not lose the box.

These lunch boxes are collected from either the respective residences, central canteen, offices, university, secretariats (mantralaya) or high courts from various places in Mumbai and Greater Mumbai and mostly distributed to various offices located in and around Churchgate, Nariman Point, Cuffe Parade, Fort, high court, secretariats (mantralaya), university, banks and Reserve Bank of India (Matani et al. 2015). Lunch boxes are placed inside the designated coded aluminum containers equipped with a handle for convenience. All lunch boxes are collected at various pickup points in the respective coded containers. As indicated earlier, the entire operation is conducted through three railways, e.g. Western, Central and Harbor Lane. Dr. Thomke reported (Thomke 2012)

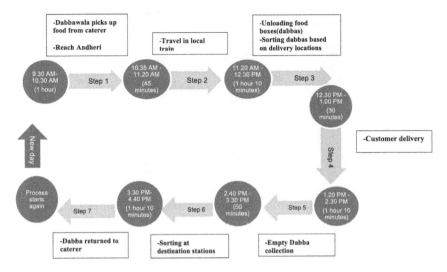

FIGURE 21.2
Activities of dabbawala team members on any particular day.

that at major train stations, Dabbawalas just get 40 seconds to load the Dabbas in the Dabbawala compartment, while at interim stations, only 20 seconds are allotted for loading and unloading at interim stations.

Furthermore, to maintain the tight time schedule, each group has 2–3 extra workers who can fill in any positions required for collecting the lunch boxes, sorting them and transporting them appropriately. At the same time, Dabbawalas manage the financial aspects and build up a professional and friendly customer relationship (Thomke 2012). For proper functioning, Dabbawalas approximately divide the total manpower in 200 units of 25 persons per unit. Figure 21.2 demonstrates the daily journey of Dabbawala team members on any particular day, which is quite a description of strict discipline, harmony and time management.

21.4.1 Process Innovation for Operational Accuracy

Every aluminum container used by the Dabbawalas are uniquely coded as follows:

1. Two color coding systems are normally used on each designated container top. The first color in the left side indicates three things: (a) area (alphabetic), e.g. area, Vile Parle is indicated as VLP, (b) road/lane/street (alphabetic), e.g. road, Hanuman Road as E, and (c) destination railway station (numerical number), e.g. railway station, Churchgate Station as 3 (Figures 21.3 and 21.4)

(a)

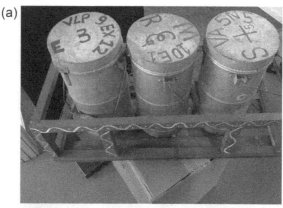

(b)

FIGURE 21.3

(a) Photos of three dabbas (boxes) with labelling on the tops depending of the delivery plan; (b) Pictures of a delivery cycle with multiple dabbas in the front and back of the cycle.

2. The other color on the right side of the container indicates three things:
   a. zone (numerical number), e.g. zone, Nariman Point denoted as 9,
   b. building name (alphabetic), e.g. Air India Building denoted as AI,
   c. the location of the office in a multistoried building is denoted by the number of floor of the office located (numerical number), viz,. an office located on the 12th floor is represented as 12 (Figure 21.4).
3. If there are more than one person to pickup and drop-off in a location, then the title is represented on the top of the container.
4. A representing coding system is shown in Figure 21.4. In early days, the coding was done using colored threads to the different areas of operations, later Dabbawala group started utilizing thrown away cotton waste from local tailors, and now, the Dabbawalas use permanent colored markers (Matani et al. 2015, Fugate et al. 2010, Thomke 2012)

VLP: Vile Parle (suburb in Mumbai)

9EX12: Dabbawala code at destination

EX: Express Towers (building name)

12: Floor Number

E: Code for Dabbawala at residential location

3: Code for destination station.
For example: Churchgate station (Nariman point)

FIGURE 21.4

A Typical Dabbawala Label with Different Numerical Codes, where E represents code for dabbawala location at the Residential Station; VLP indicates residential station Ville Parle; 3 exhibits the code for destination Station; 9 highlights the code for destination; EX represents the building name "Express Towers"; 12 denotes the floor number in the building.

21.5 CREATING A BRAND IN LOGISTICS

Over the years the Dabbawalas have created a brand which reflects discipline and accuracy. It has also publicized itself by supporting important social causes. This has translated to customer loyalty over the years. Even the customers must maintain the time schedule very critically. The customers, who are repeatedly late in keeping their lunch boxes ready for pickup and do not follow the timing norms and ethics are dropped (Erwin 2011, Thomke 2012).

All Dabbawalas are required to wear white caps and carry identity cards strictly during working hours. Drinking alcohol is not permitted during working hours. All Dabbawalas must build a "Trust" with their respective customers. All Dabbawalas must greet their customers professionally with a smiling face. Service is uninterrupted on the days of extreme weather – rain or cyclone. The organization provides white uniforms and white Gandhi Caps (called "Gandhi Topi") to each employee. Photo identity cards are provided to all employees. In addition, bicycles, hand carts and lunch box cases are provided to facilitate their daily operations including collection and/or delivery. Moreover, railway monthly/quarterly passes are provided to the employees travelling in the trains for carrying the lunch boxes. Each Dabbawala carries on average 35–40 lunch boxes in designated aluminum containers.

Dabbawala is a well-recognized organization and BBC, UTV, MTV and ZEE TV organized documentaries on Dabbawalas service efficiency and their ethics. Guinness Book of World Records also recognized and included Dabbawala in their book.

Dabbawala also has a website https://mumbaidabbawala.in, which is being viewed by more than 1 million visitors annually. Dabbawala circulates an informative newsletter, which also accepts various outside advertisements for publication and distribution with a fee. This money is used for the welfare of the Dabbawala staff members. Dabbawala organization also advertises through television, radio and newspapers.

Dabbawala Team jointly with FICCI's (Federation of Indian Chambers of Commerce & Industry, a nonprofit organization) Ladies Organization (FLO) launched a campaign "Share my Dabba" to distribute food to the homeless children. Dabbawala team requests their customers to place a red sticker on the lunch boxes, if any food is left behind. After collecting all the lunch boxes from the various offices, Dabbawala team looks for red sticker lunch boxes and hands over all these left-over lunches to FLO for distribution to the homeless children in various places in and around Mumbai. In this way, Dabbawala team distributes lunch to 80–90 homeless children per day (Bañuelas and Antony 2003).

21.6 CONCLUDING REMARKS

Dabbawala, a hundred-year-old Six Sigma concept, is one of the oldest and premium examples of food delivery supply chain developed in the late eighteenth century for efficient delivery of fresh meals from the residences to the offices at the right place, at the right time. It has been successful in implementing an optimum logistics model at the minimum cost. The most important success criteria of Dabbawala is timely and efficient delivery ensuring optimum inventory model at minimum cost.

This Dabbawala operation is primarily supported by the uneducated below poverty-line people (Mallik and Mukherjee 2007, Matani et al. 2015) who were mostly farmers and daily laborers migrated from outside. The main implication is that the uneducated labor force from the rural areas can be trained to efficiently perform logistical tasks in an urban environment under an organizational hierarchy and maintain a uniform standard of service. The Dabbawala team members are dedicated and committed persons capable for maintaining operational excellence in on-time delivery. Their key mantra is "Trust in each other, which is very important before you can build trust with customers and other stakeholders".

DISCLAIMER

Based on the available data, we have drawn Figures 21.1 and 21.2. Dr. Debasis Bagchi took the pictures of the dabbas and cycle (Figures 21.3 and 21.4) in the Exhibition Booth of The World Congress of Food Science and Technology, organized by the MM ACTIV Sci-Tech Communications during October 23–27, 2018, at the CIDCO Exhibition Centre in Mumbai, India. The conference covered diverse areas including food science and technology, food safety, food security, regulations, innovations in food processing, hygienic design and engineering, and proper permission was obtained from the Dabbawala Booth Manager on October 25, 2018.

Manashi Bagchi is currently the Chief Scientific Officer of Dr. Herbs LLC, Concord, CA. Dr. Bagchi is also a consultant for VNI, Lederach, PA. She has served as an Associate Professor in the Creighton University School of Pharmacy, Omaha, NE, and later as the Director of Research at InterHealth Nutraceuticals, Benicia, CA. She is a Fellow and currently a Board Member of the American College of Nutrition (Clearwater, FL). Dr. Bagchi has published 225 papers in peer-reviewed journals and 2 books, *Genomics, Proteomics and Metabolomics in Nutraceuticals and Functional Foods and Bio-Nanotechnology: A Revolution in Food, Biomedical and Health Sciences,* from Wiley-Blackwell.

REFERENCES

Baindur, D., & Macário, R. M. 2013. Mumbai lunch box delivery system: A transferable benchmark in urban logistics? *Research in Transportation Economics.* 38(1), 110–121. DOI: 10.1016/j.retrec.2012.06.008.

Bañuelas, R., & Antony, J. 2003. Going from six sigma to design for six sigma: an exploratory study using analytic hierarchy process. *The TQM Magazine.* 15(5), 334–344. Doi: 10.1108/09544780310487730.

Bartlett, P. A., Julien, D. M., & Baines, T. S. 2007. Improving supply chain performance through improved visibility. *The International Journal of Logistics Management,* 18(2), 294–313. DOI: 10.1108/09574090710816986.

Branch, A. E. 2010. *Global Supply Chain Management.* New York: Routledge/Taylor & Francis.

Chakraborty, A., & Hargude, A. N. 2015. Dabbawala: Introducing technology to the Dabbawalas of Mumbai. *Paper presented at the Proceedings of the 17th International Conference on Human-Computer Interaction with Mobile Devices and Services Adjunct.* ISBN 978-1-4503-3653-6.

Christopher, M. 2016. *Logistics & Supply Chain Management.* Pearson Higher Ed. ISBN: 9780273731122.

Erwin, P. M. (2011). Corporate codes of conduct: The effects of code content and quality on ethical performance. *Journal of Business Ethics.* 99(4), 535–548. DOI: 10.1007/s10551-010-0667-y.

Fugate, B. S., Mentzer, J. T., & Stank, T. P. 2010. Logistics performance: efficiency, effectiveness, and differentiation. *Journal of Business Logistics.* 31(1), 43–62. DOI: 10.1002/j.2158-1592.2010.tb00127.x.

Liu, J. J. 2012. *Supply Chain Management and Transport Logistics.* London: Routledge/Taylor & Francis.

Mallik, U., & Mukherjee, D. 2007. Sigma 6 Dabbawalas of Mumbai and their operations management-an analysis. *Management Accountant-Calcutta,* 42(5), 386.

Matani, D. A., Tripathi, M., Doifode, D. S., & Gowardhan, S. 2015. Green supply chain management in food industries. *Int. J. of Engineering and Technical Research (IJETR).* 3(7), 261–263. ISSN: 2321-0869.

Mi Dahlgaard Park, S., & Näslund, D. 2013. Lean and six sigma-critical success factors revisited. *International Journal of Quality and Service Sciences.* 5(1), 86–100. DOI: 10.1108/17566691311316266.

Pathak, G. S. 2010. Delivering the nation: The Dabbawalas of Mumbai. South Asia. *Journal of South Asian Studies.* 33(2), 235–257. ISSN: 0085-6401.

Percot, M. (2005). Dabbawalas, Tiffin Carriers of Mumbai: Answering a Need for Specific Catering. HAL ID: halshs – 00004513. https://halshs.archives–ouvertes.fr/halshs-00004513 Submitted on Aug 31, 2005.

Roncaglia, S. 2013. *Feeding the City: Work and Food Culture of the Mumbai Dabbawalas.* Cambridge, UK: Open Book Publishers. ISBN: 9781909254008.

Thomke, S. Nov 2012. Business Models. Mumbai's Models of Service Excellence. In: Harvard Business Review. https://hbr.org/2012/11/mumbais-models-of-service-excellence (accessed Feb 11, 2020).

22

Conclusion

Ishaan Das

CONTENTS

22.1 Building the Fundamentals... 425
22.2 Execute 426
22.3 Concluding Remarks.. 427
Reference 427

22.1 BUILDING THE FUNDAMENTALS

This book is not about supply chain only, but also about the people behind the supply chain. Like Marco Polo, a supply chain can succeed or fail depending on the leadership, culture, and management done by the people. As someone in that field, it is important to know how crucial and important you are in creating and impacting a successful supply chain system, whether you are an individual contributor or the head of a multibillion-dollar operation. As discussed, the supply chain industry is a large, complicated, and diverse group of networks and people where everyone must be able to demonstrate leadership and strength in their career.

Like Marco Polo, it is important to understand that you have a crucial role to play and if one cog falls apart, the entire machine can malfunction. This book has taught you the fundamental knowledge you need to succeed as a professional in the field at any level, from a fresh graduate to a seasoned executive. Joining an organization, one may find it rigid and inflexible in terms of its processes and tools, and this book tells you how to evaluate the needs and wants and goals of your department, its business' focus, and what type of supply chain organization it is. You should be able to understand the key processes it follows; identify opportunities to improve; be able to identify and collaborate with the right stakeholders; build long-term relationships; improve culture, processes, and revenue; and lead an organization to

DOI: 10.4324/9780429273155-26

success. The book teaches you to quickly ramp-up, communicate with clarity about your ideas, and move the needle of the organization

At this time, you should be able to know the various roles and responsibilities of members in a supply chain organization and other relevant functional teams. You should be able to recognize how to collaborate with members and key executives in all the different moving parts of your role and understand how to manage your daily tasks as well as going above and beyond and creating impactful goals. In order to succeed, it is crucial for you to understand everyone's expectations from you and fulfill them in a prompt and professional manner. By now, you should understand the six pillars of success and how Marco Polo was able to demonstrate it in the silk road (Haksöz et al. 2011) (product – silk; demand from the Venetian and medieval European market; supplier – Chinese weavers; value creation – to bring the best quality; relationship – with emperor Kublai khan; execution – to successfully manage through the perilous journey and then write about it) and be able to recognize and apply those tenets in your role, regardless of experience level.

In your own role, it is important to be able to recognize common problems through the examples and case studies given and be able to then solve those problems regardless of your level of experience. This book teaches you the key tools needed to create a sustainable supply chain. It enables you to identify and solve problems as well as situations to look out for in both common issues and vertical-specific knowledge, and how to apply that knowledge even outside of that situation.

Lastly, this book teaches you the transferable skills needed to succeed in any supply chain vertical. As industries create larger and more complicated supply chains, the need and demand for successful supply chain professionals will only continue to grow. With the knowledge imparted from industry leaders in the vertical-specific chapters as well as industry agnostic advice, you should now know how to enter multiple industries and be successful in their respective supply chains.

22.2 EXECUTE

Now that you have learned all the skills, it is your time and responsibility to execute what you have learned in order to become successful. Supply chain management is all about being able to execute efficiently and quickly, and success in the field comes from being able to take all the knowledge you have learned and apply it.

This book is designed to give you practical knowledge that will help you to execute at the highest functional level regardless of the size of your role or industry. The book is specifically designed to help you understand and apply how to be a successful supply chain leader at all levels, from a newcomer to the field to the chief executive, and is something that you can refer back to at all points in your career. You should now be able to successfully communicate with stakeholders and cross-functionally, influence and seek alignment with your external and internal business partners, and drive ideas and value propositions for the business.

The industry-specific knowledge as well as the pillar system is designed for you to reevaluate your own knowledge and execution throughout your career, and just like Marco Polo, in order to become a successful leader, you will continuously have to reevaluate and execute your leadership as your scope, organization, and responsibilities change. Creating a large and efficient organization and moving up in an organization takes concerted time and effort and it is of utmost importance to continue to cultivate your knowledge and skills throughout your career in order to become the best possible leader.

22.3 CONCLUDING REMARKS

Marco Polo was the first supply chain leader because he excelled at learning and executing the fundamentals of supply chain. While this book teaches you the key skills and knowledge you need to succeed in supply chain at various levels and in various industries, it is now up to you to determine how successful you can be. This book is intended to motivate and inspire you to become a great leader, and your power to execute what you have learned will be the key differentiator between you and others that will help you become the next successful supply chain leader. Good luck as you venture onward into the exciting world of supply chain management!

REFERENCE

Haksöz, C., Seshadri, S., and Iyer, A. 2011. *Managing Supply Chains on the Silk Road: Strategy, Performance, and Risk*. Boca Raton, FL. CRC Press/Taylor & Francis.

Commentary: A Path to Cognitive Leadership

Sourya Datta

Sudip Das

Ishaan Das

Debasis Bagchi
Pharmacy and Health Sciences

Supply chain is a unique organization inside an enterprise which offers amazing opportunities to demonstrate leadership skills. The organization needs and accepts talents from all walks of professional life and offers a path to leadership leveraging the diverse set of experiences that comes with it.

This book *Becoming a Supply Chain Leader: Mastering and Executing the Fundamentals* is written for academics, MBA students in supply chain management/operations research, entry-level procurement professionals, and supply chains executives. This book is also intended for executives and professionals outside of the space who are interested in understanding the workings, concerns, and nuances of a global supply chain organization from an execution point of view. Editors focused on two prime areas to make their readers successful: professional growth and leadership, and creating value and making an impact by solving real-life business challenges through supply chain.

Supply chain management is a strategic and complex process. It easily gets impacted by disruptive forces which come from all directions. It's a unique field which requires the ability to influence and build relationships based on trust with a wide spectrum of people. Simultaneously, it also requires development of business acumen through a diverse set of knowledge, ranging from market condition to unique technologies and regulations to global economic trends and financial margins and profitability to environmental disruptions and other supply base shifts. Therefore, managing a supply chain is a cognitive process which requires thinking, reasoning, acquiring factual information, and processing all for final decision making. This book focuses on this cognitive path of leadership and articulates how to put all the factors together and lead with clarity and purpose.

From a tactical operational level the supply chain could be viewed as a cluster of operations to enable procurement and delivery of goods and services. This includes management of the flow of materials, management of uninterrupted manufacturing operations, managing inventory level of finished goods, and efficient delivery and out flow of products. However, from a holistic and business perspective, supply chain touches every part of product development from concept to production to end of life. It works with all the functional verticals such as manufacturing, sales, finance, product development, and even human resources, as it's essential for producing goods and services as well as running the overall business.

In our human body there are important organs such as heart, brain, and lungs. However, blood flows everywhere and provides nutrients to every organ from head to toe. Drawing a parallel, the supply chain organization is the lifeblood of an organization which is responsible for working with all parts of the organization and making the whole business to function successfully. This requires two foundational steps: understanding the business challenges and working with people to mitigate them.

Supply chain organizations are formed to smoothen the procurement and delivery of goods and services but as an organization grows, the process owners improve efficiencies and create strategic business advantages for its customers. They also mitigate business risks. Section 1 of this book discusses that objective. It is intended to help the reader to assess the type of production strategies and processes adopted by an organization and how are they being addressed. This section covers various types of business problems and the underlying mitigating supply chain principles. Examples are provided for the reader to compare and discern with their own organizations and to identify and map out the collaborators and stakeholders.

A leader needs to have followers. Both Sections 1 and 2 in this book are dedicated to understanding the stakeholders who will follow a supply chain leader in a win-win scenario. Section 2 is intended to help supply chain professionals to understand why they have been hired and what are the day-to-day expectations of them from their partners outside their organization. It gives them a picture from the centers of their universe. It also offers them a lens to look through in the form of six knowledge pillars. These pillars offer a framework for communication and execution of their efforts.

Leaders, however, can go above and beyond in order to succeed, and Section 2 and 3 focus on how they can create value and move the needle for their organization. As the scope of roles, empowerment, and domain of influences could be different and rigid in various supply chain organizations, and between an entry-level supply and an executive, the scope of the value creating initiatives could be different. The examples are classified by the levels of effort needed to give a sense of scope as well as offer future goals. Section 3 offers a unique perspective on 10 supply chain verticals. The verticals are reviewed through the industry agnostic principles and frameworks articulated in Sections 1 and 2 while highlighting industry-specific nuances. The business problems and solutions are also articulated through case studies in Sections 3 and 4.

Overall, this book is not just an overview of a supply chain. First, it is about the people involved with all aspects of the supply chain. Second, it's a motivational book for supply chain professionals who are driven and want to demonstrate leadership and be successful. This book enables them to take complex challenges within the scopes of their roles and create impacts well beyond that boundary. Quite often, a supply chain professional feels frustrated that they have stopped learning or feel boxed in professionally, and this book is also intended for them to spark their imagination and take new challenges. The challenge could be in the form of career transition to another branch of supply chain or another industry vertical. Or it could be in the form of taking new transformational challenges for their existing business processes, including building analytics and unearthing new efficiencies. This book is intended to be a companion for reference and an inspiration to all supply chain professionals throughout their professional journey in the dynamic and exciting world of supply chain.

Index

Note: **Bold** page numbers refer to tables and *italic* page numbers refer to figures.

aging inventory, automotive industry 274–275
air traffic services flight notification (AFN) systems 337
Amazon 22–23
Amazon Web Services (AWS) strategy 62
analytics, used in supply chain 28
Anderson, D. 20
antitrust laws 183
apparel industry 361–362
    brands and the manufacturers 363
    dealing with volatility 371–372
    individual performance 372–374
    macro trends impacting the industry landscape 364
    organizational leadership 374–375
    supply chain management
        logistics 367–368
        products 365–367
        shift in perspective 364–365
        supply base 368–369
        trade regulations 369–370
application-specific integrated circuits (ASIC) 288
approval flow 162–163
ARIMA 319, 327
artificial intelligence (AI) 290
AS9100 certification 338
AS9145 certification 338
ASIC *see* application-specific integrated circuits (ASIC)
asset management, hardware 239–240
automotive industry 264
    individual performance 276–277
    macro trends impacting the industry 264–265
        global footprint 265
        remanufacturing 266–267
        service orientation 265–266
    obsolescence risk 277
    organizational leadership 275–276

stock keeping units (SKUs) 266
supply chain management
    aging inventory 274–275
    forecasting 268–269
    internal and external manufacturing 271
    price 273
    product mix 269–271
    products 267–268
    spares 273–274
    supply base 271–272
autonomous supply chain leader, six pillars for 126–127
avionics industry 332–333
    cross-functional collaboration 343–344
    individual performance 344–345
    macro trends impacting the industry 333–337
        distinction from other industries 333–334
        long life cycle 334–335
        long tail 335–336
        obsolescence 335
        quality and reliability 336–337
    organizational leadership 345–346
    regulatory requirements 344–346
    supply chain
        demand 338–339
        liability 341–343
        products 337
        reactive changes 339
        regulatory control 337–338
        relationship 341
        spares and support 339–340
AWS strategy *see* Amazon Web Services (AWS) strategy

behavior, consumer 185
big iron 200, 201
bilateral trade 370
bill of material (BOM) 23, 97, 129

biologic markets 350
biopharmaceutical industry 349
    changes brought in by pandemic
        356–357
    discovery 351–353, 355
    individual performance 358
    macro trends impacting the industry
        350–352
        drug development timeline 352
        leaders of the industry 352
    recovery and yield 354
    supply chain
        CDMO partner selection 355
        contract with CDMO 353–354
        intellectual property (IP)
            protection 355
        products and processes 353
        relationship management 356
Blockbuster LLC 26
BOM *see* bill of material (BOM)
bootstrapping supply chain
    culture 393–394
        dealing with failures 394–395
    early phase: setting up the foundation
        identify key hires 382
        laying the foundation 380–382
        stabilize 378–375
        undoing and redoing 379–380
    hyper growth dealing 392–393
    launch: rubber meets the road
        382–383
        launch blockers 383–384
        lessons learnt: evolve the supply
            base 384
        strengths and weakness of
            partners 381
    supply chain challenges
        multisourcing and slicing the
            business 383–385
        relationship management 385–386
        repositioning resources 382–383
    supply chain systems 386
        ERP system 387–388
        internal pain points prioritizing
            386–387
BPO *see* business process owner (BPO)
brand–annual goals 119–120
BRICS (Brazil, Russia, India, China and
    South Africa) 30

budget
    capital expenditure (CAPEX) 96
    operational expenditure (OPEX) 96
built to order production method 21
business consolidations 185
business knowledge and value creation
    144–147
    cost of make *versus* buy 146
    non-recurring expense analysis 146–147
    return on investment analysis 144–145
    value engineering 145
business opportunities for printers 180–181
    growth opportunities 181
    home and workplace printing practices
        181–182
business process owner (BPO) 115
butter oil 303

Capex *see* capital expenses (Capex)
capital cost savings
    in exchange of increasing inventory
        holding cost 400
        key success factors 401
        learnings 400–401
capital expenditure (CAPEX) budget 96
capital expenses (Capex) 144–145,
    282, 289
category-specific global supply chain
    team 105
CDMO *see* contract development and
    manufacturing organizations
    (CDMO)
cell line 353–354
channel assembly production method
    22–23
cheese 302–305
chief financial officer (CFO), indirect
    procurement under 95–96
chief operating officer (COO)
    indirect procurement under
        CMO/COO 97
        COO 96
    organization 96
China 14
    US-China trade tensions 30
Cisco 40–41
cloud 219, 226
Cloud service providers 279
    challenges of 282–284

individual performance 293–294
and market 281–282
organizational leadership 294–295
    execution challenges 290–293
    forecast and demand volatility
        285–287
    relationship 288–290
    suppliers 287–288
Public and Private Cloud 280
supply chain management 284–293
CNS *see* computing network systems (CNS)
collective negotiation
    for procurement efficiency 403–404
        key success factors 404
    learnings 404
commodity management team, global
        106–107
commodity manager 114–115
communicating supply chain strategy 165
communication, and information sharing
        within organization 28
company's organization structure 94–99
    direct procurement under CMO or
        COO 97
    heat map of key external influencers and
        decision-makers 98–99
    indirect and direct procurement 95
    indirect procurement
        under CFO 95–96
        under COO 96
    shadow procurement under CTO or GM
        97–98
complex opportunities 168
    excess and obsolescence (E&O)
        management and disposal 171
    logistics 170–171
    multilevel supply chain 169–170
    vendor change and consolidation 168
    vendor managed inventory (VMI) 169
computing & network systems 337
computing network systems (CNS) 332
consignment stocking system (CSS) 38
    against supply disruptions 396–397
        customer commitments 397
        key success factors 398
        learnings 397–398
        vendor commitments 397
constant replenishment production method
        21–22

consumables
    intellectual property protection of
        184–185
    printers 189
consumables ordering, predictive 194–195
consumer, behavior 185
continuous replenishment production
        method 21–22
contract development and manufacturing
        organizations (CDMO)350, 351
    contract with 353–354
    partner selection 355
contract manufacturer 383, 382, 384,
        385, 393
contract research organizations (CRO) 350
contractual agreement (CSA) 340
Conway, J.K. 43
COO *see* chief operating officer (COO)
corporate social responsibility (CSR)
        174, 368
    demands for 27
    for manufacturers 27
cost of goods (COG), printers 189–190
cost plus model 232
COVID-19 pandemic 16, 28, 201, 282, 332
    biopharmaceutical industry and
        356–357
    vaccine 357
CPM *see* critical path analysis (CPM)
critical path analysis (CPM) 23
CRO *see* contract research organizations
        (CRO)
CSA *see* contractual agreement (CSA)
CSR *see* corporate social responsibility
        (CSR)
CSS *see* consignment stocking system
        (CSS)
customer
    need 24–25
    residential 185
customers–internal and external 46–58
    engineering 52
    financial management 49–51
    information technology (IT) 51–52
    internal supply chain management
        functions 47
    legal and role in supply chain 57–58
    logistics 51
    marketing 47–49

customers–internal and external (*cont.*)
  procurement subgroups and
    collaboration with legal 53–57
  production (operations in service
    companies) 51
  supply management 52
CVS 19, 22

Dabbawala 37, 42, 414–415
  creating a brand in logistics 421–422
  long before logistics 413–414
  organizational strategic focus *416*,
    416–417, **417**
  organizational structure 416
  past and present 414–415
  process excellence in logistics
    418–419, *419*
  process innovation for operational
    accuracy 419–421, *420*, *421*
  SWOT (Strength, Weakness,
    Opportunities and Threat)
    Analysis of Dabbawala Team
    Operations **417**
Dabbawalas (Mumbai Dabbawala 2019)
  14–15
dairy market 299–300
  cheese 302–305
  dairy product producers 301
  economic trends 302
  individual performance 307
  macro trends impacting the industry
    301–302
  New Zealand 300–301, 303
  organizational leadership 308
  sterilization 306
  supply chain
    cooperative 305
    global producers 303
    logistics 304
    product 304–305
    seasonality 303–304
  ultra-heat treated (UHT) product
    306–307
data and presentation 167
data center 281–287, 291–293
  supplier managers for 232
data in supply chain 311
  challenges with managing uncertainties
    313–314

inventory management models 325–326
  logistics challenge solved through
    network optimization and linear
    programing 321–323, **323**
  machine learning-based forecasting
    326–328
  opportunities 314–315
  purchase orders (PO) 312
  risk management through probability
    simulation 323–324
  sources of 312–313
  vendor selection
    using hypothesis testing 320–321
    using TCO model 315–320, **319**
data mining 28
data privacy 225–226
data science 256, 290
data transparency
  energy industry 256–257
dead asset reduction 192–194
decision-makers 98–99
decision-making process 117
decision, transactional 98–99
Dell 21, 45
demand 6
digital supply chain, energy industry
  248–249
direct procurement manager 112–113
direct procurement team
  organizational structure of 103–109
    global commodity management
      team 106–107
    global supplier/vendor management
      team 108
    global supply chain management
      team 103–105
    supply chain planning team
      105–106
    supply chain program management
      team 108–109
direct supply chain 97, 103, 105, 118
disruption 163, 164, 166
disruptive innovation
  in product technology 29
  in supply chain 25–26
distributors 59–61
DMAIC (define, measure, analysis,
  improvement, control) steps
  67–68, 79–80

Drucker, P. 47
durable goods 60

economic price negotiation (EPN) 404
ecosystem
　　of EPC (Engineering, Procurement and
　　　　Construction) companies 246
　　supply chain 42
　　　　flexibility in 41
EDLP *see* everyday low price (EDLP)
electric vehicle (EV) manufacturer 130
end of life (EOL) 109
energy industry 245–246
　　case study 253–256
　　digital supply chain 248–249
　　individual contributor role 258–259
　　individual performance 258–259
　　macro trends impacting the industry
　　　　landscape 247–248
　　operational excellence 259–261
　　organizational leadership 257–258
　　shift in relationship management –
　　　　seller's perspective 250–253
　　shift in supply chain management
　　　　245–250
　　tribal knowledge and data transparency
　　　　256–257
energy poverty 261
energy sector supply base 246–247
energy transition 247, 248
engine blocks 266–267, 271
engineering team 52
enterprise resource planning (ERP)
　　applications 221
　　systems 16, 387–388
　　tools 115, 141, 154
entry-level SCM 124–125
　　impatient 155
　　internal demand and development
　　　　process 132–134
　　internal product knowledge 127–130, **129**
entry-level supply chain manager 93, 139,
　　142, 148
Environmental Protection Agency (EPA)
　　standards 268
E&O (excess and obsolescence) risk 208
EPC (Engineering, Procurement and
　　Construction) companies
　　ecosystem of 246, 251–252

EPN *see* economic price negotiation (EPN)
ERP *see* enterprise resource planning (ERP)
escalation management, reactive 163–164
escalation response 165–167
everyday low price (EDLP) 18
excess and obsolescence (E&O)
　　management 171
execution 8–9
existing enterprise resource planning (ERP)
　　systems 248
explore and prepare 4
external manufacturer/producers 58–59
external supply chain members 58–62
　　customers 61 62
　　distributors 59–61
　　external manufacturer or producers
　　　　58–59
　　retailers 61
　　suppliers 59
　　transporters (transportation service
　　　　providers) 59
　　wholesalers 59–61

FAR *see* Federal Acquisition Rules (FAR)
Federal Acquisition Rules (FAR) 345
FedEx 22
financial management 49–51
firmware challenge 188
flexibility
　　in supply chain ecosystem 41
　　supply chain methods and processes
　　　　41–42
flight management systems (FMS) 332
FMS *see* flight management systems (FMS)
food 60
Food and Drug Administration (FDA) 351
food delivery service, Mumbai (India) 15
Ford, H. 42
forecast 130, 132–136
'4Ps' 47–48

GAAP *see* generally accepted accounting
　　principles (GAAP)
GDPR 221, 225
Gemba way 213
General Electric 66
generally accepted accounting principles
　　(GAAP) 50
Generation Z 365

global commodity management team
106–107
global dairy trade auction 305
global supplier/vendor management
team 108
global supply chain management team
103–105
good laboratory practice (GLP) process
development 351
good manufacturing process (GMP)
production 351
goods and services 36
graphics processing unit (GPU)
manufacturers 290

hardware asset management 239–240
health monitoring systems 337
heavy-, mid-, and light-duty (HMLD)
automotives 263, 264,
266, 267
manufacturer 272, 274
supply chain team 275
vehicles 271–274
home printer 181–182, 186
HR-IT roadmap 131
human resource (HR) management 131
Humira 352
hyper growth dealing 392–393
hyperscalers 282–283, 289
hypothesis testing 320–321

IAAS 281
impatient entry-level SCM 155
impatient mid-level SCM 155
impatient senior or an SCM executive
155–156
India, food delivery service 15
indirect procurement, in information
technology 217–244
indirect procurement manager 110–112
indirect procurement team
organizational structure of 100–103
procurement to pay or a P2P team
102–103
strategic sourcing team 100–101
indirect strategic sourcing manager
110–112
indirect supply chain 95, 98, 105
industrial automotive 263, 277

industrial engineering, and operations
research 15
industry *see specific types of industry*
industry landscape, macro trends
impacting 182–183, 202–203
information technology (IT) 51–52
hardware procurement 230
direct *vs.* distributors 232–233
forecast 231–232
indirect procurement in 217–244
procurement 218–220
procurement manager 237
procurement team, individual
performance 242–243
professional service procurement
235–237
team 131
inhalers 200
inhibitor, tumor necrosis factor (TNF) 352
Institute for Supply Management (ISM) 51
institute of supply chain management
2018 16
intellectual property (IP) protection
biopharmaceutical industry 355
of consumables 184–185
internal demand and development
process
entry-level SCM 132–134
mid-level SCM 134–135
senior or an SCM executive 135–136
internal product knowledge 127–130
entry-level SCM 127–130, **129**
mid-level SCM 130–131
senior or an SCM executive 131–132
Internal Revenue Service (IRS) 49
internal supply chain management
functions 47
international traders 38
interview, preparation 118–119
inventory control, medical device
manufacturers 211–212
inventory management 399
IRS *see* Internal Revenue Service (IRS)
ISO 13485 205

John Player 60–61
just-in-time (JIT)
Japan 16
management 397

key performance indicators (KPIs) 57, 252
Khan, Kublai 4, 8–11, 426

lead compound 351, 352
leadership 425, 427
    performance and 195–196
lean, defined 68
lean supply chain 42–45
    features 44
liability, avionics industry 341–343
license-based model, traditional 221
licensing metrics 222
licensing models, user-based 220
linear programing, data in supply chain
        321–323, **323**
logistics 170–171
long term risk sharing 250
long-term strategy 39–40

machine learning (ML) 290
    based forecasting 326–328
magnetic resonance imaging (MRI)
        machine 200, 208, 212, 213
make to stock (MTS) 20–21
make-*versus*-buy and outsourcing
        decisions, supply management
        team 54–55
Mana 308
manger *see specific types of manager*
manufacture, repair, and overhaul (MRO)
        spending 339
manufacturers, corporate social
        responsibility (CSR) for 27
manufacturing resource planning (MRP)
    systems 292
    tools 134
manufacturing set up, finance role 50–51
marketing 47–49
market survey 42
mean time to failure (MTTF)317, 319, 320
mean time to repair (MTTR) 317
medical devices 200–201
    macro trends impacting the industry
        landscape 202–203
    manufacturers 201–202
    organizational leadership 209–210
    quality management 205–206
    supply chain management
        high mix low volume 207–208

pricing 208–209
technology trends 203–204
medical devices supply chain organization
    individual contributors role 210–214
        customer and product knowledge
            213–214
        inventory control 211–212
        sustaining 212–213
mental resilience 258
microprocessor space 290
mid-level SCM 125
    impatient 155
    internal demand and development
        process 134–135
    internal product knowledge 130–131
mid-level supplier manager 144, 148
mid-level supply chain manager 93
milk curve 303, 304, 309
milk solids 305
mindset, self-centered 26
mining, data 28
mitigating risk 208
MLCC (multilayer ceramic chip)
        capacitors 139
modern lean supply chain system,
        features 44
modern supply chain, partners in 45–46
Moore's law 283, 286
MRO spending *see* manufacture, repair,
        and overhaul (MRO) spending
MTS *see* make to stock (MTS)
MTTF *see* mean time to failure (MTTF)
MTTR *see* mean time to repair (MTTR)
multilevel supply chain 169–170
multisourcing and slicing the business
        383–385
multiyear roadmap 40
    methods and processes 40

natural disaster 30
natural products 5
NDA *see* nondisclosure agreement (NDA)
NetApp 21
Netflix 25
network optimization, data in supply chain
        321–323, **323**
new product introduction (NPI)
        186–187, 381
New Zealand, dairy market 300–301, 303

Nike 43, 44
nondisclosure agreement (NDA) 220
nondurable goods 60
nongovernmental organizations (NGO) 368
non-recurring expense (NRE) analysis
        146–147
NPI *see* new product introduction (NPI)

ODM *see* original design manufacturer
        (ODM)
OEM *see* original equipment manufacturer
        (OEM)
open source software 226
operational expenditure (OPEX) budget 96
operational expenses (Opex) 144–145
operational features
    of printer business 186–188
        firmware challenge 188
        products as generally repairable 187
        selling to VARs and distributors 188
        SKU proliferation 186
        unit volumes as not large 186–187
        value-added reseller (VAD) 187–188
operations, defined 51
operations research, industrial engineering
        and 15
opportunities
    approval flow 162–163
    communicating supply chain
        strategy 165
    complex 168
        excess and obsolescence (E&O)
            management and disposal 171
        logistics 170–171
        multilevel supply chain 169–170
        vendor change and consolidation 168
        vendor managed inventory
            (VMI) 169
    dashboard, white paper and spend
        summary 159–160
    data and presentation 167
    escalation response 165–167
    price 161–162
    reactive escalation management 163–164
order to make production method 23–24
organizational leadership
    apparel industry 374–375
    automotive industry 275–276
    avionics industry 344–346

energy industry 257–258
    IT supply chain leadership 240–241
    medical devices 209–210
organizational structure
    Dabbawala 416
    of direct procurement team 103–109
        global commodity management
            team 106–107
        global supplier/vendor management
            team 108
        global supply chain management
            team 103–105
        supply chain planning team 105–106
        supply chain program management
            team 108–109
    of indirect procurement team 100–103
        procurement to pay or a P2P team
            102–103
        strategic sourcing team 100–101
organization, communication and
        information sharing within 28
organization structure, company 94–99
original design manufacturer (ODM) 104
original equipment manufacturer
        (OEM) 104
    supplier 181
outsourcing
    decisions, make-*versus*-buy and 54–55
    of manufacturing for low value-added
        products 402–403

PAAS *see* platform as a service (PAAS)
        providers
packaging, retail-ready 48
Packowski, J. 42
partners, in modern supply chain 45–46
performance and leadership 195–196
personal protective equipment (PPE) 200
platform as a service (PAAS) providers 281
POC *see* proof of concept (POC)
point of sale (POS) 28
Polo, Marco 3, 9–11, 14, 37, 425, 427
POS *see* point of sale (POS)
P2P *see* procurement to pay (P2P)
PPE *see* personal protective equipment
        (PPE)
Practice of Management (Drucker) 47
predictive consumables ordering 194–195
preferred supplier list (PSL) 108

preferred suppliers (strategic suppliers) *vs.*
        transactional suppliers 62–63, **63**
preparation 117–120
    brand–annual goals 119–120
    interview 118–119
    resume 118
price 161–162
    medical devices 208–209
The Principles of Scientific Management
        (Hyde 2019) 15
print engine 190
printer business
    operational features of 186–188
        firmware challenge 188
        products as generally repairable 187
        selling to VARs and distributors 188
        SKU proliferation 186
        unit volumes as not large 186–187
        value-added reseller (VAD) 187–188
printer manufacturing
    supply chain implications in 188–191
        consumables 189
        cost of goods (COG) 189–190
        print engine 190
        printers 188
        RIP controller 190
        spare parts 189, 191
printers 180
    business opportunities for 180–181
        growth opportunities 181
        home and workplace printing
            practices 181–182
    manufacturers 180
    supply chain, relationship management
        191–192
printing into service, convert 183–184
Private Cloud 280
private labels 363
procurement manager 222–223
procurement process 162
procurement team *see specific types of
        procurement team*
procurement to pay (P2P)
    manager 115–116
    team 102–103, 156
production, operations in service
        companies 51
production strategies, supply chain
        management 20–24

built to order 21
channel assembly 22–23
constant replenishment/continuous
        replenishment 21–22
make to stock (MTS) 20–21
real-time change of supply sources 23
3D printing process for prototype
        development 29
Turnkey contract or order to make
        23–24
warehouse management for storing 29
product portfolio 129, 137
products 5–6
    natural 5
product SKUs (stock keeping unit) 185
product technology, disruptive innovations
        in 29
professional service procurement
    Information Technology (IT) 235–237
proof of concept (POC) 220, 352
prototype development, 3D printing
        process for 29
PRs *see* purchase requests (PRs)
PSL *see* preferred supplier list (PSL)
Public Cloud 280
purchase approvals 102
purchase order (PO) 102
purchase requests (PRs) 102

quality management, medical devices
        205–206
quarterly business reviews (QBR) 144

reactive escalation management 163–164
reading guidance 157–158
real-time change of supply sources 23
recreation vehicles (RV) customers 268
Red Hat 219
regulations, trade 369–370
regulatory compliance, in supply chain
        204–205
relationship 7–8
relationship management,
        biopharmaceutical industry 356
remanufacturing, automotive industry
        266–267
request for information (RFI) 54
request for proposal (RFP) 111
request for quote (RFQ) 54, 111

request ID (RID 193
residential customers 185
resume, preparation 118
retailers 61
retail-ready packaging 48
return merchandise authorization
    (RMA) 317
return on investment (ROI) 132, 387,
    399, 415
    analysis 144–145
RFI *see* request for information (RFI)
RFID 29
RFP *see* request for proposal (RFP)
RFQ *see* request for quote (RFQ)
RIP controller 190
risk management, data in supply chain
    323–324
risk mitigation 154, 174, 182
RMA *see* return merchandise authorization
    (RMA)
ROI *see* return on investment (ROI)
Roylance, D. 47

SAAS *see* software as a service (SAAS)
SAS *see* Statistical Analysis System (SAS)
SCM *see* supply chain management (SCM);
    supply chain manager (SCM)
SCOR model *see* supply chain operations
    reference (SCOR) model
Securities and Exchange Commission
    (SEC) 49–50
self-centered mindset 26
senior manager 148
senior or an SCM executive 125
    impatient 155–156
    internal demand and development
        process 135–136
    internal product knowledge 131–132
service level agreement (SLA) 233–234, 265
    optimizing 234
shift in supply chain management 183–185
    business consolidations 185
    convert printing into service 183–184
    energy industry 249–250
    intellectual property protection of
        consumables 184–185
Silk Road 3–4, 10, 15
    demand 6
    execution 8–9

product 5–6
relationship 7–8
significance of 5–9
supplier 6–7
value creation 7
Silk Route 6, 7, 37, 39
Silk Route Romeo 11
SIPOC 72, *72*
Six Sigma 66, 417, 423
    current performance measurement **83,**
        83–85
    customer requirements define **82,** 82–83
    defined 66–67
    expand and integration 85–87, **86**
    identify key customers *81,* 81–82
    prioritize, analyze and implement
        improvements 85, **85**
    process details of 79–80
    projects 67–69
Six Sigma concepts, implementation 69–78
    analyze phase 76–77, *77*
        actions/analysis 76
        return back fishbone diagram to
            determine actual causes 76, *77*
        ship and invoice errors 76
        why analysis 77, **77**
    case problem statement 69–70
    conclusions 77–78
    define phase
        action/analysis (voice of customer:
            data analysis) 72
        action/analysis: supplier–input–
            process–output–customer
            diagram 72, *72*
        current status determination 71–72
        outcome/results key findings 73
        voice of customer phase
            determination **71**
    measure phase *73,* 73–75, *74, 75*
        data collection plan and key
            inputs 74
        defect types determination 74–75
        purpose/objectives (data collection)
            74, *74*
    next steps/lessons learned 78
    solution steps 70–78
Six Sigma roadmap 66
skim milk 303, 304
SLA *see* service level agreement (SLA)

Smith, B. 66
software as a service (SAAS) 281
    applications 218, 219, 221
    model 282
software license distribution 237–239
software procurement 220–221
    compliance 221
    data privacy 225–226
    deliverables 221–222
    open source 226
    preparation 222–224
    removing the bells and whistles 224–225
SOP *see* standard operating procedure
        (SOP)
sourcing manager 112–113
SOW contract *see* statement of work
        (SOW)-based contract
spare parts, printer 191
spares, automotive industry 273–274
SRM *see* supplier relationship management
        (SRM)
stakeholders 112
standard operating procedure (SOP) 78
statement of work (SOW)-based
        contract 236
Statistical Analysis System (SAS) 28
sterilization, dairy market 306
stocking system, consignment 38
stock keeping units (SKUs) 185
    automotive industry 266
    proliferation 186
storage drive 287
storing, warehouse management for 29
strategic outsourcing
    of manufacturing for low value-added
        products 402–403
        key success factors 403
        learnings 403
strategic sourcing
    manager 112–113
    team 100–101
strategic suppliers (preferred suppliers) *vs.*
        transactional suppliers 62–63, **63**
supplier 6–7
supplier landscape 136–144
    business review of strategic and
        operational performance history
        and metrics 143–144
    competitive landscape 138–139

financial health of the supply base
        139–141
    spend, dependencies, and risk 141–143
    technology 136–138
supplier managers 149
    for data centers 232
    mid-level 144, 148
supplier relationship management (SRM)
        52, 56–57
suppliers, external supply chain
        members 59
supplier/vendor management team,
        global 108
supply chain
    aspects of the ancient 37–38
    dashboard 159
    flexibility of 26–27
    goal of 7
    history of 15–16
    lean 42–45
    legal and role in 57–58
    methods and processes 39–42
        flexibility 41–42
        long-term strategy creation–business
            problems and opportunities
            39–40
        make trade-offs 40–41
        multiyear roadmap 40
    modern, partners in 45–46
    process and systems, global perspectives
        29–30
    regulatory compliance in 204–205
    root of 15
    strategy 21–22
    strategy, communication 165
    streamlined multiple components in
        37–38
    system, significance 37–38
    timeline of (mid-1900s–2000s) 16
    value creation within 32
supply chain ecosystem 42
    flexibility in 41
supply chain implications, in printer
        manufacturing 188–191
supply chain leader 140–141
    pillars for 126–127
        business knowledge and value
            creation 144–147
        execution 149–151

supply chain leader (*cont.*)
>   internal demand and development
>   process 132–136
>   internal product knowledge 127–132
>   relationship management 147–149
>   supplier landscape 136–144
supply chain management (SCM) 396 *see
>   also specific types of supply chain
>   management (SCM)*
>   executive 93
>   functions, internal 47
>   responsibilities 56
>   shift in 183–185
>   business consolidations 185
>   convert printing into service 183–184
>   intellectual property protection of
>   consumables 184–185
>   student 93
>   team, global 103–105
supply chain management principles 24–32
>   analytics used 28
>   collaboration between verticals within
>   organization 26
>   communication and information
>   sharing within organization 28
>   corporate social responsibility for
>   manufacturers 27
>   customer need 24–25
>   disruptive innovations 25–26
>   in product technology 29
>   flexibility 26–27
>   global perspectives 29–30
>   impact of economic conditions 30–31
>   political conditions and navigating
>   through political changes 31
>   risk management and mitigation
>   strategies for buyers 31
>   system thinking 25
>   technology-based transformation 28
>   value creation 32
>   visibility of end-to-end supply chain and
>   catching problems early 32
supply chain management–production
>   methods 20–24
>   built to order 21
>   channel assembly 22–23
>   constant replenishment/continuous
>   replenishment 21–22
>   make to stock (MTS) 20–21

real-time change of supply sources 23
3D printing process for prototype
>   development 29
Turnkey contract or order to make 23–24
warehouse management for storing 29
supply chain manager (SCM) 29–30, 99,
>   154, 164, 185
>   entry-level 93, 139, 142, 148
>   mid-level 93
>   role 165
>   works 53
supply chain members, external 58–62
supply chain need 17–20
>   accuracy and quality focus 18
>   all round efficiency focus 19–20
>   cost focus 18
>   customer focus 18–19
>   lead time focus 17–18
supply chain operations reference (SCOR)
>   model 55
supply chain organization 154, 163
>   maturity and rigidity in 156–157
>   roles in 93
supply chain planning team 105–106
supply chain program management team
>   108–109
supply chain program manager 116–117
supply chain roles 109–117
>   commodity manager 114–115
>   direct procurement manager/sourcing
>   manager/strategic sourcing
>   manager 112–113
>   indirect procurement manager/indirect
>   strategic sourcing manager
>   110–112
>   P2P manager 115–116
>   supply chain program manager 116–117
>   vendor managers 113
supply management
>   defined 52
>   primary function of 53
>   supports 53
>   team 53
>   global sourcing 55–56
>   make-*versus*-buy and outsourcing
>   decisions 54–55
>   supplier relationship management
>   (SRM) 56–57
>   supplier selection 53–54

supply sources, real-time change of 23
SWOT (Strength, Weakness, Opportunities
    and Threat) analysis of
    Dabbawala team operations **417**
system thinking, in supply chain 25

Target 23
Taylor, F. 15
TCO *see* total cost of ownership (TCO)
technical bid evaluation 249
technical support officers 399
technological disruption in supply chain
    398–399
  key success factors 400
  learnings 399
technology-based transformation, in supply
    chain 28
technology transition 105, 291, 323
technology trends, medical devices
    203–204
Tesla 22, 25
3D printing process, for prototype
    development 29
time and material (T&M) contract 185
timeline of (mid-1900s–2000s) supply
    chain 16
total cost of ownership (TCO) 56, 161, 162,
    185, 221, 283
  elements of 56
  model 315–320, **319**
Toyota Production system 42
trade, bilateral 370
trade-offs, make 40–41
trade regulations
  apparel industry 369–370
  US 40
traders, international 38
trade tensions, US-China 30
traditional license-based model 221
transactional decision 98–99
transactional suppliers *vs.* preferred
    suppliers (strategic suppliers)
    62–63, **63**
transporters (transportation service
    providers) 59
tribal knowledge and data transparency,
    energy industry 256–257
tumor necrosis factor (TNF) inhibitor 352
Turnkey contract 23–24

ultra-heat treated (UHT) product, dairy
    market 306–307
unchecked cost reduction
  causing supply chain disruption
    404–405
  key success factors 405–406
  learnings 405
UPS 22
US-China trade tensions 30
user-based licensing models 220
US Securities and Exchange Commission
    (SEC) 140
US trade regulation 40

value-added reseller (VAD) 187–188
  selling to VARs and distributors 188
value creation 7
  within supply chain 32
value engineering 145
VARs 191
vendor-managed inventory (VMI) 22, 169
  hub location 321
vendor managers 113
vertical integration 401
  key success factors 402
  learnings 401
video streaming services 25
VMI *see* vendor-managed inventory (VMI)

Walmart 18, 23, 38, 45
warehouse management, for storing 29
warranty support 233–235
web-based companies 218
WEF *see* World Economic Forum (WEF)
Welch, J. 66
whole milk 303–305
wholesalers 59–61
WIFI router manufacturer 129, **129**
working capital 41–42
workplace printer 181–182
World Economic Forum (WEF) 31
worries of a leader 171–175
  automation 172
  decision support and data analytics
    172–173
  lost opportunities 173–174
  whole lot of issues 174–175

Xerox Corp. 183–184

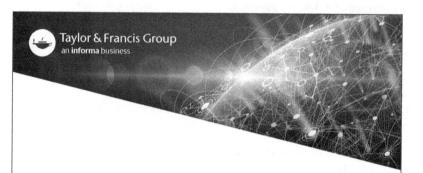